Memoirs of The New York Botanical Garden

Volume 123

Launched in 1900 by The New York Botanical Garden (NYBG), and with 120 volumes published as of 2019, Memoirs has traditionally focused on taxonomy and floristics, with an emphasis on plants of the Western Hemisphere. Most titles comprised a single volume, but quite a few were multi-volume titles and constitute some of the most iconic and substantial monographic publications of NYBG. In recent years, the Memoirs series has expanded to encompass field guides and botanical history as well. All in-print volumes of Memoirs through volume 120 are available exclusively through NYBG Shop: https://nybgshop.org/nybg-press/. Volumes 121 and going forward are available exclusively through Springer.

Roy Osborne • Dennis Wm. Stevenson

The Cycad Botanists

Springer

Roy Osborne
Private Researcher
Oxenford, QLD, Australia

Dennis Wm. Stevenson
Science
The New York Botanical Garden
Bronx, NY, USA

ISSN 0077-8931 ISSN 2662-2858 (electronic)
Memoirs of The New York Botanical Garden
ISBN 978-3-031-79081-2 ISBN 978-3-031-79082-9 (eBook)
https://doi.org/10.1007/978-3-031-79082-9

Disclaimer: This book is intended as a reference work, in which the information provided herein is meant only for educational purposes. This information has been recorded in an attempt to preserve and return traditional knowledge about the uses of medicinal plants, or bush medicines. As such, this information is not to be used as a medical manual for self-treatment with plants, or as a substitute for medical advice, diagnosis, or treatment of any health condition or problem. In matters of personal healthcare, a qualified medical specialist should be consulted. Delay in seeking appropriate medical care may worsen illness. The editors do not make any claims regarding the accuracy, completeness, timeliness, controversial nature, or usefulness of any information contained or referenced in this book since scientific knowledge changes frequently.

This Springer imprint is published by the registered company Springer Nature Switzerland AG
The registered company address is: Gewerbestrasse 11, 6330 Cham, Switzerland

Plants have determined man's history and they will determine his future (Ayres, P. "The Aliveness of Plants", 2008).

Acknowledgments

Many people were kind enough to provide material assistance over the four years we spent in the preparation of this volume; some supplied photographs, some provided data, and some gave leads to other often quite obscure sources of information. For this help, we are indebted to the following people and their associated institutions: Charles Allsopp, William Archer, Duke Benadom, Doug Benson, Rodrigo Bernal, Edwin Brown, Luigi Califano, Michael Calonje, Andrew Cameron, David Cantrill, Sharon Chirgwin, James Clugston, Bret and Yvonne Dalziel, Jacques De Leuze, John Donaldson, Wynand van Eeden, Kate Edgar, Lesley Elkan, Tim Entwisle, Wes Field, Marg Friedel, Greg Ginsburg, Favio González, Patrick Griffith, Hanneke and Nat Grobbelaar Jr., Dennis Haines, Bill Hayes, Jody Haynes, Horace Hobbs, Roderick Home, Bruce Ironmonger, Chip Jones, Jiban Khuraijam, Carl Lewis, Anders Lindström, David Mabberley, Santiago Madriñán, Lufuno Makwarela, Lilí Martínez-Domínguez, Kristine McColl, Hannah McPherson, Ed Medlock, Alan Meerow, Aldo Moretti, Ian Morris, Alex and Des Nelson, Hiêp Nguyen, Fernando Nicolalde-Morejón, Alexandra Nylström, Rolf Oberprieler, Angela Osborne, Hans-Helmut Poppendieck, Peter Rae, Peter Raven, Nathalie Rasolofo, Shayla Salzman, Uwe Schippmann, Matthias Schultz, Giancarlo Sibilio, Rita Singh, Leonie Stanberg, Fred Stauffer, Sally Steward, Wendy Strahm, Brett Summerell, Hiroyuki Takanashi, Craig Thompson, Kirill Tkachenko, Andreas Trepte, Gerhard van Deventer, Piet Vorster, Andrew Vovides, and Iris Wenderholm. The contributions of the four reviewers are appreciated in making this work more accurate and more readable. A special thank you to Bill Buck, Editor of the Memoirs of the New York Botanical Garden series, for his many useful and thoughtful suggestions as well as his rigorous attention to detail that vastly improved the text and its accuracy.

Contents

The original version of the book has been revised. A correction to this book can be found at https://doi.org/10.1007/978-3-031-79082-9_8

About the Authors

Roy Osborne with *Bowenia spectabilis* in Far North Queensland, Australia. (Photo: Angela Osborne)

Roy Osborne was born in London, England, in 1944. He was schooled in Port Elizabeth, South Africa, before studying botany and chemistry at Rhodes University, Grahamstown, during which time he first became acquainted with cycads in theory and in habitat. He continued with postgraduate studies in natural products chemistry at the University of kwaZulu-Natal in Durban, and in botany at the Pietermaritzburg campus of the same institution. In 1984, he discovered, characterized, and named the triterpenoid compound, jessic acid, from the Zimbabwean tree *Combretum eleagnoides*. His Ph.D. thesis, completed in 1988, relates to cycad culture and phytochemistry. From 1979 to 1995, Roy lectured in the Faculties of Architecture, Engineering, Medicine and Science at the University of kwaZulu-Natal. He served on the Board of Durban Botanic Gardens and was founder and first President of the Cycad Society of South Africa. Roy and his family emigrated to Australia in 1996 where he joined a scientific arm of Queensland Government. He has published over 200 papers in scientific and popular journals and is co-author of *Cycads of Australia* (2001) and *Cycads of Vietnam* (2007). He is a long-serving member of the IUCN's Cycad Specialist Group and has been involved with many of the International Conferences on Cycad Biology.

Dennis Wm. Stevenson was born in rural Ohio in 1942 and raised on a farm. He received his B.S. and M.S. degrees from Ohio State University. As a senior student, he met Dr. Knut Norstog who introduced him to the field of cycad biology. In 1975, he completed his Ph.D. at the University of California at Davis. From there he spent 4 years at Fairchild Tropical Botanical Garden as a Harvard University Fellow, interspersed with a year as NATO Postdoctoral Fellow at the Royal Botanic Gardens at Kew. From 1980 to 1987, he was on the faculty at Barnard College of Columbia University, after which he joined the staff of the New York Botanical Garden, a position he held for over 30 years. He has published extensively in the areas of cycad paleobotany, reproductive biology, anatomy, cytology, molecular systematics and genomics, as well as compiling taxonomic monographs on cycads for various floras. He has been responsible for raising major funding for projects in plant molecular systematics and plant genomics and for the construction of a new laboratory at the NYBG. He has conducted extensive field work studies, has been a member of the IUCN's Cycad Specialist Group since its inception, and has been actively involved with each of the International Conferences on Cycad Biology.

Dennis Wm. Stevenson with *Zamia lindleyi* in Fortuna, Chiriquí, Panama. (Photo: Shayla Salzman)

Chapter 1
Introduction

This is not a book about cycads; it is a book about people—the 57 men and women who, in our opinion, have been most influential in the sphere of cycad biology. These are people of past years whose work we acknowledge with gratitude; without them, our appreciation and understanding of cycads would be woefully inadequate. They are as diverse as the plants they studied: they are from the seventeenth through to the twenty-first centuries, they are of many different nationalities, they speak numerous languages, they have varying beliefs, and they often have opposing political viewpoints. But they have all made a significant contribution to our fundamental knowledge of the living cycads. For that, we owe them. We have arranged the sequence of chapters in a chronological sequence to reflect periods of activity and within each alphabetically by surname.

Many early civilizations were aware of cycads; they knew of the plants' toxicity and, at some cost to human life, had independently derived various ways to render the cycad material, usually the seeds, but sometimes stems or roots, safe for human consumption. These people were from Central and South America, Florida in the U.S.A., the Dominican Republic, Japan, India and tropical eastern Asia, Southern and Central Africa, New Guinea, Australia, the islands of the Western Pacific, and elsewhere. The value of cycads to these people is part of a rich folklore that has been handed down through generations, but there are no written records.

We do have evidence to support this early awareness of cycads. Archaeological explorations of cave dwellings in Western Australia (Fig. 1.1) and in Queensland have uncovered remains of *Macrozamia* seeds that date back 13,000 and 4300 years, respectively, while representations of *Cycas* plants and cones are seen in undated aboriginal rock paintings in the Northern Territory of Australia. At Angkor Thom in Cambodia, there is part of a large mural depicting the Khmer people in battle with the Cham, who controlled the city from 1177–1181, so the carving can probably be assigned to the twelfth or thirteenth century and appears clearly to be a large *Cycas* plant (Fig. 1.2). In South Africa, the use of cycads by the *khoikhoi* tribespeople led to the colonists' descriptive term *broodboom* ("breadtree") and the subsequent

R. Osborne, D. Wm. Stevenson, *The Cycad Botanists*, Memoirs of The New York Botanical Garden 123, https://doi.org/10.1007/978-3-031-79082-9_1

Fig. 1.1 An Australian Aboriginal rock art representation of *Cycas arnhemica* from an old archaeological site in Arnhem Land, Northern Territory. (Photo: Ian Morris)

derivation of the genus name *Encephalartos*. In Florida, it was the Calusa, Seminole, and Tequesta Indians who used the fleshy roots of the "coontie," *Zamia integrifolia*, to prepare starch—an observation that led to a short-term industry at the turn of the twentieth century.

More recent evidence of human interactions with cycads originates in the Orient. For example, representations of *Cycas revoluta* are common in Japanese *imari* porcelainware, *ukiyo-e* woodblock prints of the seventeenth and eighteenth centuries, and the *jūbako* ornately lacquered boxes of more recent vintage (Fig. 1.3). While the details above are clear indications of man's association with cycads, it is at best a cursory introduction to what would become the field of cycad biology—and the platform on which we introduce the people whose lives are the substance of this volume.

Fig. 1.2 A stone carving of what we believe to be a *Cycas* plant on the outer gallery wall of Angkor Thom in Cambodia. (Photo: Horace Hobbs)

The first cycad to be described (after the Linnean system of classification was established) was Linnaeus' 1753 description of *Cycas circinalis* based on the Indian *Todda Panna* of Rheede nearly a hundred years earlier. Our readers will find details of the adventures, discoveries, and writings of seventeenth- and eighteenth-century

Fig. 1.3 These Japanese wooden boxes, called *jūbako*, are used to hold special dishes during rituals, festivals, and similar events. In the production technique, called *maki-e*, the polished wood is treated with *urushi* lacquer (the pigmented sap of *Toxicodendron vernicifluum*). This series of boxes is inlaid with golden images of *Cycas revoluta*. (Photo: Hiroyuki Takanashi)

botanists with cycad interests—Georg Rumphius, Hendrik van Rheede, Carl Linnaeus, Nicolaus Jacquin, Joseph Banks, Carl Thunberg, Carl Willdenow, Robert Brown, and Johan Lehmann—in the pages that follow. Later in our volume, we provide summaries of the nineteenth-, twentieth-, and twenty-first-century cycad biologists who have built on and extended from those early foundations.

Not all the nuances are positive, diversity is paramount, and our interpretation of "cycad biologists" is broad. Linnaeus senior was justifiably accused of nepotism in the appointment of his son to a professorship.

Julius Schuster seems to have had bizarre sexual interests and extremist political affiliations, while his 1932 publication confounded advances in cycad taxonomy for decades. Eduard von Regel, Johan Lehmann, William Thiselton-Dyer, and Ronald Melville were considered "difficult" by their colleagues. Joseph Banks' prodigious wealth contrasts with the limited finances of, for example, Carl Thunberg. Benedikt Roezl, Józef Warszewicz, Gustav Wallis, and other professional plant hunters of the Victorian Era, some of whom plundered such large numbers of specimens from habitat that the viability of several plant populations was compromised. Usually, these living collections were done deliberately and often with false locality records, for the purpose of inflating the value of the plant harvests. We have added Kew Directors William Hooker, Joseph Hooker, and William Thiselton-Dyer for their work in editing and contributing to the journals in which so many cycad descriptions and some quite magnificent lithographic illustrations were portrayed. Two non-botanists, Cynthia Giddy and Loran Whitelock, are in our treatment because of their vigorous promotion of cycad collecting as a pastime, but with some questions about the consequences to cycad populations in the wild. Also included is the highly respected but somewhat controversial clinical neurologist, Oliver Sacks. Among his many works was a book that highlighted the subject of cycad toxicity in his account of the cycad-associated Lytico-Bodig neurodegenerative disease complex of the island of Guam in the Western Pacific.

Our selection of candidates has been carefully considered. We are also aware that research in cycad biology during the seventeenth to the twentieth centuries was a male-dominated affair, and the ratio of women to men is pitifully small. The twenty-first century presents a refreshing change with a significant increase in the number of female cycad biologists in all continents. We should also point out that the candidates in this volume have all passed away and that, too, has limited the number of persons we have included. There is an encouraging number of extant workers, both male and female, who are active in cycad projects worldwide and who will undoubtedly be worthy of inclusion in a future edition of this book.

In general, our candidate botanists demand respect. We have to admire Rumphius for so valiantly struggling to complete his *Herbarium Amboinense* after losing his sight. We can only express astonishment at the prodigious taxonomic productivity of botanists like Friedrich Miquel and Ken Hill. On another positive note, we have been honored to know several of the cycad biologists in this volume personally; in that respect, one or both of us has been especially privileged to call as friends Paolo De Luca, Nat Grobbelaar, Ken Hill, Knut Norstog, Sergio Sabato, and Oliver Sacks.

Our treatment is chronological, focusing on those people who we believe have most greatly influenced the present generation of cycad scientists and plant lovers. Space constraints have not allowed us to include those scientists who have been mostly concerned with palaeobotanical aspects of cycadology—apart from Adolphe Théodore Brongniart in view of his founding of the genus *Ceratozamia*. Our decisions of inclusion and exclusion have been personal; we apologize to any readers who feel that we have not given sufficient credit to someone whose work they particularly admire.

We have used "Taxonomic Literature: A selective guide to botanical publications and collections with dates, commentaries and types, Second edition" by Stafleu and Cowan (1976–1988) and the supplements by Stafleu and Mennega (1992–2000) and Dorr and Nicolson (2008–2009) as our standard reference sources for the names and dates of the botanists in this work as well as publication dates for the literature cited therein. Those works also known as TL2 can be accessed at the following public Biodiversity Heritage Library (BHL) site: https://www.biodiversitylibrary.org/item/103414#page/11/mode/1up.

We have used the entries in Brummitt and Powell (1992) for the abbreviations of authors' names.

Chapter 2
The Seventeenth Century

Hendrik Adriaan van Rheede tot Draakestein (1636–1691)

The first cycad to be described in the Western botanical literature as a binomial was the Indian species, *Cycas circinalis*, and that very brief description was published by Linnaeus in 1753 in the section *Palmae* of his *Species plantarum*. But the basis for this description goes back almost a century earlier to the work of two Dutch naturalists, Hendrik Adriaan van Rheede tot Draakestein and George Rumphius.

Hendrik Adriaan van Rheede tot Draakestein (Fig. 2.1) was born the youngest of seven children in Utrecht, on April 13, 1636. At the age of 20, he joined the Dutch East India Company serving in the campaign against the Portuguese on the Malabar (west) coast of India. After several promotions, he was appointed Commander of Dutch Malabar in 1670.

Fortunately for Rheede, the Dutch East India Company (*Verenigde Oost-Indische Compagnie*, VOC) encouraged and facilitated his interest in natural history. In 1674, he commenced three decades of work on his *Hortus Indicus Malabaricus* (Fig. 2.2). This huge work was ultimately published between 1678 and 1693 as a series of twelve volumes, each of about 500 pages, with 794 copper engravings. It is written in Latin, Sanskrit, and Arabic with later translations into Malaysian, Dutch, and English. In this work, 740 plants of the Malabar region are described. Rheede was assisted editorially by botanists Arnold Syen, Johannes Munnicks, and Jan Commelin, supported by a huge team of technical assistants and advisors, and with instructions from the publishers Johannes van Someren and Jan van Dyck. Rheede was indeed a pioneer of Indian botany.

In Volume 3 of the *Hortus*, published in 1682, Rheede has eight engravings (Plates 13–20) of a cycad from the Malabar coast where he cited local names *todda panna* and *mouta panna* which appear to mean "plant with a thick trunk." It is on this basis that Linnaeus published his brief description of *Cycas circinalis* in section *Palmae* of *Species plantarum* of 1753. Much later, Dennis Stevenson (1993)

R. Osborne, D. Wm. Stevenson, *The Cycad Botanists*, Memoirs of The New York Botanical Garden 123, https://doi.org/10.1007/978-3-031-79082-9_2

Fig. 2.1 Hendricus Adrianus van Rheede tot Draakestein, from *Hortus Indicus Malabaricus* of 1678

Fig. 2.2 The elaborate frontispiece to the first volume of Hendrik van Rheede tot Draakestein's *Hortus Indicus Malabaricus* published in 1678

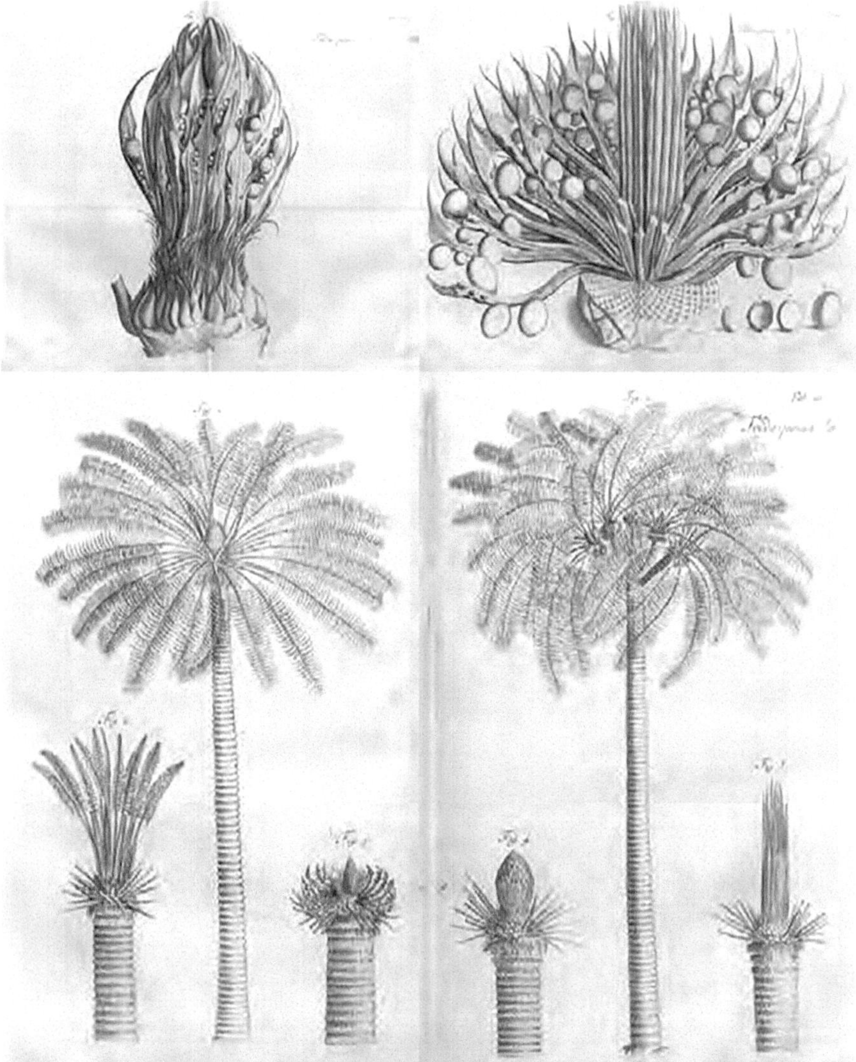

Fig. 2.3 Plates from four of the eight engravings of *todda panna* (the original Indian name for *Cycas* and meaning "plant with a thick trunk") published in Rheede tot Draakestein's *Hortus Indicus Malabaricus*, Part 3, of 1682. In the absence of any holotype, Dennis Stevenson designated Plate 19 (upper, right) as the lectotype for the name *Cycas circinalis*

addressed the absence of any holotype and designated Plate 19 (Fig. 2.3, the cluster of megasporophylls) as the lectotype for the name *Cycas circinalis*. The lack of aerenchyma in the sections of seeds in the plate was significant in that decision. A plant representing the current concept of *C. circinalis* (Hill 1995; Calonje et al. 2024) is shown in Fig. 2.4.

Fig. 2.4 *Cycas circinalis* in forest habitat in Andhra Province, India. (Photo: Keystone Foundation)

In 1684, after the publication of the *Hortus*, the VOC sent Rheede to inspect and report on their international company operations. As Lord of Mydrecht, he also played a role in the governance of the Cape Colony, so it was logical that he spent time with the then Governor, Simon van der Stel. Thus, in 1685, he visited the growing "refreshment station" at the Cape of Good Hope (now Cape Town) where the valley *Groot Drakenstein* was named for him. Rheede considered preparing a volume on African plants of economic importance, *Hortus Africanus*, but his idea never came to fruition.

After his visit to the Cape, Rheede traveled to various trading posts in Ceylon (now Sri Lanka) and India but died at sea on December 15, 1691. He was buried in Surat, India, on January 3, 1692.

Notes on Cycas circinalis vis-à-vis C. rumphii

When Linnaeus established the genus *Cycas* in 1753, he based some of his conclusions on earlier works where different species had not been correctly identified. In subsequent commentaries on the plant, the names *Cycas rumphii* Miq. and *Cycas circinalis* L. were greatly confused; even today, many of the older living *Cycas* specimens in international botanical gardens are incorrectly labeled. Hill (1995) sought to bring some clarity to the issue, stating that "almost all plants in cultivation that have been known as *C. circinalis* ...belong to the *C. rumphii* complex." He pointed out that the seeds of the *Cycas rumphii* complex have a well-defined spongy flotation layer comprising aerenchyma within the sclerotesta, a feature absent in *C. circinalis*. Geographically, *Cycas circinalis* is an Indian endemic restricted to the Western Ghats and hills of the southern peninsula, whereas *Cycas rumphii* ranges over a large territorial area encompassing much of Indonesia and some adjacent areas.

Rheede's Botanical Publication

Rheede tot Draakestein, H. A. van. 1678–1693. *Hortus Indicus Malabaricus, Continens regni Malabarici apud Indos celeberrimi omnis generis plantas rariores.* 12 vols. J. van Someren & J. van Dyck, Amsterdam.

Georg Everhard Rumphius (1627–1702)

Georg Everhard (or Eberhard) Rumphius (Fig. 2.5) was born Rumpf in Wölfersheim, Germany, on November 1, 1627, during the devastating Thirty Years' War (Veldkamp 2011). After schooling at the gymnasium in Hanau, he was recruited at age 20 by the West India Company to serve in Venice, but his ship was oddly diverted to Brazil and later Portugal, after which he returned to Germany. In 1652, he joined the powerful Dutch East India Company (*Verenigde Oost-Indische Compagnie*, VOC), traveling first to Batavia (now Jakarta, Indonesia) and then being sent in 1654 to undertake administrative duties in the Dutch trading post of Ambon where he had a residence (Fig. 2.6) in the Moluccas (the "Spice Islands," now the Maluku archipelago) situated between Borneo and New Guinea.

With the support of Joan Maetsuycker, Governor-General of Batavia, Rumphius was given dispensation from his ordinary duties to explore the flora and fauna of the 775 km^2 mountainous Ambon and neighboring islands. He also undertook

Fig. 2.5 One of the few portraits of Georg Eberhard Rumphius, this one made by his son, Paul Augustus, around 1695–1696. (From the *Herbarium Amboinense*)

Fig. 2.6 Rumphius' residence in Ambon, photographed ca. 1910. (From the archives of the Tropenmuseum, Amsterdam)

excursions to different parts of the East Indies. Of particular significance to us, he spent time exploring the Malabar (west coast) of India where, in 1658, he recorded details of the cycad that was named *Cycas circinalis* by Linnaeus nearly a century later.

Having the unusual and useful ability to establish excellent rapport with the local inhabitants of the areas he visited, Rumphius gathered an enormous amount of ethnobotanical data from them. As an example, he learned of the toxicity of cycads and wrote, rather gruesomely, that in the Celebes (now Sulawesi):

> the sap from the kernels…was given to children to drink in order to kill them so that the parents would not be hampered when they went to follow their roving life in the forest.

Rumphius' diligence in gathering material of this kind led ultimately to the preparation of his *Herbarium Amboinense* (Figs. 2.7 and 2.8), a huge multi-volume compendium written initially in Latin. In it, he describes more than 1200 plants, 930 with definite species names and another 140 identified to genus level and illustrated with numerous plates. Despite its name, the work also included informative notes on plants from India, China, Japan, New Guinea, Madagascar, Mexico, and South America. Rumphius also worked on several other projects, including a folio on the natural history of Ambon entitled *D'Amboinsche Rariteitkamer* ("The Ambonese Curiosity Cabinet").

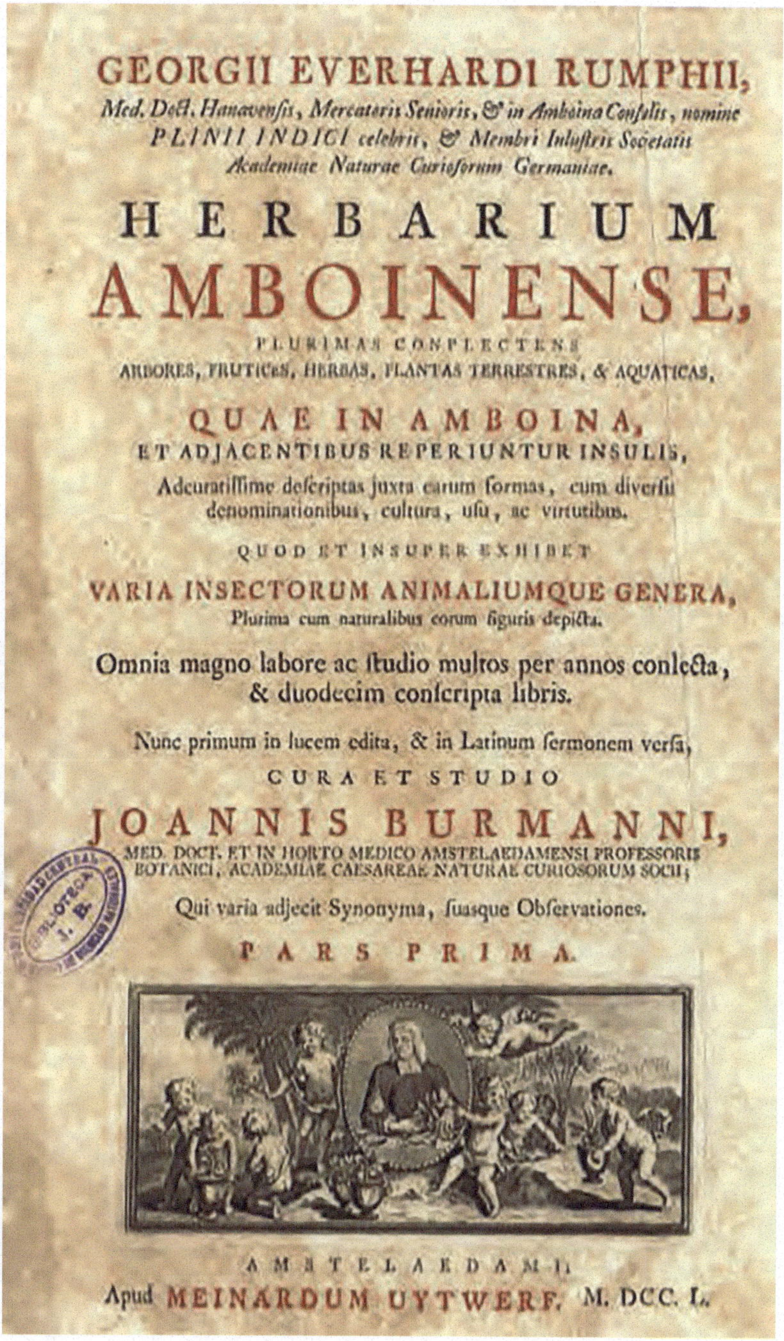

GEORGII EVERHARDI RUMPHII,
Med. Doct. Hanavenſis, Mercatoris Senioris, & in Amboina Conſilis, nomine
PLINII INDICI celebris, & Membri Inluſtris Societatis
Academiae Naturae Curioſorum Germaniae.

HERBARIUM
AMBOINENSE,
PLURIMAS COMPLECTENS
ARBORES, FRUTICES, HERBAS, PLANTAS TERRESTRES, & AQUATICAS,
QUAE IN AMBOINA,
ET ADJACENTIBUS REPERIUNTUR INSULIS,
Adcuratiſſime deſcriptas juxta earum formas, cum diverſis
denominationibus, cultura, uſu, ac virtutibus.
QUOD ET INSUPER EXHIBET
VARIA INSECTORUM ANIMALIUMQUE GENERA,
Plurima cum naturalibus eorum figuris depicta.
Omnia magno labore ac ſtudio multos per annos conlecta,
& duodecim conſcripta libris.
Nunc primum in lucem edita, & in Latinum ſermonem verſa,
CURA ET STUDIO
JOANNIS BURMANNI,
MED. DOCT. ET IN HORTO MEDICO AMSTELAEDAMENSI PROFESSORIS
BOTANICI, ACADEMIAE CAESAREAE NATURAE CURIOSORUM SOCII;
Qui varia adjecit Synonyma, ſuasque Obſervationes.
PARS PRIMA.

AMSTELAEDAMI,
Apud MEINARDUM UYTWERF. M. DCC. L.

Fig. 2.7 The title page of Rumphius' *Herbarium Amboinense*

Fig. 2.8 The elaborate frontispiece to Rumphius' *Herbarium Amboinense*

But Rumphius' life soon turned into a series of tragic events (Margulis and Raven 2009; Peters 2021). In 1670, he lost his sight from glaucoma, but he continued work by dictating a translation of his *Herbarium* from Latin to Dutch. Then, during the Chinese New Year celebrations of 1674, his wife and youngest daughter were killed when a stone building collapsed upon them in an earthquake. Further hardship forced him to sell, to the Medici family in Tuscany, his collection of shells and natural curiosities:

> treasure[s] which I have brought together over many years with much cost and labour and which, in the future, it will be impossible to acquire again, especially since I am now old and blind.

Adding further to his woes, in 1687, a fire destroyed his library containing writings and other items from his collection, including the illustrations for the *Herbarium*. The drawings were redone, with some then stolen and needing to be redrawn yet again, the mammoth task finally reaching completion in the early 1690s. Compounding Rumphius' difficulties, the first part of the publication was lost during passage to Amsterdam when the vessel was sunk by the French. Only in 1696 did the entire compendium reach the Netherlands, the production then comprising six books in the first part and six books in the second. Rumphius (1705) also completed an *Auctuarium*, a supplement dispatched to the Netherlands a month before his death on June 15, 1702.

The saga does not end there; the VOC Executive Council locked the entire manuscript away in the Amsterdam vaults as it contained information that could prejudice their hold over the lucrative spice trade. Ultimately, the ban was lifted, and the *Herbarium Amboinense* was printed through the efforts of the botanist Johannes Burman in 1742, 40 years after the author's death. The final work contained nearly 1700 pages of text and 700 plates, including six detailed illustrations of cycads. English translations of Rumphius' works have been recently published by the late Professor Eric Beekman (2003) of the University of Massachusetts. In recognition of his huge contribution to natural history, in 1681, Rumphius was made a Member by the *Academia Naturae Curiosorum* of Vienna and given the title *Plinius Indicus* (Pliny of the Indies).

More than a century after Rumphius' death, Friedrich Miquel (1839) referred to the cycad that Rumphius called *Olus Calappoides* and named it *Cycas rumphii* in 1839, citing Plates 22 and 23 as illustrations. Ken Hill (1994) discussed the validity of this typification in detail, while Lindstrom (2002) excludes discordant elements from the protologue and confirms Rumphius' Plate 23 (the plant with a pollen cone in Fig. 2.9) as the lectotype for the name *Cycas rumphii* (Fig. 2.9). This species, as now understood, is shown in Fig. 2.10.

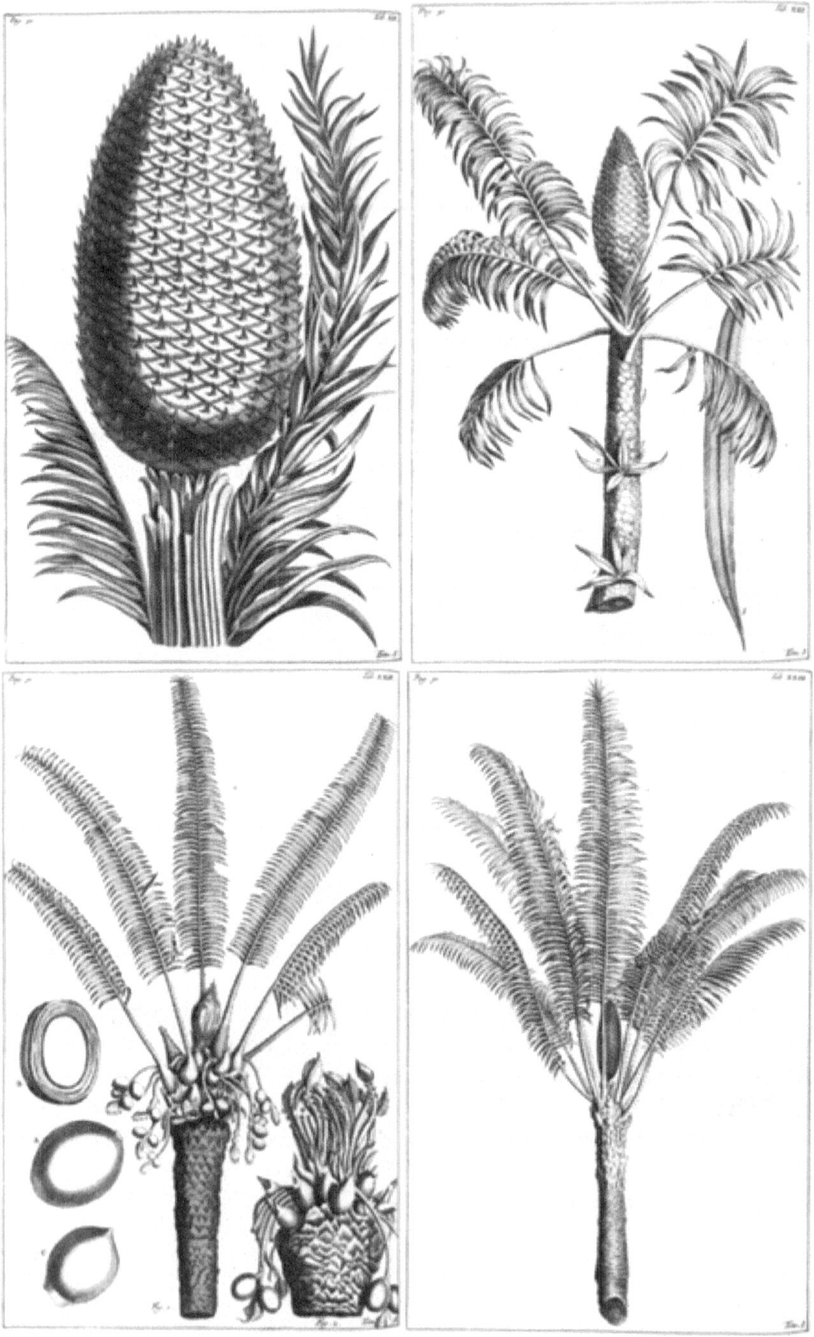

Fig. 2.9 Four engravings of a cycad in Rumphius' *Herbarium Amboinense* of 1742 that Friedrich Miquel would name *Cycas rumphii.* In the absence of any holotype, Ken Hill later designated Rumphius' Plate 23 (bottom, right) as the lectotype for the name *Cycas rumphii*

Fig. 2.10 *Cycas rumphii* on Halmahera, the largest island in the Maluku Archipelago, some 400 km north of Rumphius' home on Ambon. (Photo: Anders Lindström)

Rumphius' Biology Publications

Rumphius, G. E. 1705. *D'Amboinsche Rariteitkamer*, Halma, Amsterdam. "The Ambonese Curiosity Cabinet". English translation 1999 by E. M. Beekman, Yale University Press, New Haven.

Rumphius, G. E. 1742. *Herbarium Amboinense* (Latin), *Amboinsche Kruidboek* (Dutch). J. Burman, Amsterdam. English translation 2011 by E. M. Beekman, 6 vols. Yale University Press, New Haven.

Chapter 3
The Eighteenth Century

Joseph Banks (1743–1820)

Born in London on February 13, 1743, Joseph Banks (Figs. 3.1 and 3.2) was the son of a wealthy landowner, William Banks, and he appreciated the wonders of the British countryside at an early age. He was schooled at Harrow and later at Eton, before being admitted to Oxford's Christchurch College at the age of 17. Three years later, he matriculated with an extensive knowledge of natural history and a passion for botany. Then, at age 21, he inherited the Revesby Estate in Lincolnshire with an income of £6000 per year, instantly becoming one of England's wealthiest men.

With his new status, Banks vigorously embraced the Age of Enlightenment, gathering a large circle of influential and learned friends and spending much time studying the flora and fauna of the British countryside. But the lure of the exotic was strong, and soon, Banks was devising ways to explore fields anew. In 1766, Banks and an Etonian classmate, Constantine Phipps, sailed on the frigate HMS *Niger*, under the command of Sir Thomas Adams, to Newfoundland and Labrador. There they spent four months studying the local natural history and collecting numerous specimens, calling in at Lisbon for six weeks on the return trip.

That voyage was the forerunner to Banks' participation in the Royal Society's historic expedition to observe the Transit of Venus across the Sun from a site in the South Pacific (so as to facilitate the calculation of the distance from the Earth to the Sun). However, there were sealed orders from the Admiralty for a subsequent British geopolitical quest and one that coincided happily with Banks' personal scientific interests. The vessel of choice was a Whitby collier, the *Earl of Pembroke*, greatly modified (much at Banks' personal expense) and renamed HMS *Endeavour*. The ship was given to the command of the phlegmatic Lieutenant James Cook, RN, accompanied by Charles Green of the Royal Observatory, and the expedition was to run from August 1768 to July 1771.

R. Osborne, D. Wm. Stevenson, *The Cycad Botanists*, Memoirs of The New York Botanical Garden 123, https://doi.org/10.1007/978-3-031-79082-9_3

Fig. 3.1 The 29-year-old Joseph Banks in a giant (230 × 157 cm) oil painting by Benjamin West soon after his monumental voyage of astronomical, geographical, and biological discovery. He stands wearing a Māori cloak made from New Zealand flax (*Phormium tenax*) and is surrounded by treasures from his expedition. (From the State Library of New South Wales)

Fig. 3.2 Sir Joseph Banks at age 71 wearing decorations bestowed by King George III and sitting amidst the ceremonial trappings of the Royal Society; an engraving by Nicholas Schiavonetti after a painting by Thomas Phillips. (From the State Library of New South Wales)

Banks arranged that he should be a supernumerary on the voyage, paying his own way, and somewhat to Cook's chagrin, taking with him an eight-man team and 20 tonnes of equipment, including a library of over 150 books, and a miscellany of microscopes, telescopes, and other items. The team comprised Banks' lifelong friend, Swedish naturalist, and one of Linnaeus' 17 "apostles," Daniel Solander; his assistant Herman Spöring; the talented Scottish botanical artist Sydney Parkinson; the landscape artist Alexander Buchan; four personal servants; and Banks' two greyhounds.

The three-year expedition made stops at Madeira, Rio de Janeiro, Tierra del Fuego, Otaheite (Tahiti, where there are fairly well-substantiated reports of Banks' dalliance with a local lady of high rank), the Society Islands, New Zealand's north and south islands, the east coast of Australia, New Guinea, Batavia (now Jakarta) and several Indonesian islands, Cape Town, and St. Helena. Banks and his team collected, preserved, and drew botanical and zoological specimens at every opportunity. Sadly, at the end of the hazardous voyage, only 41 of the 94 in the company that had set out with Cook were to return alive to England. Of the nine members of the scientific party, only Banks himself, Solander, and two servants survived. In their final major voyage in 1772, Banks and Solander toured Iceland and again collected many natural history specimens.

The vast amount of herbarium vouchers, pickled plant and animal specimens, stuffed birds, skins, skeletons, anthropological artifacts, and numerous drawings from these various expeditions were all carefully cataloged and stored in Banks' large working laboratory at his home in Soho Square (Fig. 3.3), readily available to

Fig. 3.3 The herbarium and library in Banks' home at 32 Soho Square, London, an active workshop and meeting venue for research workers and members of the Royal Society. After Banks' death, the material was bequeathed to the British Museum and the Museum of Natural History

the many visitors anxious to inspect, discuss, and learn from the material. Ultimately, these items became a core part of the collection of the British Museum (Chambers 2016).

Despite his profound knowledge of and enthusiasm for all things scientific, Banks' own written publications are limited, the most extensive being his records later edited into "The *Endeavour* Journal" in two volumes (Banks 1962). His patronage and encouragement inspired and facilitated many scientific projects in an enduring legacy, particularly as he put politics aside and continued vigorous correspondence with "foreign" scientists throughout the years of the French Revolution, the Napoleonic Wars, and the American War of Independence. His efforts also contributed significantly to the establishment of the wool industry of Australia, the tea plantations of Assam, and mango and breadfruit farming in the Caribbean. Importantly, for the scientific records, Banks supervised the production of over seven hundred copper plate engravings based mainly on Sydney Parkinson's botanical sketches, these eventually being collated into the work *Florilegium* (Banks et al. 2019), although that complete volume was not published until 160 years after Banks' death.

There are only limited references to cycads in the various accounts of Banks' expeditions. When HMS *Endeavour* anchored near Botany Bay in New South Wales in late April 1770, the naturalists would have seen plants of *Macrozamia communis* that occur in the area, but it is possible that these were confused with palms (Benson and Eldershaw 2007). A voucher (*NSW 43060*), originally deposited in the British Museum under the name *M. spiralis* (Fig. 3.4), is presently filed at the New South Wales herbarium and annotated with Lawrie Johnson's (1959) determination of the name *M. communis* and illustrated here in Fig. 3.5. This may be the very first cycad voucher known from Australia. We have been unable to establish further details nor can we locate any drawings of this species from the 1770 landing. One of us (RO) has seen a population of cycads near Botany Bay and estimates that the plants are about 2.5 km from the expedition's landing site; we believe that Banks and Solander would have passed near or by this stand of plants during the 8 days they spent in the area.

The HMS *Endeavour* made landfall at another ten sites as it progressed northwards along the Australian east coast, but these were simply one-day visits to collect fresh water (Benson 2020). However, somewhat greater detail emerged from much further up the coast when the vessel ran aground on the Great Barrier Reef in June 1770. In difficult circumstances, Cook and his crew managed to get the vessel ashore near the present site of Cooktown, where the crew spent 7 weeks making the necessary repairs before being able to resume their journey. During this time Cook, Banks, Solander, and Parkinson explored the surrounding areas, meeting with the Indigenous people and collecting flora and fauna. It is here that we find references to the cycad that Ken Hill (1996) later and appropriately named as *Cycas media* subsp. *banksii* (Fig. 3.6). Parkinson's notes (from Mabberley 2019, 2020) read:

> of vegetables, we found.... *Cicas circinalis* [sic], the kernels of which roasted, tasted like parched pease; but it made some of our people sick, who ate of it; of this fruit they make a kind of sago in the East-Indies.

Fig. 3.4 The voucher *NSW 43060* originally deposited in the British Museum and presently filed at the New South Wales herbarium. The specimen was at one time recorded under the name *Macrozamia spiralis* but was determined by Lawrie Johnson in 1958 as *M. communis*. This may be the very first cycad voucher known from Australia. (Image kindly supplied by the New South Wales Herbarium, Royal Botanic Gardens, Sydney, Australia)

Fig. 3.5 A specimen of *Macrozamia communis* approximately 2.5 km from the site where James Cook and the botanists on HMS *Endeavour* first set foot in Australia in April 1770 and quite probably part of the population where Banks and Solander collected the herbarium specimen *NSW 43060*. (Photo: Roy Osborne)

Fig. 3.6 A plant of *Cycas media* subsp. *banksii* near the site where HMS *Endeavour* was repaired after running aground on the Great Barrier Reef and quite probably in the area visited by Cook, Banks, and Solander during that seven-week period in mid-1770. Interestingly, this particular plant is a good example of pseudo-dichotomous branching. (Photo: Bruce Ironmonger)

In Banks' Journal, we find this comment on the same plant mentioned:

> ….a low [palm], seldom ten feet in hight, with small pennated leaves resembling those of some kind of fern. Cabbage it had none but [it] generally bore a plentiful Crop of nuts about the size of a large chestnut and rounder. By the hulls of these which we found plentifully near the Indian fires, we were assurd that these people eat them, and some of our gentlemen tried to do the same but were deterrd…by a hearty fit of vomiting and purging.

Banks goes on to note that the ship's pigs also succumbed to the effects of the cycad toxins but then very astutely concluded that the "nuts" contained a toxic element that required various boiling, steaming, and drying to render them edible. Herbarium specimens of this cycad under the name *Banks and Solander 1770* are filed in the British Museum (BM), the Northern Territory (*DNA A0085190*), and the Melbourne (*MEL 78133*) herbaria (Fig. 3.7).

The second cycad-related note refers to the travelers' stop at Prince's Island (Panaitan Island) off the Indonesian coast near Jakarta. Here, Banks tells of local people eating…

> …. nuts of the Palm called *Cycas circinalis* [probably *Cycas edentata*] with which on the Coast of New Holland some of our people were made ill and some of our hogs Poisoned outright. Their method of preparing them to get out their deleterious qualities they told me were first to cut the nuts into thin slices and dry them in the sun, then to steep them in fresh water for three months, afterwards pressing the water from them and drying them in the sun once more; they however were so far from being a delicious food that they never usd them but in time of scarcity when they mixt the preparation with their rice.

A peripheral cycad-related matter is this: subsequent to his own voyages of exploration, Banks was responsible for promoting the scientific career of his assistant Robert Brown (1773–1858), who was to become another of Britain's leading botanists. Banks facilitated Brown's passage with Matthew Flinders on HMS *Investigator*, a historic voyage of 1801–1803 circumnavigating Australia, during which Brown collected the types for the first *Cycas* species to be named from Australia, *C. media* and *C. angulata*.

Joseph Banks was elected a Fellow of the Royal Society in 1766, becoming its longest-serving President from 1777 until his death in 1820. In 1781, he was made a baronet and in so became Sir Joseph Banks. In 1795, he received the order of Knight Commander of the Bath. Two years later, he was admitted to the privy council. Among other activities, Banks found time to serve as High Sheriff of Lincolnshire, Advisor on Colonial affairs to King George III and the British Government, an enthusiastic patron and advisor to Kew Gardens, and a trustee of the British Museum.

Further afield geographically, Banks was elected a Member of the American Philosophical Society in 1787, a Foreign Honorary Member of the American Academy of Arts and Sciences in 1788, and an Associate of the Institute of France in 1802. Perhaps, one of the best accolades is reflected in the title of one biographies of Banks, *The Multifarious Mr. Banks. From Botany Bay to Kew (*Musgrave 2020*). The Natural Historian Who Shaped the World.*

Banks' name is best known in the botanical world through his commemoration in the genus *Banksia*, although more than 80 species of other plants were named for

Fig. 3.7 The voucher *MEL 78133* represents the cycad presently known as *Cycas media* subsp. *banksii* and was collected by Joseph Banks and Daniel Solander after the HMS *Endeavour* ran aground on the Great Barrier Reef in June 1770. (Image kindly supplied by Melbourne Herbarium (MEL), Royal Botanic Gardens, Victoria, Australia)

Fig. 3.8 Left: Joseph Banks as depicted on an Australia 90¢ postage stamp, part of the New Holland–Cook's voyage 1986 bicentennial series. Right: Banks depicted on part of a British presentation postage stamp cover issued on the 250th anniversary of Cook's setting sail on HMS *Endeavour*'s voyage of discovery. The background drawings are from Sydney Parkinson's sketches of *Phaethon rubricauda* (red-tailed tropic bird) and *Passiflora aurantia*

him. Australian examples are *Grevillea banksii*, *Eucalyptus banksii*, *Piper banksii*, the sundew *Drosera banksii*, and a seaweed *Hormosira banksii*. In 1996, Ken Hill appropriately created the taxon *Cycas media* subspecies *banksii* for the cycad population that Banks and Solander would have seen while repair work was carried out on HMS *Endeavour* in 1770.

In Lincoln, England, the Sir Joseph Banks Conservatory was constructed in 1989 holding numerous plants related to Banks' voyages. The conservatory was moved to Woodside Wildlife Park in 2016 and has been renamed "Endeavour." A plaque was installed in Lincoln Cathedral in Banks' honor. The Sir Joseph Banks Centre is located in Horncastle, Lincolnshire, and housed in a building recently restored by the Heritage Trust of Lincolnshire.

Captain Cook named the north headland of Botany Bay as Cape Banks, and the Sir Joseph Banks Group comprises an archipelago off the coast of South Australia. In New South Wales, the Sydney suburbs of Bankstown, Banksia, and Banksmeadow are named after him as is the Joseph Banks High School in Revesby and the suburb of Banks in Canberra. A monument to Banks stands in Kurnell, New South Wales. An image of Banks was featured on the older style $5 Australian banknote, and his portrait was depicted on Australian (1986) and British (2018) postage stamps (Fig. 3.8). Banks' name features in the South Pacific in the Banks Peninsula in the South Island, New Zealand, the Banks Islands near Vanuatu, the Banks Strait between Tasmania and the Furneaux Islands, and the Banks Island in the Northwest Territories in Canada.

In March 1779, Banks had married Dorothea Hugessen, the eldest daughter of a wealthy landowner in Kent. They lived in a house in London's Soho Square, with Dorothea's unmarried sister Sophia. Described by David Attenborough as "the great panjandrum of British Science," Banks died in London in 1820 at the age of 77, survived only by his wife. He was buried at St. Leonard's Church, Heston.

Banks' Major Publications

Banks, J.1962. The *Endeavour* Journal 1768–1771. J. C. Beaglehole (ed.). Angus & Robertson Ltd., Sydney.

Banks, J., M. Gooding, D. Mabberley & J. Studholme. 2019. Florilegium. Compact Edition. Thames & Hudson, London.

Nicolaus Joseph von Jacquin (1727–1817)

Nicolaus (Nikolaus, Nicolaas) Joseph von Jacquin (Fig. 3.9) was born into a French Catholic family in Leiden, the Netherlands, on February 16, 1727. After being schooled at a Jesuit college in Antwerp, he returned to Leiden in 1744 to study medicine and botany, meeting the leading Dutch botanists Jan Gronovius, Laurens Theodoor, and Adriaan van Royen, all keen proponents of the new novel Linnaean classification system. He then spent a further two years studying medicine in Paris.

In his mid-twenties, Jacquin moved to Vienna at the invitation of Gerard van Swieten, Professor of Medicine at the University of Vienna and physician to Empress Maria Theresa, last Monarch of the House of Hapsburg and who married Emperor Franz Joseph I, first Holy Roman Emperor of the House of Lorraine. Van Swieten was greatly involved with the development of the gardens at *Schloss Schönbrunn* and persuaded the Emperor to send Jacquin on a four-year expedition to the West Indies, which eventually took him to Venezuela and New Granada so that he could collect tropical plant material for the newly completed conservatories and hothouses, as well as animals for the *Tiergarten* (zoo) and mineral samples for the palace museum. A full account of Jacquin's expedition to the Caribbean can be found in Madriñán (2013).

After returning to Vienna, Jacquin published a report of his New World expedition in his *Enumeratio systematica plantarum* (1760). Linnaeus greeted this work with enthusiasm, and the two then commenced a vigorous correspondence. Three years later, Jacquin (1762–1763) published a more detailed illustrated description of his Caribbean discoveries in *Selectarum stirpium Americanarum historia* (Howard 1973; Stafeu,1971). He then took a six-year appointment as Professor of Chemistry and Metallurgy at the *Bergakademie Schemnitz* in Hungary (now Slovakia).

In 1769, Jacquin returned to Austria to take over as Head of Botany and Chemistry at the University of Vienna, remaining in that position for 28 years. He also became Director of the University Botanic Gardens and supervised the Palace Gardens at Schönbrunn. In 1785, Maria Theresa's son, Emperor Joseph II, sent two Austrian plant collectors, Franz Boos and Georg Scholl, to the Cape of Good Hope and Mauritius to collect natural history material for Schönbrunn. They sent various consignments of living and dried plants, bulbs, seeds, birds, and other animals back to Vienna, Scholl himself returning home only in 1799 (Garside 1942). In the interim, Jacquin had met with and assisted Alexander von Humboldt in preparation for von Humboldt's journey to the Americas.

Fig. 3.9 Nicolaus Joseph von Jacquin (1727–1817). (Image from Wikipedia)

Complete details of the origin of the cycads in the hothouses at Schönbrunn are somewhat sketchy. We do know that some plants (and a menagerie of animals) were transferred from the nearby Belvedere residence of Prince Eugene of Savoy when he died in 1736; the plants had been originally sourced from Kew but is unlikely that cycads were included in that acquisition. Many of the exotics, probably including some cycad specimens, were sourced later from European nurseries. Jacquin (1791) described *Zamia angustifolia* (Fig. 3.10) from male and female plants grown at Schönbrunn from seed originating in 1784 from the Bahamas. He described *Z.*

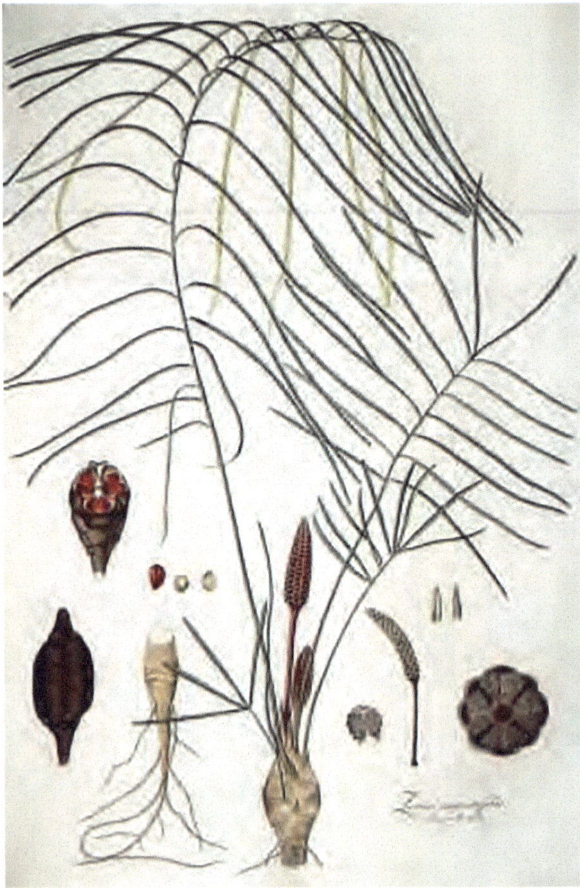

Fig. 3.10 A color plate of Jacquin's *Zamia angustifolia*. Plate 636 from *Icones plantarum rariorum*, Vol. 3 of 1792. Based on male and female plants raised in the hothouses at Schönbrunn from seeds collected in the Bahamas in 1784

integrifolia (Fig. 3.11) as a plant originally introduced from Florida to Kew in 1768 but later cultivated in the hothouses at Schönbrunn.

Georg Scholl brought several Eastern Cape *Encephalartos* plant specimens and their cones back to Vienna in 1799 after his visit to the Cape of Good Hope (Dyer,1965), and this was the material used in the drawings in Jacquin's (1801) *Fragmenta botanica* (e.g., *Encephalartos horridus* in Fig. 3.12). We cannot establish for certain which artist/s drew these superb but unsigned cycad illustrations; the style is such that Ferdinand Bauer may have been involved. But Jacquin himself was a skilled artist, and it is possible that student illustrators may have contributed. There is some discussion of this in the reports by Garside (1942) and Madriñán (2013).

In recognition of his contribution to Austrian botany, Jacquin received the title of *Edler* in 1774 and *Freiherr* (Baron) in 1806. He has been described as "the Austrian

Fig. 3.11 A color plate of Jacquin's *Zamia integrifolia*. Plate 635 from *Icones plantarum rariorum* Vol. 3 of 1792. Based on a plant originally introduced from Florida to Kew in 1768 but later cultivated in the hothouses at Schönbrunn

Linnaeus" (Erickson 2011). He was a Member of the Royal Swedish Academy of Sciences and the Royal Netherlands Academy of Arts and Sciences. He passed away in Vienna on October 26, 1817. His name is commemorated in the genera *Jacquinia* (Theophrastaceae) and *Jacquiniella* (Orchidaceae). In 2011, the Austrian

Fig. 3.12 Hand-colored copper engravings of Jacquin's *Zamia horrida* (now *Encephalartos horridus*). Plates 27 and 28 from *Fragmenta Botanica* of 1801. The leaf is atypical for *E. horridus* and may better represent *E. trispinosus*

Mint issued silver 20€ coins to mark his Caribbean expedition. His elder son, Joseph Franz, succeeded him as Professor of Botany and Chemistry at the University of Vienna; his younger son, Emil Gottfried, and his daughter, Franziska, were friends of Mozart.

Names of Cycads with Jacquin's Authorship
Data from the *World List of Cycads*

Zamia angustifolia Jacq. (1791).
Zamia cycadifolia Jacq. (1801) ≡ *Encephalartos cycadifolius* (Jacq.) Lehm.
Zamia horrida Jacq. (1801) ≡ *Encephalartos horridus* (Jacq.) Lehm.
Zamia lanuginosa Jacq. (1801) ≡ *Encephalartos longifolius* (Jacq.) Lehm.
Zamia longifolia Jacq. (1801) ≡ *Encephalartos longifolius* (Jacq.) Lehm.

Jacquin's Major Publications

Jacquin, N. J. 1760. *Enumeratio systematica plantarum.*

Jacquin, N. J. 1762–1763. *Selectarum stirpium Americanarum historia.*

Jacquin, N. J. 1789–1793. *Icones plantarum rariorum.* 3 vols.

Jacquin, N. J. 1791. *Collectanea ad botanicam, chemiam, et historiam naturalem spectantia.* Vol. 3: 261–265. [*Zamia angustifolia* Jacq.].

Jacquin, N. J. 1798. *Plantarum rariorum horti Caesarei Schoenbrunnensis*, Vol. 3, Nos 397–398: 77–78. [*Zamia media* Jacq.].

Jacquin, N. J. 1801. *Fragmenta botanica, figures coloratis illustrata,* Nos 91–94: 27–28 & Plates 25–31. [*Zamia cycadifolia* Jacq., *Z. horrida* Jacq., *Z. lanuginosa* Jacq., *Z. longifolia* Jacq.].

Carl Linnaeus (1707–1778) and Carl Linnaeus *filius* (1741–1783)

Carl Linnaeus (Carolus Linnæus, Carl von Linné, Fig. 3.13a) was born in Råshult, Småland, in southern Sweden, on May 23, 1707. He was the first of five children of Nils Ingemarsson (an amateur botanist and Lutheran pastor who later adopted the family name Linnaeus) and Christina Brodersonia.

Linnaeus was not regarded by his school teachers as a star pupil, being more interested in plants and nature than conventional subjects. Familiar with the existing botanical literature at the age of 17, he entered the Växjö Katedralskola to study courses designed for boys preparing for the priesthood. However, botanist, teacher, and state doctor Johan Rothman suggested that Linnaeus would be better suited to a different career and offered to tutor him in physiology and botany.

Fig. 3.13 Left: Carl Linnaeus against a Natural History background. Image reproduced from https://openclipart.org. Right: Carl Linnaeus the younger. (From a portrait by Jonas Forsslund)

In 1727, Linnaeus, aged 21, enrolled in Lund University in Skåne but, on the advice of Rothman, moved after one year to the more prestigious Uppsala University, where he met a new benefactor, Olof Celsius, Professor of Theology and an amateur botanist. Celsius received Linnaeus into his home and allowed him the use of his library, then one of the richest botanical libraries in Sweden.

In 1729, Linnaeus wrote a thesis on plant sexual reproduction, his *Praeludia sponsaliorum plantarum*. In this work, he drew some controversial parallels between sexuality in plants and in humans, writing:

> The flowers' leaves … serve as bridal beds which the Creator has so gloriously arranged, adorned with such noble bed curtains, and perfumed with so many soft scents that the bride-groom with his bride might there celebrate their nuptials with so much the greater solemnity.

Linnaeus' thesis drew the attention of Professor Olof Rudbeck, who then selected Linnaeus—a mere second-year student—to teach courses at the university. His lectures were popular, often attracting an audience of several hundred. As a senior student, Linnaeus began to question the then-entrenched Tournefort system of classification of plants and decided to create one of his own, based on floral sexual parts. He thus began work that would lead to his *Systema naturae* (Fig. 3.14a), *Genera plantarum* (Fig. 3.14b), and other volumes.

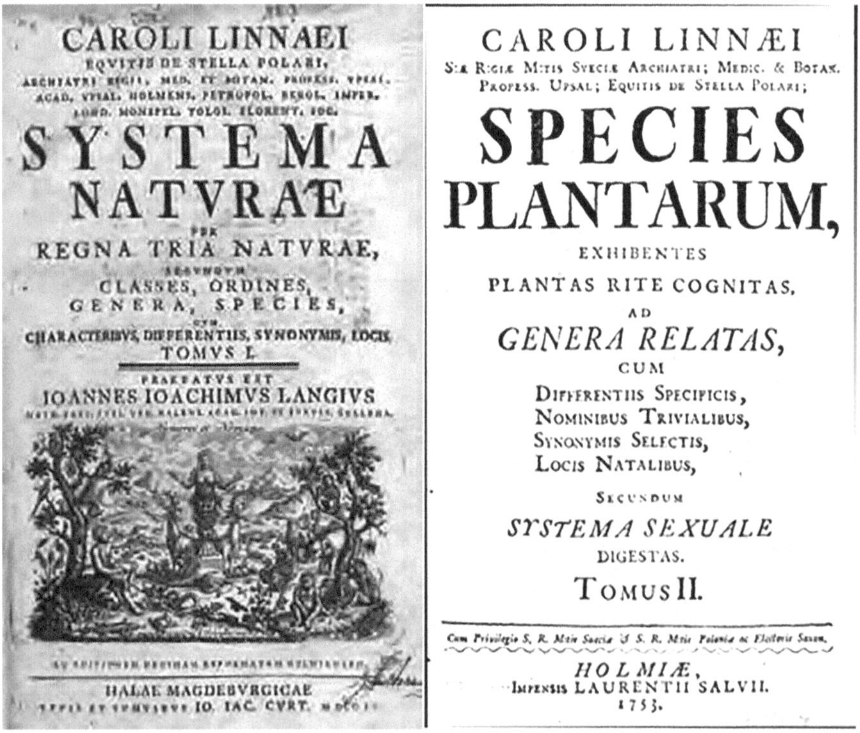
CAROLI LINNAEI
SYSTEMA NATVRAE
PER REGNA TRIA NATVRAE,
CLASSES, ORDINES, GENERA, SPECIES,
CVM CHARACTERIBVS, DIFFERENTIIS, SYNONYMIS, LOCIS.
TOMVS I.
IOANNES IOACHIMVS LANGIVS
HALAE MAGDEBVRGICAE
IO. IAC. CVRT.

CAROLI LINNÆI
S:æ R:giæ M:tis Sveciæ Archiatri; Medic. & Botan. Profess. Upsal; Equitis de Stella Polari;
SPECIES PLANTARUM,
EXHIBENTES
PLANTAS RITE COGNITAS,
AD
GENERA RELATAS,
CUM
DIFFERENTIIS SPECIFICIS,
NOMINIBUS TRIVIALIBUS,
SYNONYMIS SELECTIS,
LOCIS NATALIBUS,
SECUNDUM
SYSTEMA SEXUALE
DIGESTAS.
TOMUS II.
HOLMIÆ,
Impensis LAURENTII SALVII.
1753.

Fig. 3.14 Left: The title page of *Systema naturae* (first published in 1735) in which Linnaeus enumerated his binomial system of plant taxonomy based on the sexual parts of flowers. From Wikipedia. Right: The title page of Linnaeus' *Species plantarum* (1753) in which *Cycas circinalis* was the first cycad given an accepted botanical description. *Zamia pumila* followed in the second volume (1763)

In April 1732, the Royal Society of Sciences in Uppsala awarded Linnaeus a grant to explore Lapland, then the northernmost province of Sweden, for plants, animals, and possibly valuable minerals. During the six-month expedition, Linnaeus traveled over 2000 kilometers on foot and by horse and discovered about 100 previously unidentified plants. This became the basis of his *Flora Lapponicam.*

In April 1735, Linnaeus enrolled to study medicine at the University of Harderwijk in the Netherlands. Within two weeks, he had completed his oral and practical examinations, submitted a paper that he had prepared in advance on the subject of intermittent fevers, and was awarded a doctorate. During his stay in the Netherlands, Linnaeus met influential physicians and botanists such as Johan Gronovius, Herman Boerhaave, Johannes Burman, and George Clifford III, all of whom encouraged him in his research and publications. It was during this period when Linnaeus published his *Systema naturae*, the *Hortus Cliffortianus*, and several other texts. He also had the opportunity to visit England where he met Sir Hans Sloane, Phillip Miller, Johann Dillenius, and also Antoine de Jussieu in Paris.

On his return to Sweden, Linnaeus first took a position as a physician, helped initiate the Royal Swedish Academy of Sciences, and married Sara Moræa. The first of their children was Carl Linnaeus *filius*, the subject of the latter part of this text. In May 1741, Linnaeus was appointed to Professorship at Uppsala University, responsible for courses in medicine and natural history and for supervision of the Botanical Garden (Fig. 3.15). He continued teaching and writing but also found time to lead field trips to Öland, Gotland, Västergötland, and Scania.

Fig. 3.15 Founded in 1655 by Olof Rudbeck the Elder, the Linnaean Garden is one of two botanical gardens belonging to the University of Uppsala. The more modern garden, with some 7000 plant taxa, is situated near Uppsala Castle. (Photo: Andreas Trepte)

In 1750, Linnaeus became Rector of Uppsala University. He then relied on his senior students to travel and collect botanical samples; he called the best of these students his "apostles"; 17 such protégés included Carl Thunberg and Daniel Solander, both of whom collected data on cycads for him. In 1753 and 1763, he published his two-volume *Species plantarum*, now internationally accepted as the starting point of modern botanical nomenclature. The work contained 1200 pages and described over 7300 plant species including the cycads *Cycas circinalis* and *Zamia pumila*.

Linnaeus' last years were troubled by illness. He suffered from a disease called the Uppsala fever and sciatica and had endured several paralyzing strokes, the third of which led to his death on January 10, 1778. He was buried in Uppsala Cathedral.

Among Linnaeus' many accolades, he was given the title *archiater* or chief physician by King Adolf Frederick, elected a Member of the Academy of Sciences of Berlin and the American Philosophical Society, and dubbed a knight of the Order of the Polar Star, an ennoblement which entitled him to take the name Carl von Linné in 1761. A 100-kronor note was issued in his honor in 1986 (Fig. 3.16). The philosopher Jean-Jacques Rousseau said of him "I know no greater man on earth," while Goethe wrote, "With the exception of Shakespeare and Spinoza, I know no one among the no longer living who has influenced me more strongly." Swedish author August Strindberg added, "Linnaeus was in reality a poet who happened to become a naturalist." Linnaeus has been called *Princeps botanicorum* ("Prince of Botanists") and "The Pliny of the North."

Carl Linnaeus the Younger (Carolus Linnaeus the Younger, Carl von Linné d.y. or Linnaeus *filius,* Fig. 3.13b) was born on January 20, 1741.He was enrolled at the University of Uppsala at the age of nine and was taught science by his father's students. In 1763, aged just 22, he succeeded his father as the head of Medicine at Uppsala, and his promotion to professorship without taking exams or defending a thesis led to resentment among his colleagues.

Linnaeus the Younger's work was modest in comparison to that of his father. His best-known publication is the *Supplementum plantarum systematis vegetabilium* (1781–1782); it contains mainly botanical descriptions by the elder Linnaeus and his colleagues, as edited by the son. However, that work included descriptions of the cycads *Zamia cycadis* and *Z. debilis* and also led to his descriptions in Aiton's Hortus Kewensis (1789) of *Zamia integrifolia*, *Z. furfuracea*, and *Z. pungens*.

Between 1781 and 1783, Linnaeus the Younger took a two-year trip to England, France, the Netherlands, and Denmark but suffered from fever and a stroke and died shortly after his return to Sweden, on November 1, 1783, at age 42. He was entombed in the family grave at Uppsala Cathedral.

Linnaeus the Younger had inherited his father's extensive scientific collections of books, specimens, and correspondence. After his passing, his mother Sara sold the library and herbarium to the English botanist Sir James Smith, Founder of the Linnean Society of London, which ultimately acquired all the Linnean material.

Names of Cycads with Linnaeus and L.f.'s Authorship
Data from the *World List of Cycads*

Cycas circinalis L. (1753).
Zamia pumila L. (1763).
Zamia cycadis L.*f.* (1782) ≡ *Encephalartos affer* (Thunb.) Lehm.
Zamia debilis L.*f.* (1789 in Aiton) = Z.*pumila* L.
Zamia integrifolia L.*f.* (1789 in Aiton).
Zamia furfuracea L.*f.* (1789 in Aiton).
Zamia pungens L.*f.* (1789 in Aiton) ≡ *Encephalartos pungens* (L.*f.*) Lehm., *nomen dubium*.

Linnaeus' and L.f.'s Major Publications

Linnaeus, C. 1729. *Praeludia sponsaliorum plantarum*. Thesis, Uppsala University.

Linnaeus, C. 1735. *Systema naturae, sive regna tria naturae systematice proposita per classes, ordines, genera, & species*. See also Linnaeus 1758 below.

Linnaeus, C. 1737. *Flora Lapponica*. Salomonem Schouten, Amsterdam.

Linnaeus, C. 1737. *Hortus Cliffortianus*. Amstelaedami, Amsterdam.

Linnaeus, C. 1751. *Philosophia botanica: in qua explicantur fundamenta botanica* cum *definitionibus partium, exemplister minorum, observationibus rariorum, adiectis figuris aeneis*. Published simultaneously by R. Kiesewetter, Stockholm and Z. Chatelain, Amsterdam.

Linnaeus, C.1753. *Species plantarum: exhibentes plantas ritecognitas, ad genera relatas,* cum *differentiis specificis, nominibus trivialibus, synonymis selectis, locis natalibus, secundum systema sexuale digestas*. Laurentius Salvius, Stockholm. [*Cycas circinalis* L.] Second Edition 1763. [*Zamia pumila* L.].

Linnaeus, C. 1758. *Systema naturæ perregnatria naturæ, secundum classes, ordines, genera, species,* cum *characteribus, differentiis, synonymis, locis*. tenth edition, Laurentius Salvius, Stockholm.

José Celestino Bruno Mutis (1732–1808)

José Mutis (Figs. 3.17 and 3.18), in full José Celestino Bruno Mutis y Bosio, was born in Cádiz, Spanish Andalucia, on April 6, 1732. He studied medicine, physics, chemistry, and botany at Seville University, receiving his bachelor's degree in 1753 and a doctorate in medicine in 1757. He then became a physician to the royal household of King Ferdinand VI in Madrid while simultaneously teaching anatomy; extending his studies into astronomy, philosophy, and mathematics; and learning botanical skills at the Migas Calientes Botanical Gardens (now the Real Jardín Botánico de Madrid).

After three years, he emigrated to South America, taking an appointment in Bogotá in 1761 as a physician to Pedro Messía Corea de la Cerda, then Viceroy of New Grenada (present-day Colombia, Venezuela, Panama, and Ecuador). Mutis

Fig. 3.16 The 100-kronor note issued in 1986 to honor Linnaeus. Top: Carl von Linné with drawings from his garden in Uppsala. To the right of his portrait is his motto *Omnia mirari etiam tristissima* ("Find wonder in all things, even the most common place"). Below: The reverse side of the note depicts a bee harvesting pollen from a flower

immediately started exploring the local flora and commenced an investigation into the ethnobotanical use of the bark of *Cinchona*, a Rubiaceous tree in which the bark contains the alkaloid quinine and which was considered by the locals as a panacea for the treatment of all kinds of diseases.

In 1763, Mutis proposed to the new king, Charles III, that Spain sponsor an expedition to study the flora and fauna of New Grenada, but he had to wait nearly 20 years before the Bourbon king granted his consent! Meanwhile, Mutis continued a range of activities that embraced various mineralogical, medicinal, social, and economic projects. Deeply religious, he was ordained a priest in 1774. He kept up a regular correspondence with scientists in Europe and had a particularly strong interaction with Carl Linnaeus, who referred to Mutis as *Phytigorum Americanorum*

Fig. 3.17 José Celestino Bruno Mutis. Oil on canvas painting by R. Cristobal, ca. 1930. (From the Wellcome Collection)

Princeps ("Prince of American Botanists") and later named the genus *Mutisia* (Asteraceae) for him.

Shortly after the arrival of a new viceroy in 1782, Mutis was named botanist and astronomer of the *Real Expedición Botánica del Nuevo Reyno de Granada*, a task

Fig. 3.18 José Celestino Bruno Mutis. A statue in the grounds of the Department of Biology, National University of Colombia, Bogotá. (Photo: Favio González)

that was to occupy the next 25 years of his life (Fernández and Tamaro 2004). Together with a team of associated botanists, zoologists, and artists, he conducted the operation with meticulous thoroughness, keeping detailed notes and drawings on all observations and materials gathered. The expedition headquarters moved on three occasions. Initially, in 1782, it was based in the municipality of La Mesa in the Department of Cundinamarca, and then, in late 1783, it moved to Mariquita in the Department of Tolima. It was there that Mutis founded a botanical garden (which is still in existence and has a small library and museum with some memorabilia from Mutis' time). Finally, in 1791, the expedition operated from Bogotá. In 1801, the explorer Alexander von Humboldt and his botanist companion Aimé Bonpland stayed with Mutis for two months and were said to have greatly admired the work of the expedition.

The expedition statistics are astonishing: Between 1782 and 1808, the team covered some 8000 km^2 of territory; gathered more than 24,000 plant vouchers; prepared 8000 illustrative plates, numerous maps, notes, and manuscripts; and assembled collections of wood, shells, resins, minerals, and skins. Eight years after the conclusion of the expedition, the Spanish King ordered that all the materials were to be packed in crates and returned to Spain. Unfortunately, some of the collation between the field notes and the plates was lost during the passage to Madrid, and much of the work was archived without being edited and analyzed until very much later. In a similar vein, Mutis' grand work, *Flora de la Real Expedición Botánica del Nuevo Reino de Granada* (1783–1816), a multi-volume tome with 6000 illustrations, was not published because of a lack of funds. Then, in 1929, the Mutis materials were rediscovered in the archives of the Madrid Botanical Garden by American botanist Ellsworth Killip, prompting a resurgence of interest in the Royal Expedition. At the end of the Spanish Civil War and World War II, the governments of Spain and Colombia joined forces to publish the expedition's botanical descriptions and illustrations. The first volume of this work was published in 1953, and at present, 23 volumes, with more than 500 illustrations by nearly 40 artists along with scientific descriptions, have been printed.

During his exploration into the interior of Colombia, the Mutis expedition came across several cycads, two of which were illustrated as plates. The first was of a seed cone shown in its entirety and in a longitudinal section. Mutis described the seeds as being white in color in his illustration (Fig. 3.19), an uncommon feature for *Zamia.* It was over 200 years later that Stevenson (2001) described *Zamia encephalartoides* from a population near the Río Chicamocha in the Department of Santander, a species that matched Mutis' original description that probably was collected by Eloy Valenzuela. The second cycad illustration in the Mutis work was more enigmatic. It showed a seed cone partially exposed to show a reddish internal color, with features somewhat intermediate between those of a *Zamia* and an *Encephalartos*. No plant corresponding to that image has since been located, and the challenging possibility remains that this may represent an entirely novel cycad genus (Sabato 1990) or simply be an inaccurate representation. Although the veracity of the unknown cycad has been questioned by many, the fact that the illustration of *Z. encephalartoides* is so accurate makes one believe that the image of the unknown cycad is also accurate. Moreover, this has been true for all the other plant illustrations in the Mutis volumes (Fig. 3.20).

Fig. 3.19 The illustration of the *Zamia encephalartoides* seed cone prepared by José Mutis during the Royal Spanish Expedition to New Grenada

Mutis died in Bogotá on September 2, 1808, at age 76, a victim of apoplexy. Although much of his work remains lost or as yet unpublished, he is known to history as an extraordinary explorer and leader and one of the most influential authors of the Spanish Universalist School of the eighteenth century. He is commemorated on the Colombian 200 pesos banknote of 1983–1992 and also the Spanish 2000 peseta banknote of 1992–2002 (Fig. 3.21).

An important acknowledgment to Mutis' botanical work is the 40-acre *Jardín Botánico de Bogotá José Celestino Mutis*, established in 1955 by Enrique Arbeláez and Teresa Bueno. Located near the University in central Bogotá, the garden serves both as a place of recreation and a research center, with an emphasis on Andean and

Fig. 3.20 Another of Mutis' plates is of a seed cone partially exposed to reveal a reddish internal color, with features somewhat intermediate between those of a *Zamia* and an *Encephalartos*. No plant corresponding to this has since been located, and the challenging possibility remains that this may represent an entirely novel cycad genus

Páramo ecosystems, and a fine collection of native orchids. In addition to his work in biology, Mutis was also a knowledgeable student of mathematics, astronomy, and physics; the José Celestino Mutis Award is a recently instituted biennial prize presented by the Colombian Mathematical Society to teachers of mathematics who

Fig. 3.21 Mutis' image on the Colombian 200 pesos banknote of 1983–1992 and also the Spanish 2000 peseta banknote of 1992–2002; on the reverse of the latter is a representation of a *Mutisia clematis* flower, a genus named by Linnaeus in his honor

have excelled in their professional work. Mutis is also remembered for founding an astronomical observatory in Bogotá. The town of Puerto Mutis on the Río San Juan in the Department of Chocó is named in his honor as well as over 100 specific epithets.

Mutis's Major Publication

Mutis, J. C. 1783–1816. *Flora de la Real Expedición Botánica del Nuevo Reino de Granada.* Volumes published sequentially since 1953 by Ediciones Cultural Hispánica, Madrid.

Carl Peter Thunberg (1743–1828)

Carl Peter Thunberg (Karl Peter von Thunberg, Carl Pehr Thunberg, or Carl Per Thunberg, Fig. 3.22) was born on November 11, 1743, in the city of Jönköping in southern Sweden. His father, Johan, a local shopkeeper, passed away when Carl was seven, leaving Carl's mother, Margaretha, to take over the family shop. After attending the local schools, he entered Uppsala University, enrolling for courses in theology, philosophy, chemistry, medicine, and natural history. His enthusiasm for the latter subject brought him to the attention of Carl Linnaeus, then Rector of the university during a period when natural sciences were esteemed. Thunberg graduated with a Bachelor of Medicine in 1767, after 6 years of study. He was awarded a Doctorate in Medicine in absentia some years later.

As one of Linnaeus' seventeen "apostles," Thunberg applied for a scholarship to further his studies in Amsterdam and Paris. In Amsterdam, he met and worked with the influential Dutch botanists Johannes and Nicolaas Burman and also met Leiden

Fig. 3.22 Portrait of Carl Peter Thunberg by Jacob Ek, ca.1820–1825. From the portrait collection of the Swedish Academy of Sciences

councilor Laurens Gronovius. With their sponsorship, Thunberg joined the Dutch East India Company (the VOC) in December 1771 and set sail as surgeon on board the vessel *Schoonzicht* to the Cape of Good Hope—and later onwards to Japan and other countries in the Orient.

During his three-year stay in the Cape, Thunberg mastered the Dutch language, studied the culture of the native *khoikhoi* (Hottentot) people, and set about exploring the natural history of the area. Between 1772 and 1775, he made three trips by ox-wagon into the eastern and central-eastern parts of what was to become the Cape Province, each time collecting numerous flora and fauna specimens, many of which were then unknown to science. On the latter two expeditions, he was accompanied by Francis Masson, who had come to the Cape to collect plants for the Royal Botanic Gardens at Kew (Gunn and Codd 1981).

Thunberg encountered the cycads that we now know as *Encephalartos longifolius* in the Langkloof area and *E. afer* near present-day Humansdorp during the first of his trips, in 1772. At the time, he apparently considered them adults and juveniles of one species, which he named *Cycas afra* in 1775. This name went through a confusing number of revisions until the two species were recognized as being distinct and when Lehmann placed them in the genus *Encephalartos* in 1834. Lehmann was certainly aware of Thunberg's "discovery," and his *en-cephal-artos* ("bread-in-head") name construction was based largely on the following extract from Thunberg's field notes, published in 1793:

> The Bread tree (*Zamia afra*) is a species of palm, which grows on the hills, below the mountains, in these tracts. It was of the height and thickness of a man at most, very much spread, and single. I have sometimes seen from one root, two or three stems spring. It is out of the pith (medulla) of this tree that the Hottentots contrive to prepare their bread.

There may be some doubt as to whether it was the *khoikhoi* (Hottentot) people who used the cycad material because it was the Xhosa (Bantu) folk who lived near the cycad localities. We also have reservations as to whether the material for making the crude bread was the pith from the cycad stems; the use and processing description cited by Thunberg is more consistent with it being the seeds that were harvested and detoxified.

It is interesting to note that, during his field inspections of the cycad(s) that he had named *Zamia afra* in his *Prodromus plantarum capensium* (Fig. 3.23), Thunberg (1784a) also collected insects that were closely associated with the cycad cones (Fig. 3.24), describing and illustrating the long-snouted weevil *Curculio zamiae* (now *Antliarhinus zamiae*). Thunberg (1784a) commented that its larvae feed solely in the seeds of the ovulate cones and marveled at the exceptionally long and thin rostrum of the female weevils, noting that it exceeded the body length several times, making it unique among the weevils.

Thunberg left the Cape in March 1775 for Japan, via Java, and arrived at the VOC administrative base on the small island of Deshima in Nagasaki Bay in August 1775. The ruling Japanese *shōgun*, Tokugawa Ieyasu, unimpressed with the earlier activities of Portuguese missionaries, was wary of foreigners, and Thunberg was at first rarely allowed to leave the island. However, he gradually gained credibility with the locals, largely by offering his medical expertise to help the local doctors cure syphilis, known in Japan as the "Dutch disease." As a result, he was granted more visits to Nagasaki and was eventually able to explore the surrounding area and even make a trip to the *shōgun*'s court in Edo (now Tokyo).

Once again, he collected numerous specimens of plants, animals, and Japanese artifacts to take back to Europe. Thunberg would have seen cycad specimens,

CLASSIS XIII.

POLYANDRIA.

Monogynia.

ZAMIA.

caffra. Z. foliis pinnatis: pinnis lanceolatis, petiolis inmeribus.

CAPPARIS.

triphylla. C. foliis ternatis.

capenſis. C. foliis oblongis mucronatis, floribus umbellatis, ſtipulis aculeatis.

PAPAVER.

aculeatum. P. capſulis oblongis glabris, caule multifloro aculeato folioſo, foliis pinnatifidis ſerratis.

NYMPHAEA.

capenſis. N. foliis ſubpeltatis cordatis orbiculatis integris, petalis lanceolatis calyci æqualibus.

CHRYSITRIX.

capenſis. C.

SPARRMANNIA.

africana. S.

GREWIA.

occidentalis. G. foliis ſubovatis.

MIMOSA.

nilotica. M. ſpinis ſtipularibus patentibus, foliis bipinnatis: partialibus extimis glandula interſtinctis, ſpicis globoſis pedunculatis.

caffra. M. aculeis binis oppoſitis, foliis bipinnatis: pinnulis lanceolatis glabris, ſpicis elongatis.

Fig. 3.23 The page in Thunberg's *Prodromus plantarum capensium*, Vol. 2, where he named *Zamia afra* in 1800

probably grown as ornamentals in temple gardens, and he sent at least two living specimens to Amsterdam. Noting the characteristic revolute leaflet margins, he named the species *Cycas revoluta* in 1782 after his return to Europe.

In November 1776, he left for Java and then traveled to Ceylon (now Sri Lanka), and after a stopover at the Cape of Good Hope and a visit to London, where he met Joseph Banks at the British Museum, he ultimately reached Sweden in 1779. After

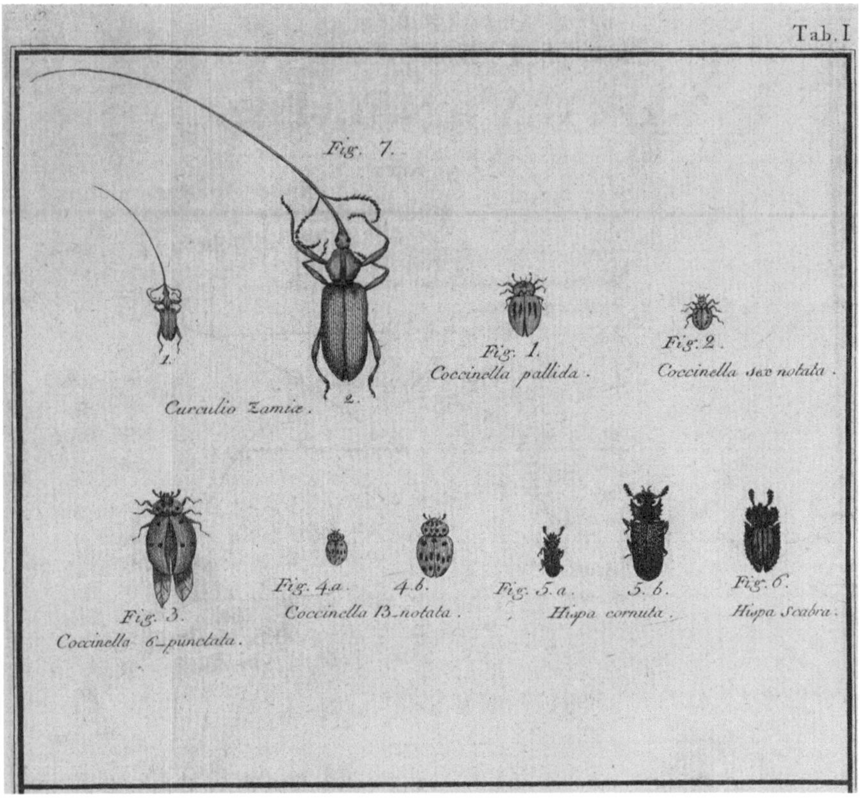

Fig. 3.24 Drawings of selected insects that Thunberg found in the Cape area of South Africa, including *Curculio zamiae* (Fig. 7, later *Anthliarhinus zamiae*) that he observed in the seed-bearing cones of his *Zamia afra* (probably *Encephalartos longifolius*). (From Thunberg 1784a)

his arrival in Sweden, he rejoined the University of Uppsala as a Professor of Medicine and Natural Philosophy.

There he continued working on the numerous monographs, which he published in the journals of various Swedish and international scientific societies. The three major works were his "Travels in Europe, Africa and Asia," the *Flora Capensis* (1801–1817), and the Thunberg (1784b). In addition to his academic and writing tasks, Thunberg established a new botanical garden in Uppsala with the support of King Gustav III, and he also founded a museum at the University to house his extensive herbarium and collection of insects, shells, birds, and other animals.

Thunberg was an Honorary Member of sixty-six scientific societies, an elected Member of the Royal Swedish Academy of Sciences and the American Philosophical Society, and a Correspondent and an Associate Member of the Royal Institute of the Netherlands (later the Royal Netherlands Academy of Arts and Sciences). As one of Linnaeus' seventeen "apostles," Thunberg stands out as one of the most productive scientists of his era, collecting and describing more plants and animals than all his contemporaries. The number of plants that he described has been estimated as nearly 1900; of these, some 1200 are still listed in *Index Kewensis* as valid names (Svedelius 1944). He has been called "Father of South African Botany," "Pioneer of Occidental Medicine in Japan," and the "Japanese Linnaeus." Other botanists referencing his name have created the genus *Thunbergia* (Acanthaceae) and the species *Allium thunbergii*, *Amaranthus thunbergii*, *Arisaema thunbergii*, *Berberis thunbergii*, *Fritillaria thunbergii*, *Geranium thunbergii*, *Lespedeza thunbergii*, *Pinus thunbergii*, and *Spiraea thunbergii*. Carl Thunberg died at Thunberg, near Uppsala, on August 8, 1828.

Names of Cycads with Thunberg's Authorship
Data from the *World List of Cycads*

Cycas afra Thunb. (1775) ≡ *Encephalartos afer* (Thunb.) Lehm.
Cycas revoluta Thunb. (1782).
Zamia afra (Thunb.) Thunb. (1800) ≡ *Encephalartos afer* (Thunb.) Lehm.

Thunberg's Cycad Publications

Thunberg, C. P. 1775.*Cycas afra, nova palmae species, descripta. Nova acta regiae societatis scientiarum Upsaliensis* 2: 285–290. [*Cycas afra* Thunb.]

Thunberg, C. P. 1782. Beschryving van twee nieuwe soorte van palmboomachtige gewassen, uit Japan en van de Kaap der Goede Hoope/Holland.*Verhandelingen van de Hollandse maatschappij der wetenschappen* 20(2): 419–433. [*Cycas revoluta* Thunb.].

Thunberg, C. P. 1784. *Curculio zamiae, nova et singularis insecti species e Capite Bonae Spei, descripta. Nova acta regiae societatis scientiarum Upsaliensis* 4: 29–30.

Thunberg, C. P. 1784. *Flora Japonica: sistens plantas insularum Japonicarum*. Leipzig.

Thunberg, C. P. 1788–1796. *Resa uti Europa, Africa, Asia förrättad åren 1770–1779*. Four volumes. English edition "Travels in Europe, Africa and Asia performed between the years 1770 and 1779" published by W. Richardson & J. Egerton, London, in 1793.

Thunberg, C. P. 1794–1800. *Prodromus plantarum capensium*. Vols1–2. Uppsala. [*Zamia caffra* (Thunb.) Thunb.].

Thunberg, C. P. 1807–1813. *Flora Capensis: Sistens plantas promontorii Bonae Spei Africes*. Vols 1 & 2. Uppsala. Complete edition by J. A. Schultes, 1823. Stuttgart.

Chapter 4
The Early Nineteenth Century

Adolphe Théodore Brongniart (1801–1876)

Adolphe Théodore Brongniart (Fig. 4.1) was born in Paris on January 14, 1801, son to geologist Alexandre Brongniart and grandson to architect Alexandre Théodore Brongniart. He was schooled by his father until entering the University of Paris where he graduated in medicine in 1816 and in botany three years later. In 1833, he succeeded René Desfontaines as Professor of Botany at the Paris Museum of Natural History, where he worked until his death in 1876.

Brongniart was a passionate researcher and a prolific writer. At the age of 21, he (1822) wrote a paper on the classification and distribution of fossil plants. That was the first of a great many such works (Brongniart 1828) culminating in his two-volume *Histoire des végétaux fossiles* (1828–1837), with 199 illustrations, which earned him the title of "Father of Paleobotany." That publication was of special interest to cycad biologists as Brongniart, supporting the view of Robert Brown, was the first to place the Cycadeae in the gymnosperms.

Apart from his palaeobotanical interests, Brongniart was also respected as a leading plant anatomist, morphologist, taxonomist, and biogeographer, with a large and eclectic range of publications to his credit. His systematic work led to papers on algae, fungi, mosses, and lycopods through to many families of higher plants. In terms of biogeography, he was involved in writing up the botanical section of the multi-volume report on Louis Duperrey's 1822–1825 circumnavigation of the world in the vessel "La Coquille," a joint collaboration with the expedition's botanist, Jules d'Urville. The expedition returned to France with more than 3000 plant specimens, about 400 of which were new to science, many of which were deposited into the herbarium of the Natural History Museum in Paris under Brongniart's supervision.

R. Osborne, D. Wm. Stevenson, *The Cycad Botanists*, Memoirs of The New York Botanical Garden 123, https://doi.org/10.1007/978-3-031-79082-9_4

Fig. 4.1 Adolphe-Théodore Brongniart (1801–1876). (From J.-B. Dumas 1881)

Brongniart's *Énumération des genres de plantes cultivées au muséum d'histoire naturelle de Paris* of 1843 represented a novel classification system later modified by the German workers Alexander Braun, August Eichler, and Adolf Engler, which was not superseded until the development of molecular technologies.

Brongniart had a special interest in cycads, both fossils and extant representatives. His *Recherches sur l'organisation des tiges des cycadées* (1829) and *Recherches sur la structure et les fonctions des feuilles* (1830) paved the way for the detailed text, and numerous impressive illustrations (e.g., Fig. 4.2) in his *Description des grains fossiles silicifiées du terrain Houiller de Saint-Étienne* (1881) are testimony to Brongniart's rigour. The micropyle and pollen chamber arrangement seen in the fossil material from La Grand-Croix (near St. Étienne, France) relates very clearly to the structures in *Note sur un noveau genre de cycadées du Mexique* (1846) where he named the genus *Ceratozamia*.

Brongniart's (1846) description of *Ceratozamia mexicana* (Figs. 4.3 and 4.4) was based on material growing in the garden of the Natural History Museum in Paris. That material had been gathered in Veracruz, Mexico, by the Belgian plant and animal collector Auguste Ghiesbrecht who spent many years collecting specimens (Schätti et al. 2018).

Possibly because of the influence of his father, Brongniart also had a remarkable insight into the evolution of our terrestrial climate. He realized that there was a correlation between the proliferation of plants and the emergence of terrestrial vertebrates; as plants increased and diversified, they removed carbon dioxide from the global atmosphere and produced oxygen, thus providing an environment to support the first air-breathing reptiles and mammals.

Brongniart founded the journal *Annales des sciences naturelles* in his early 20 s. He was the first President of the *Société Botanique de France*, a Member of the *Académie des Sciences* of Paris, and a Foreign Member of the Royal Swedish Academy of Sciences. Survived by his wife Agathe Françoise Boitel and two sons, he passed away in Paris on February 18, 1876, and was buried in the Père Lachaise cemetery. His name is commemorated in the cycad initially called *Zamia brongniartii* (Fig. 4.5), a name given by Weddell (1849) but revised to *Z. boliviana* by Alphonse de Candolle (1868) as a new combination of the name published earlier in 1846 by Brongniart as *Ceratozamia boliviana*. Brongniart is also commemorated by the names of such diverse taxa as *Adolphia* (Rhamnaceae), *Brongniartia* (Fabaceae), the palm *Brongniartikentia*, the marine alga *Brongniartella*, and a Russian fossil genus named *Brongniartites*.

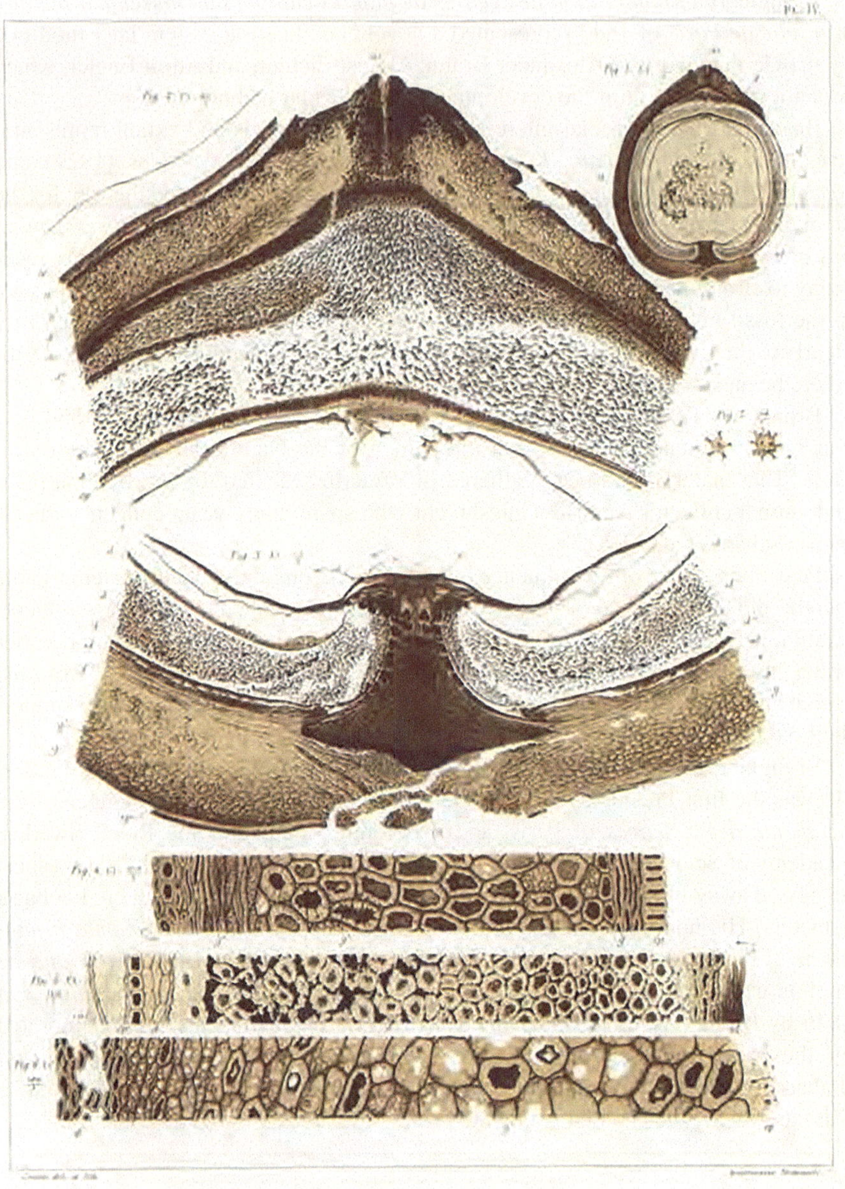

Fig. 4.2 Drawings of sections of a fossil cycad seed named *Cyclocarpus nummularis*, found near Saint-Étienne in France. Plate 4 from Brongniart (1881). The micropyle and pollen chamber arrangement seen in the fossil material relates very clearly to the structures in extant cycads

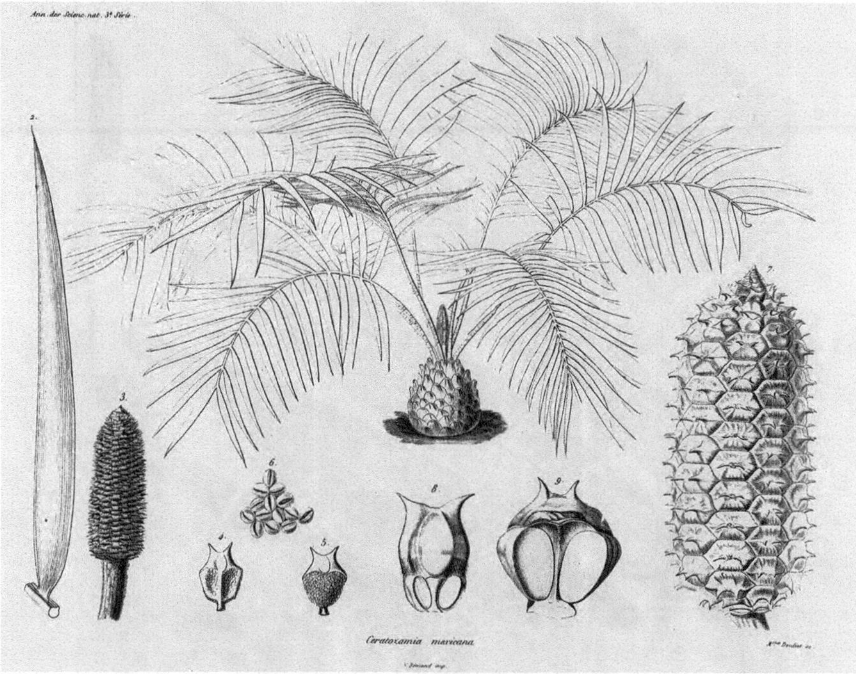

Fig. 4.3 The illustration of *Ceratozamia mexicana* published by Brongniart when he named the genus in 1846

Fig. 4.4 The 1845 holotype for *Ceratozamia mexicana* from the herbarium of the Museum of Natural History in Paris. The specimen was taken from a plant collected by Auguste Ghiesbrecht in Veracruz, Mexico, and cultivated in Paris

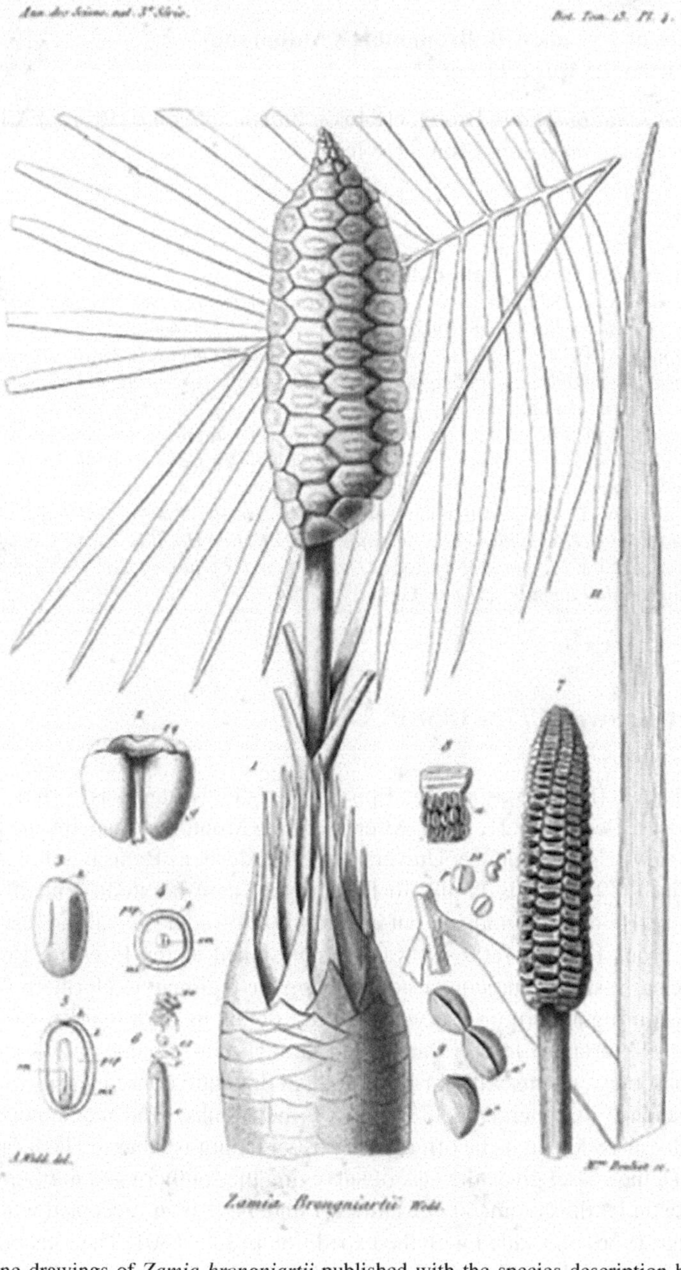

Fig. 4.5 Line drawings of *Zamia brongniartii* published with the species description by Hugh Weddell in 1849. The name was revised to *Z. boliviana* by Alphonse de Candolle in 1868

Names of Cycads with Brongniart's Authorship
Data from the *World List of Cycads*

Ceratozamia boliviana Brong. (1846) ≡ *Zamia boliviana* (Brong.) A.DC.
Ceratozamia mexicana Brong. (1846).

Brongniart's Cycad Publications

Brongniart, A. T. 1829. *Recherches sur l'organisation des tiges des cycadées. Annales des sciences naturelles* 16: 389–402.

Brongniart, A. T. 1846. *Note sur un nouveau genre de cycadées du Mexique. Annales des sciences naturelles, botanique, série* 3, 5: 5–9, plate 1. [*Ceratozamia boliviana* Brong., *C. mexicana* Brong.].

Brongniart, A. T. 1875. *Sur la structure de l'ovule et de la graine des cycadées, comparée à celle de diverses graines fossiles du terrain houiller. Comptes rendus de l'académie des sciences* 81: 305–307.

Brongniart, A. T. 1881, posthumous. *Structure de l'ovule et de la graine des cycadées, comparée à celle de diverses graines fossiles du terrain Houiller.* Chap. 3, pp. 31–33 in M. Renault & M. Grand'Eury, (eds), *Description des graines fossiles silicifiées du terrain Houiller de Saint-Étienne.* G. Masson, Paris.

Robert Brown (1773–1858)

Robert Brown (Fig. 4.6), son of an Episcopalian clergyman, was born in Montrose, Scotland, on December 21, 1773. After attending Montrose Academy, he proceeded in 1787 to Marischal College, University of Aberdeen, as Ramsay scholar, and then moved in 1789 with his family to Edinburgh where he studied medicine at the University. He did not graduate but apparently showed a special interest in natural history. From 1795 to 1800, he was commissioned to the Fifeshire Regiment of Fencibles as assistant surgeon, accompanying the regiment to Northern Ireland.

It was during a recruiting service visit to London in 1798 that Brown met Jonas Dryander, librarian to Joseph Banks, and through him became an Associate of the Linnean Society, events with far-reaching and dramatic consequences to his career and to science in general. In 1799, Brown met Banks, who recommended to the Admiralty in 1800 that he be offered the post of naturalist aboard HMS *Investigator* (Fig. 4.7), the vessel given the task of surveying the southern and northern coasts of Australia under the command of Matthew Flinders. Brown "accepted with alacrity" (Burbidge 1966) and sailed with the expedition in July 1801. The total compliment of 88 on the vessel also included botanical artist Ferdinand Bauer, landscape artist William Westall, horticulturist Peter Good, astronomer John Crosley (who fell ill en route and was replaced by James Inman), Miner John Allen, and their four servants.

The vessel made brief stops at Bugio Islands and at Madeira and then anchored at the Cape of Good Hope for a little over two weeks, during which Brown climbed

Fig. 4.6 Robert Brown (1773–1858), explorer, naturalist, and botanist of the British Museum and President of the Linnean Society, at the age of 50. (From a painting displayed in the Linnean Society lecture room)

Table Mountain at least twice and collected and named two species of *Serruria*: *S. foeniculacea* and *S. flagellaris*.

In December 1801, HMS *Investigator* reached Cape Leeuwin in Western Australia, an area of great floral richness and diversity. For the next two and a half years, Brown, Bauer, Good, and Westall inspected, collected, and drew natural history specimens as the vessel circumnavigated Australia on a counter-clockwise course—until the vessel's leaks and the deteriorating state of the crew forced a return to Sydney. While Flinders sailed home to procure another ship, Brown and Bauer (and possibly also Good and Westall) remained in the colony and continued explorations of Port Phillip and the Kent Group islands in Bass Strait and then further south to Tasmania (then Van Diemen's Land), spending six months in the

Fig. 4.7 Landscape artist William Westall's painting of HMS *Investigator* anchored in Lucky Bay on the southern coast of Western Australia in December 1801. The coning cycad in the foreground is a representation of *Macrozamia dyeri*. (From Taylor 2018)

vicinity of Hobart, visiting Mount Wellington and the Derwent and Huon River valleys. In New South Wales, again during 1804, Brown and Bauer collected further in the Sydney area and explored the Hunter River before returning to England in May 1805 on a refitted HMS *Investigator* under the command of William Kent.

Although much of the material gathered along the southern coast was lost in the wreck of HMS *Porpoise* in August 1803, Brown's work nevertheless resulted in a collection of between 3500 and 3900 plants, more than 2000 of which were previously unknown to science, as well as numerous zoological and geological specimens. As the material required much time in classifying on Brown's return, Banks arranged for him to be appointed to the post of clerk, librarian, and housekeeper to the Linnean Society.

This posting and his collections allowed him to publish his major work in 1810, the first flora of Australia (then New Holland), *Prodromus florae Novae Hollandiae et insulae Van Diemen*. The fame of Brown's *Prodromus* rests both on its quality and on its support for the "natural system" of classification of Jussieu (as against rigid Linnaean practice); this approach revitalizes botanical science. Brown (1814) also contributed an Appendix *General remark, geographical and systematical, on the botany of Terra Australis* in Flinders' account of the voyage of the *Investigator*. His analytical treatment in this paper led to him being regarded as a pioneer in the science of plant geography.

Fig. 4.8 The microscope used by Robert Brown in his study of plant cells. His observations of the constant random movement of small particles such as pollen grains became known as Brownian motion. The microscope, made by Bancks and Son of London, is a favored piece of the Linnean Society's memorabilia

When Banks died in 1820, Brown inherited an annuity of £200, a lease of the house in Soho Square, and a life interest in Banks' magnificent collections, which were subsequently to pass to the Trustees of the British Museum, but under Brown's direction as Keeper of their newly formed botanical department. Throughout this time until his passing in 1858, Brown maintained close relations with the Linnean Society, assisting them financially and administratively as fellow, vice president, and president.

During this time, Brown vigorously continued his own research projects. He resolved taxonomic matters with respect to families such as Asclepiadaceae, Compositae, Dasypogonaceae, Gesneriaceae, Orchidaceae, and Proteaceae. His eclectic work also contributed substantially to the understanding of plant morphology, embryology, and palynology. He was the first to recognize the nucleus as a constant feature of plant cells (Brown 1828), the plurality of embryos in conifers (Brown 1844), and the fundamental differences between angiosperms and gymnosperms, the latter work revolutionizing plant systematics (Tomlinson 2012). Brown was an acute observer and the epitome of a scientist. Complimenting his botanical work, Brown is also known to physicists and mathematicians as the first person to describe what is known as "Brownian motion," the random motion of small particles, observed microscopically (Fig. 4.8) in living cells but demonstrated as a physical phenomenon (Brown 1828).

As an early botanist studying plant anatomy and development, Brown had a profound influence on Matthias Jakob Schleiden both personally and because of Brown's discovery of the plant cell nuclei in 1831. More so when Brown visited Berlin in 1836, he encouraged the young Schleiden's interest in microscopy and anatomy (Mendelson 2024). This eventually led to Schleiden (1839) and Schwann (1839) developing their landmark concept of cell theory in plants and animals.

Brown's team certainly saw cycads of various kinds during the voyage on HMS *Investigator*, although references to them are limited. Landscape artist William Westall captured a coning cycad in his painting of Lucky Bay (near present-day Esperance) (Taylor 2018), and we can assume this is a representation of *Macrozamia dyeri*. Westall also has a cycad in his painting of the Bay of Pines (Port Bowen), and we speculate that this may represent *Macrozamia miquelii*.

Of greater botanical relevance is Brown's description of the first two Australian *Cycas* species; there are two Latinized descriptions on page 348 of his *Prodromus*. The first of these refers to their visit to the Cumberland Island group, some 60 km north-east of Mackay on the Queensland central coast, where on Calder Island, Brown collected material he later named *Cycas media*. The date of that collection is thought to be October 16, 1802 (Hill 1992), and the type (Fig. 4.9) is filed in the British Museum (*BM 0000630122*) with an isotype at Kew. An excellent color illustration of the habit and pollen-bearing structures was produced by Ferdinand Bauer (Fig. 4.10). Why Brown coined the epithet is uncertain; Hill and Osborne (2001) speculated that it refers to the taxon as being intermediate in morphology between *C. revoluta* and *C. circinalis*, the only other *Cycas* species known at the time.

Three months further into the counterclockwise voyage circumnavigating Australia, Brown collected samples (Fig. 4.11) of a second cycad that he named *Cycas angulata*. The epithet is from the word *angulatus*, angled, and refers to the sharp V insertion of the leaflets on the rachis. Although Brown's herbarium sheets do not record the detailed position for the type, it is believed that the material was collected in December 1802 from one of the two Bountiful Islands in the Gulf of Carpentaria, thus technically within the State of Queensland (Osborne et al. 2019). The holotype (BM *0000630115*) is filed in the British Museum (BM) with an isotype at Kew. An older annotation of *Cycas media* on these sheets was corrected by John Maconochie to *C. angulata*, but similar confusion applies to some of the early drawings of these two species.

A third *Cycas* involving Brown is an enigmatic species from Madagascar, incorrectly named in 1804 as *Cycas circinalis* by Louis Marie du Petit-Thouars. Brown studied vouchers at Kew and proposed the name *Cycas thuarsii* [sic] but did not provide a description. *Cycas thouarsii* R.Br. *ex* Gaudich. was later described in 1829 by French explorer and botanist Charles Gaudichaud-Beaupré in Louis de Freycinet's account of the voyage of the S.M. *Uranie*; he provided a minimal but valid description and attributed the name to Robert Brown (Hill 1998).

Brown was elected a Fellow of the Royal Society in 1810 and a Fellow of the Linnean Society in 1822 (serving as President from 1849 through 1853). In 1822, he also became a Foreign Member of the Royal Swedish Academy of Sciences. In 1827, he was a Correspondent of the Royal Institute of the Netherlands (later the

Fig. 4.9 The type for *Cycas media* as collected by Robert Brown on Calder Island, off the coast of Queensland, during Matthew Flinders' circumnavigation of Australia. The date of that collection is thought to be October 16, 1802 (Hill 1992), and the voucher is filed in the British Museum (BM *0000630122*)

Fig. 4.10 The elegant and accurate painting of *Cycas media* as drawn by botanical artist Ferdinand Bauer who accompanied Brown on Matthew Flinders' circumnavigation of Australia. (From Plate 4 in Mabberley 1985)

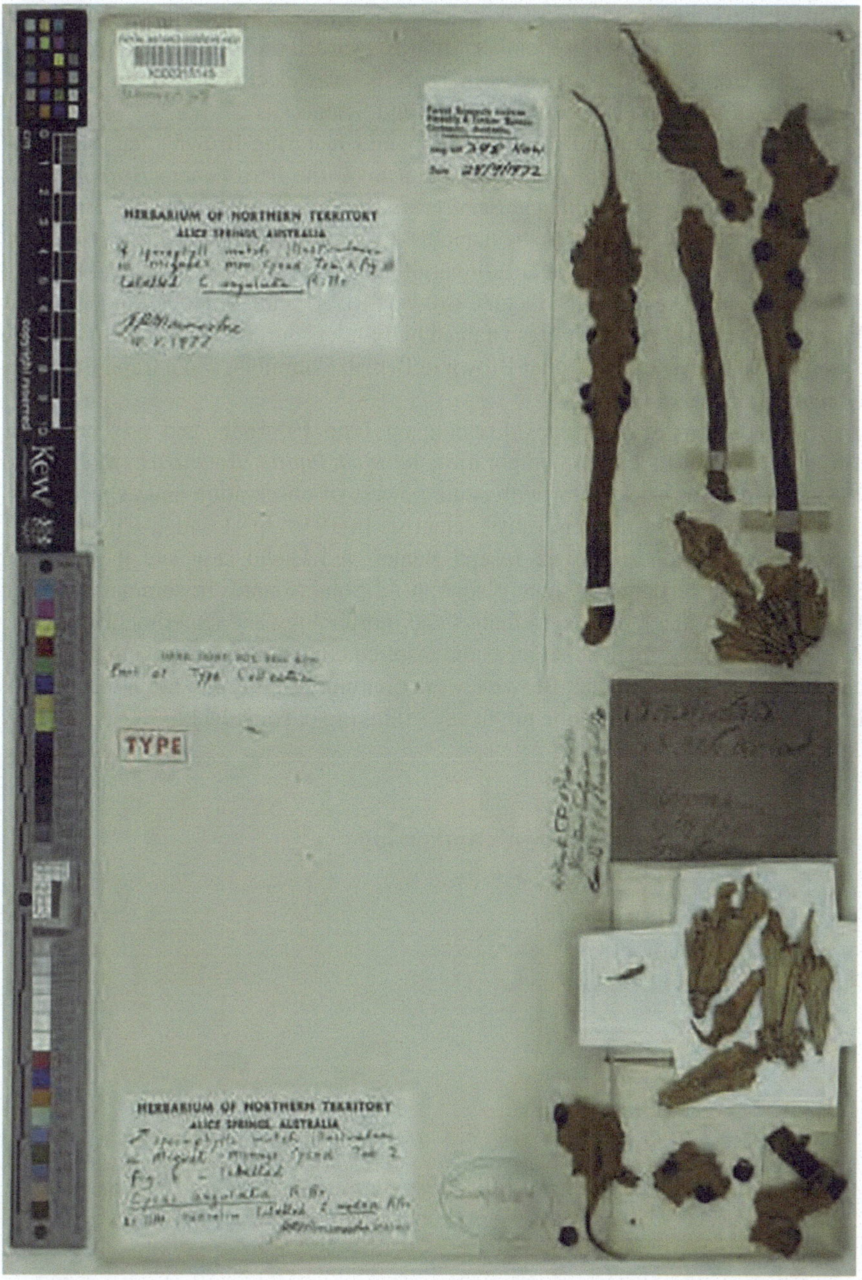

Fig. 4.11 Part of the Kew isotype of *Cycas angulata* of Brown's collection in December 1802 from one of the two Bountiful Islands in the Gulf of Carpentaria. The holotype (BM *0000630115*) is filed in the British Museum. An older annotation of *Cycas media* on these sheets was corrected by John Maconochie to *C. angulata*, but similar confusion applies to some of the early drawings of these two species

Royal Netherlands Academy of Arts and Sciences), subsequently becoming an Associate and then a Foreign Member. He was awarded the Royal Society's highest honor, the Copley Medal, in 1839. In 1848, he was elected a Foreign Honorary Member of the American Academy of Arts and Sciences.

Recognition in plant and geographical names featuring Brown's name is commemorated in a number of Australian plant taxa: the herb genera *Brunonia* and *Brunoniella*, *Banksia brownii*, and *Eucalyptus brownii*. The small New Zealand tree *Pisonia brunoniana* also takes his name as does the moss *Tetrodontium brownianum*, a species he discovered near Edinburgh while a student. Geographical features associated with Brown's name include Brown's River, south of Hobart in Tasmania, Mount Brown and Point Brown (so named by Flinders during the *Investigator* expedition) in South Australia, Mount Brown in British Columbia, and a Cape Brown in Greenland.

Brown died in Soho Square, London, on June 10, 1858, and was buried at Turnham Green. To his contemporaries, he was *Jupiter Botanicus* (Mabberley, 1985). He was an investigator with acute powers of observation and experimentation, the hallmark of the innovative scientist. In 1938, the London City Council commemorated Robert Brown, Joseph Banks, and David Don and the Linnean Society, with a rectangular stone plaque at 32 Soho Square. In summary, Robert Brown revolutionized plant systematics and morphology and experimental botany. As one of Britain's greatest botanists, he inspired contemporaries, young botanists, and future generations alike. Brown's work continues to this day as an example of innovative thinking and, as one might say, "thinking outside the box."

Names of Cycads with Brown's Authorship
Data from the *World List of Cycads*

Cycas angulata R.Br. (1810).
Cycas media R.Br. (1810).
Cycas thouarsii R.Br. *ex* Gaudich. (1826).

Brown's Cycad Publications

Brown, R. 1810. *Prodromus florae Novae Hollandiae et insulae Van- Diemen.* 1830. *Supplementum primum prodromiflorae Novae Hollandiae.* Richard Taylor, London. Reprinted in 1960 by J. Cramer, Weinheim, Germany. [*Cycas angulata* R.Br., *C. media* R.Br.]

Brown, R. 1814. General remarks, geographical and systematical, on the botany of Terra Australis. In: M. Flinders, Voyage to Terra Australis. Two vols, G. & W. Nicol, London.

Johann Georg Christian Lehmann (1792–1860)

Johann Georg Christian Lehmann (Fig. 4.12), son of Pastor Johann Gottlieb Lehmann and his wife Maria Elisabeth, was born in the town of Haselau in Schleswig-Holstein, Germany, on February 25, 1792. He had an elder brother, lawyer and naturalist Martin Christian Gottlieb Lehmann (1775–1856) and one other sibling.

Lehmann studied medicine in Copenhagen and Göttingen, graduating as a medical doctor in 1813, after which he obtained a Ph.D. from Jena University a year later. In February 1818, he was appointed Professor of Physics and Natural Sciences at the Academic Gymnasium in Hamburg, also taking on the role of Chief Librarian, a position he held for the rest of his life. Lehmann was an active member of some 26 learned organizations including the *Academy Leopoldina* (the Royal Society of Natural Historians); the Prussian Society (Berlin); *Academy Lipsiae* (Leipzig), where he held the post of Rector; and the Russian Academy of Sciences in Saint Petersburg.

In 1821, Lehmann established the Botanical Garden of Hamburg (*Alter Botanischer Garten*). He immediately began publishing notes on the garden and its collections (Lehmann 1821–1824). He also instituted an annual seed list from 1822–1840 as well as special lists for exchanges in 1830–1840 and 1849–1852. The garden and its structures were greatly damaged during WW II, and the present greenhouses (*Schaugewächshäuser* or *Tropengewächshäuser*) were erected only in 1963. Remarkably, at least one of the original plants brought to Hamburg by the Danish plant collector Christian Ecklon in 1832, a large *Encephalartos altensteinii*, remains an impressive specimen in the greenhouse complex.

One of the quite interesting aspects of Lehmann's early career is his mentoring of Matthias Jakob Schleiden (1804–1881). An interesting account of this is provided by Wallace Mendelson (2024) in a new biography of Schleiden entitled "From Despair to Discovery: The Botanical Odyssey of Matthias Jakob Schleiden and the Dawn of Cell Theory." Briefly, Schleiden was born and spent his youth in Hamburg. From 1821–1824, as a secondary school student, he studied with Lehmann, the new director of the Botanical Garden of Hamburg. Lehmann had an influence that resulted in Schleiden abandoning a career in law and becoming a botanist with further training in the center of botany at that time, the University of Berlin. Botanist Schleiden (1839) and zoologist Theodor Schwann (1839) first proposed a unified cell theory in 1838. The unified cell theory states that all living things are composed of one or more cells, the cell is the basic unit of life, and new cells arise from existing cells. This idea was a major conceptual advance in biology and led to renewed interest in the living processes that take place in cells. Thus, Schleiden (1839) and Schwann (1839) became the first to formulate what was then an informal belief as a principle of biology. Lehmann's lasting influence is that Schleiden developed his love for botany and his combative academic style from his mentor as well as Robert Brown.

Fig. 4.12 "D. Jo. Christian Lehmann." Image of a copper engraving, artist with initials "M.B.," housed in the Diocesan and County Library, Skara, Sweden

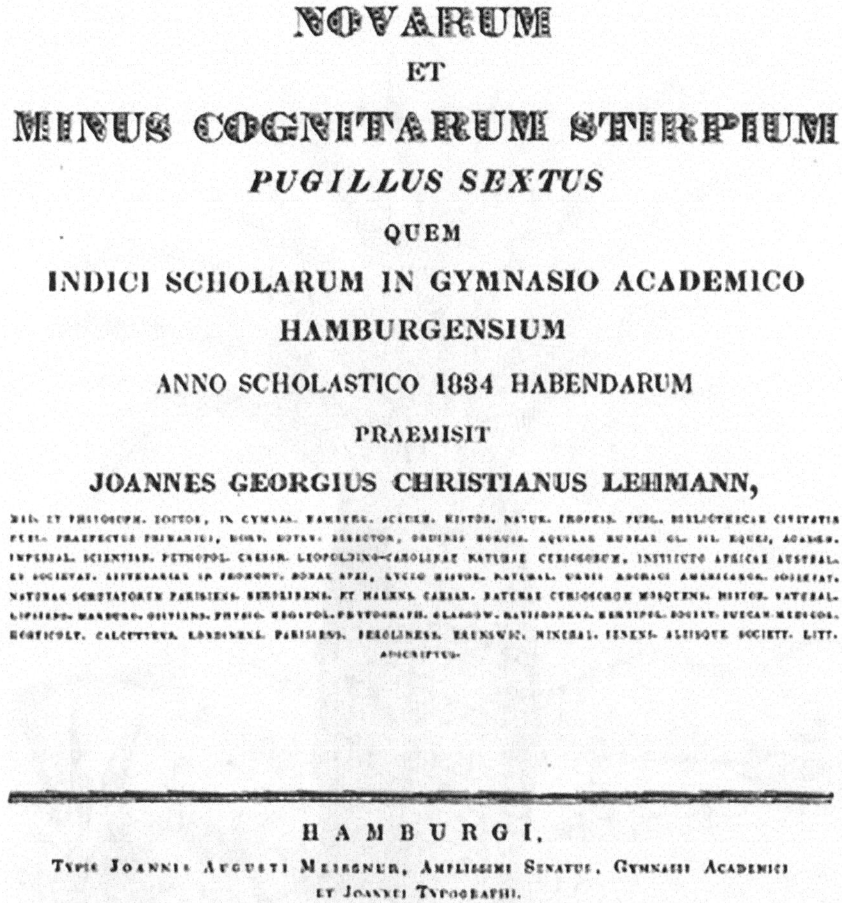

NOVARUM
ET
MINUS COGNITARUM STIRPIUM
PUGILLUS SEXTUS
QUEM
INDICI SCHOLARUM IN GYMNASIO ACADEMICO
HAMBURGENSIUM
ANNO SCHOLASTICO 1834 HABENDARUM
PRAEMISIT
JOANNES GEORGIUS CHRISTIANUS LEHMANN,
[illegible]

HAMBURGI,
Typis Joannis Augusti Meissner, Amplissimi Senatus, Gymnasii Academici et Joannei Typographi.

Fig. 4.13 The title page of Lehmann's *Pugillus* 6 (1834) in which he named the genus *Encephalartos*

Lehmann wrote widely on botanical matters over the period 1818–1856. His interests and publications prior to 1820 addressed such diverse groups as the family Boraginaceae; the genera *Nymphaea*, *Nicotiana*, *Potentilla*, and *Primula*; and the liverworts. However, his most extensive, multi-volume work was his *Novarum et minus cognitarum stirpium*, *Pugillus*, Vols. 1–10 (1828–1857). It was in Vol. 6, Part 1 (Fig. 4.13) that Lehmann (1834) devoted his attention to cycads in a section entitled *De plantis cycadeis praesertim Africae Australis*. Some copies of this work were illustrated with beautiful hand-painted color plates, but it seems that the only

Fig. 4.14 *Encephalartos friderici-guiliemi*. Color plate 1 from Part 1 of *Pugillus* 6 of 1834

extant copy is now under safekeeping at the New York Botanical Garden (Figs. 4.14, 4.15, 4.16, 4.17, and 4.18).

Lehmann was not the first botanist to write about cycads from the Southern Hemisphere, and he never visited South Africa or Australia; instead, he relied on information from Thunberg, Ecklon, Zeyher, Preiss, and other botanical collectors. Early and somewhat confused names were Thunberg's *Cycas afra* of 1775; Linnaeus' *Zamia cycadis* of 1782; Jacquin's *Zamia cycadifolia*, *Z. horrida*, *Z. lanuginosa*, and *Z. longifolia* of 1801; and Ecklon's *Zamia lehmanniana* of 1833. But it

Fig. 4.15 *Encephalartos friderici-guiliemi*. Color plate 2 from Part 1 of *Pugillus* 6 of 1834

was Lehmann who, in his 1834 *Pugillus,* reviewed and amended these names when he coined the genus name *Encephalartos*. He derived the name from the Greek phrase anglicized as "en-cephal-artos" meaning "in head, bread" and referring to the natives' habit of harvesting starch from the cycads in order to make a crude bread. The derivation is similarly related to the Afrikaans word *broodboom.* Lehmann described two new species in the *Pugillus*: *Encephalartos friderici-guilielmi* (in honor of King Friedrich Wilhelm III of Prussia) and *E. altensteinii* (to honor the King's statesman, Karl Altenstein). At the same time, he transferred

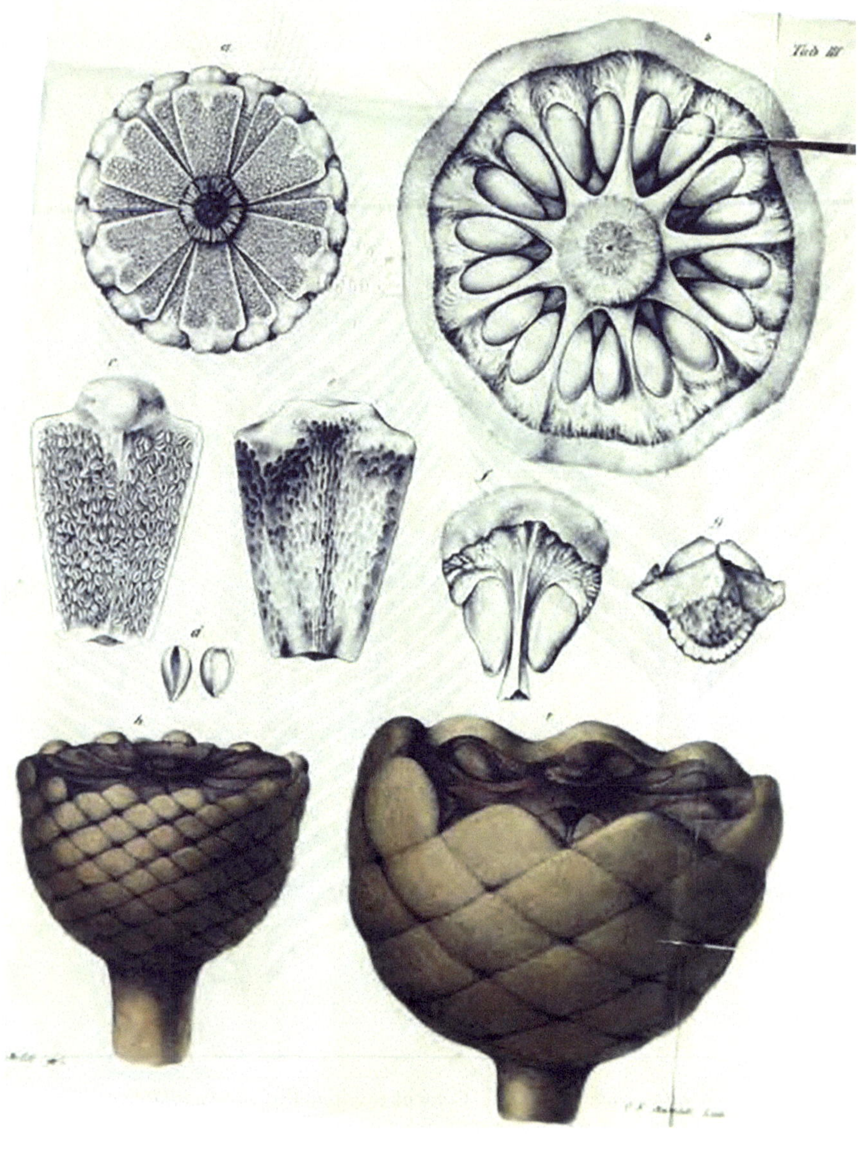

Fig. 4.16 *Encephalartos friderici-guiliemi*. Color plate 3 from Part 1 of *Pugillus* 6 of 1834

earlier names from *Cycas* and *Zamia* to *Encephalartos caffer*, *E. cycadifolius*, *E. horridus*, *E. lehmannii*, and *E. longifolius*.

When Lehmann erected the genus *Encephalartos*, he did so without indicating a type; it was Eicher (1926) who nominated *Encephalartos afer* (Thunb.) Lehm. as the type for the genus. Much later, Stevenson (1992) designated *E. friderici-guilielmi* Lehm. as the lectotype (Fig. 4.19) in view of the accurate illustrations in the *Pugillus*—although that recommendation was not supported by Vorster (2004).

Fig. 4.17 *Encephalartos altensteinii*. Color plate 4 from Part 1 of *Pugillus* 6 of 1834

Of relevance to the Australian readers, Lehmann's concept of *Encephalartos* also included taxa we now know as species of *Macrozamia* Lehmann (1844) and *Lepidozamia*. Lehmann collaborated with several other workers to publish *Plantae Preissianae* (1844–1847), an extensive compilation of the botanical work of Ludwig Preiss, a German-born British botanist and zoologist who explored and collected some 200,000 plant specimens in Western Australia during 1838–1842.

Some of Lehmann's colleagues seem to have found him difficult, perhaps quarrelsome, but he was clearly a person of letters, literature, learned societies, and

Fig. 4.18 *Encephalartos altensteinii*. Color plate 5 from Part 1 of *Pugillus* 6 of 1834

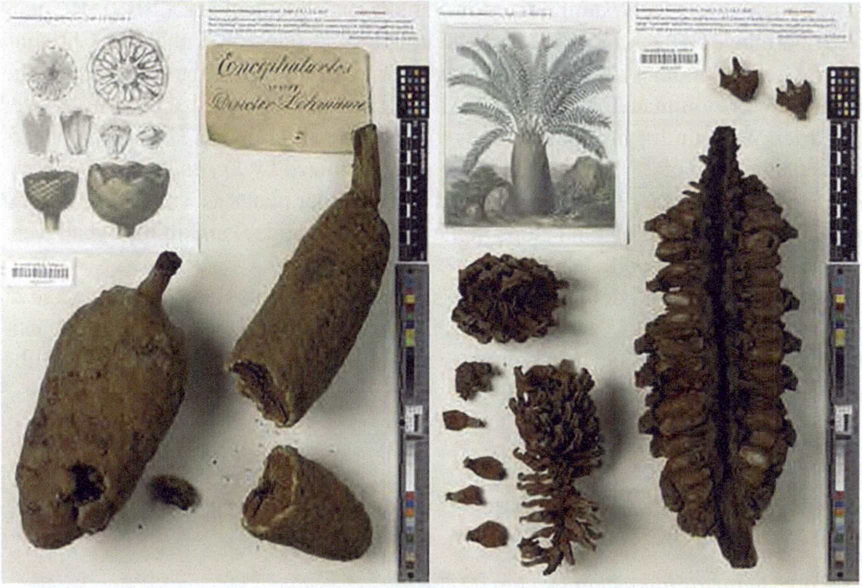

Fig. 4.19 Cone material of *Encephalartos friderici-guiliemi* (left) and *E. altensteinii* (right) collected by Ecklon and Zeyher in South Africa and stored by Lehmann in a leather-bound carton since 1832. These are still present in the Hamburg Herbarium and the images are reproduced by permission

Fig. 4.20 *Encephalartos lehmannii* on a steep granite cliff in the Suurberg, near Addo Elephant Park, Eastern Cape, South Africa. (Photo: Gerhard van Deventer)

science which is nicely summarized in Schultz and Poppendieck (2019). He married Dorothea Baltzer in 1824, and the couple had one son, Johannes Christian Eugen Lehmann, later a senator and mayoral candidate of Hamburg. Lehmann the botanist was nominated Knight of the Red Eagle Order, third Class, by Kaiser Friedrich Wilhelm for his contributions to medicine and science. His name appears in the authorship of 25 cycads. His name is eponymized by *Encephalartos lehmannii* (Fig. 4.20), *Eucalyptus lehmannii*, and several other plants as well as *Lehmannia* (a genus of air-breathing land slugs in the family Limacidae) and in the cycad weevil *Platymerus lehmannii*.

After a lengthy and debilitating illness, Lehmann died in Hamburg at the age of 68 on February 12, 1860. His herbarium material was distributed to botanical institutions in Stockholm and Melbourne, with the algae going to Hamburg and the mosses to Leiden, Stockholm, and Paris.

Names of Cycads with Lehmann's Authorship
Data from the *World List of Cycads*

Encephalartos altensteinii Lehm. (1834).
Encephalartos brachyphyllus Lehm. & de Vriese (1838) = *E. afer* (Thunb.) Lehm.
Encephalartos afer (Thunb.) Lehm. (1834).
Encephalartos cycadifolius (Jacq.) Lehm. (1834).
Encephalartos elongatus Lehm. (1838) = *E. lehmannii* Lehm.
Encephalartos friderici-guilielmi Lehm. (1834).
Encephalartos horridus (Jacq.) Lehm. (1834).
Encephalartos lanuginosus (Jacq.) Lehm. (1834) = *E. longifolius* (Jacq.) Lehm.
Encephalartos latifrons Lehm. (1834).
Encephalartos lehmannii Lehm. (1834).
Encephalartos longifolius (Jacq.) Lehm. (1834).
Encephalartos nanus Lehm. (1838) = *E. horridus* (Jacq.) Lehm.
Encephalartos pungens (L.*fil.*) Lehm. (1834), *nomen dubium.*
Encephalartos spinulosus Lehm. (1838), *nomen dubium.*
Encephalartos spiralis (Salisb.) Lehm. (1834) ≡ *Macrozamia spiralis* (Salisb.) Miq.
Encephalartos tridentatus (Willd.) Lehm. (1834) = *Macrozamia* sp.
Macrozamia preissii Lehm. (1844) = *M. riedlei* (Gaudich.) C.A.Gardner.

Lehmann's Cycad Publications

Lehmann, J. G. C. 1821–1824. *Icones et descriptions novarum et minus cognitarum stirpium.* Five parts of 10 plates each.

Lehmann, J. G. C. 1822–1840. *Semina in Horto Botanico Hamburgensi.*

Lehmann, J. G. C. 1830–1840, 1849–1852. *Delectus seminum quae in Horto Hamburgensium Botanico e collectione anni 1834 mutuae commutation.*

Lehmann, J. G. C.1834. *De plantis cycadeis praesertim Africae Australis. Novarum et minus cognitarum stirpium. Pugillus* 6: 1–66, Tables 1–5. Meissner, Hamburg. [*Encephalartos altensteinii* Lehm., *E. afer* (Thunb.) Lehm., *E. cycadifolius* (Jacq.) Lehm., *E. friderici-guilielmi* Lehm., *E. horridus* (Jacq.) Lehm., *E. latifrons* Lehm., *E. lehmannii* Lehm., *E. lanuginosus* (Jacq.) Lehm., *E. longifolius* (Jacq.) Lehm., *E. pungens* (L.*fil.*) Lehm., *E. spiralis* (Salisb.) Lehm., *E. tridentatus* (Willd.) Lehm.].

Lehmann, J. G. C. 1838. *Tijdschrift voor natuurlijke geschiedenis en physiologie* 4(4): 414–425. [*Encephalartos brachyphyllus* Lehm. & de Vriese, *E. elongatus* Lehm., *E. nanus* Lehm., *E. spinulosus* Lehm.].

Lehmann, J. G. C. 1844. *Novarum et minus cognitarum stirpium. Pugillus* 8: 31. Meissner, Hamburg. [*Macrozamia preissii* Lehm.].

Lehmann, J. G. C. 1844–1847. *Plantae preissianae sive enumeratio plantarum quas in Australasia occidental et meridionale-occidentali annis 1838–1841 collegit Ludovicus Preiss.*

Lehmann, J. G. C. 1851–1855. *Index seminum in Horto Botanico Hamburgensi collectorum.*

Friedrich Anton Wilhelm Miquel (1811–1871)

Friedrich Anton Wilhelm Miquel (Fig. 4.21) was born on October 24, 1811, in Neuenhaus, Lower Saxony, the son of a rural physician. His youth was spent in the village of his birth under the tutelage of his classically oriented father. This was followed by a period at a gymnasium in the nearby town of Lingen (1828–1829) where the emphasis was on classical languages and history. His teachers commented on young Miquel's "punctuality, orderliness, tremendous industry, and a remarkable proficiency in Latin and Greek" (Stafleu 1966). In 1828, he enrolled at the University of Groningen, where medical students had first to obtain a primary degree in the natural sciences (mathematics, chemistry, botany, and zoology). During his studies, it was Professor of Botany Herman van Hall (1801–1874), only ten years older than Miquel, who became his mentor and undoubtedly influenced his career.

Following his graduation from Groningen University in 1833, he was employed for two years as a resident physician at the *Buitengasthaus*, an Amsterdam hospital (1834–1835). Botanical interests were to become an important relief during his extramural hours.

In June 1835, Miquel took his first academic position: that of lecturer in general and pharmaceutical botany at the clinical school at Rotterdam. This post gave him access to the well-developed Rotterdam Botanical Garden; his interests now extend to plant morphology, physiology, and taxonomy. By necessity, he had to supplement the meager academic salary by running a practice as a medical doctor.

Fig. 4.21 Friedrich Miquel at Utrecht University in 1871. (From an official portrait by Dutch artist Johan Heinrich Neuman)

It was during his time in Rotterdam that Miquel commenced a vigorous interaction with botanists in Europe. His local contacts were with Pieter Korthals, Carl Blume, and staff of the *Rijksherbarium* at Leiden, while his German associations were with Diederich Schlechtendal (Editor of the journal *Linnaea*), Gustav Kunze, and the two Nees von Esenbecks, Gottfried Treviranus and Carl Martius. His most important French correspondents were Adolphe Brongniart, Adrien de Jussieu, Joseph Decaisne, and Charles-François Mirbel. In England, he befriended the young Joseph Hooker and his father, Sir William Hooker, later also meeting Robert Brown on a visit to Kew. Although his travels were limited, he also corresponded widely with plant collectors in the Dutch East Indies, Suriname, India, and Australia and facilitated a significant number of herbarium deposits.

However, Miquel never traveled outside his home country and Germany, other than to visit Kew and the British Museum; his descriptions of plants were based on material sent to him by others. Much of his material came from importing nurseries,

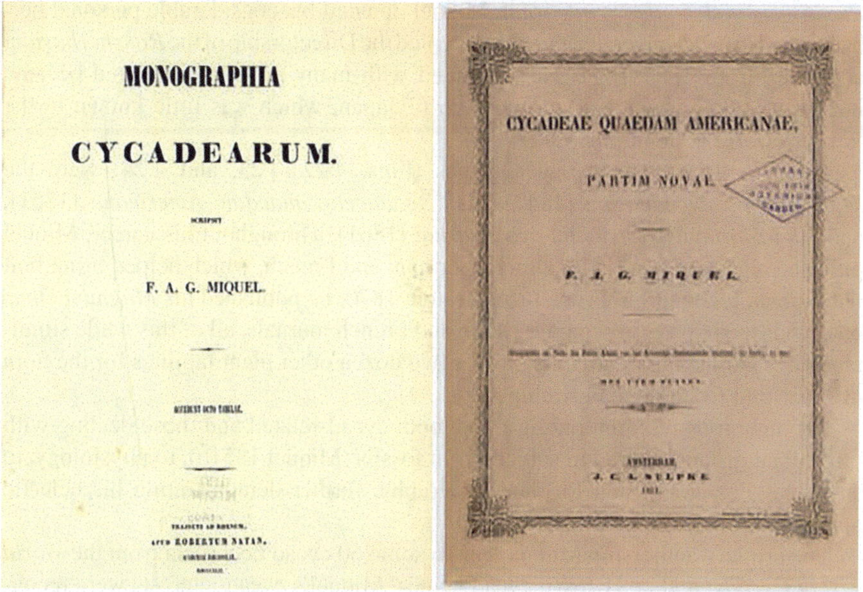

Fig. 4.22 Left: The frontispiece of the 1842 Latin version of Miquel's *Monographia cycadearum*. Right: The frontispiece of Miquel's *Cycadaceae quedam Americanae*, Part 9 of 1851

such as that of the Loddiges family, which Kew Director William Hooker described as "the largest and finest in Europe" (Hooker 1830).

It was also during the Rotterdam period that Miquel began publishing in earnest, his *Monographia cycadearum* of 1842 (Fig. 4.22a) being but one example of a hugely impressive number of works printed in multiple languages.

In 1846, Miquel accepted an appointment as Professor at the Athenaeum in Amsterdam, where his teaching duties encompassed botany, pharmacology, paleontology, and geology. The enhanced salary meant he no longer had to continue his medical practice and could focus wholly on teaching and research work. Another positive was his access to the *Hortus Botanicus*, undoubtedly the richest botanic garden in the Netherlands. He became President of the Royal Institute of Sciences and was actively involved with other scientific and charitable organizations.

His botanical interests extended internationally. Connections to the Dutch East Indies allowed him to commence a four-year project on the *Flora Indiae Batavae*, a tome of 3700 pages and 41 plates (Stafleu 1978). Batavia was then the capital of the Dutch East Indies, an area that corresponds to present-day Jakarta, Indonesia. In the New World, he worked on South and Central American botanical material, publishing the *Cycadaceae quaedam Americanae* in 1851 (Fig. 4.22b) and writing details for the Flora of Suriname.

In 1859, Miquel's succeeded his long-term rival, Cornelis Bergsma, as Professor of Botany at Utrecht University, where the sciences were taught per se and not simply as an adjunct to medical studies. In 1861, Miquel completed the first edition of what was probably the most important cycad text of the nineteenth century, his

Prodromus systematis cycadearum. Miquel donated his considerable personal herbarium to Utrecht University and also accepted the Directorship of the *Rijksherbarium* at Leiden from 1862–1871. He continued with many other projects and became increasingly knowledgeable on the flora of Japan, which was little known to the Western world at that time.

Miquel's three main cycad volumes (Figs. 4.22, 4.23, and 4.24) were the *Monographia cycadearum* (1842), his *Cycadaceae quaedam Americana* (1851), and his *Prodromus systematis cycadearum* (1861). Throughout his career, Miquel published his works in Latin, Dutch, German, and French, which helped bring him widespread recognition. From 1868 through 1870, he published his *Magnum Opus* on cycads in six parts in separate Dutch and French journals, all of this while simultaneously publishing treatments of over two dozen other plant families for the flora of Japan and for the Indian Archipelago.

His numerous research publications, both cycad-related and those dealing with other plant groups, ranged in subject from fossils (Miquel 1851b), to physiology, to structure (Fig. 4.25), and to plant geography, further demonstrating his eclectic approach.

The primary authorship Miq. is seen in some 60 cycad taxa (data from the *World List of Cycads*) and in 31 fossil cycad names. Miquel's cycad interests were recognized by colleagues in the specific epithets of *Macrozamia miquelii* (Fig. 4.26), *Ceratozamia miqueliana* (Fig. 4.27), and the now obsolete *Cycas miquellii*, *Epicycas miquellii*, and *Zamia miquellii*.

Fig. 4.23 The frontispiece pages of the 1861 Latin version of Miquel's *Prodromus systematis cycadearum*

Fig. 4.24 Left: *Zamia angustifolia*. Right: *Zamia ottonis*. From Miquel's *Cycadaceae quaedam Americanae*, Part 9 of 1851

Fig. 4.25 Plates 10 and 11 on cycad seed morphology and embryology from Miquel's *Noveaux matériaux pour server à la connaissance des cycadées* of 1868

Fig. 4.26 *Macrozamia miquelii* in habitat in the Byfield National Park, Queensland, Australia. (Photo: Yvonne Dalziel)

Fig. 4.27 *Ceratozamia miqueliana* in habitat near Santiago Tuxtla, Veracruz, Mexico. (Photo: Chip Jones)

Little is known of Miquel's family life other than that in 1840 he married Catherina Madry, daughter of a leading Rotterdam banker, a step that materially advanced his financial status. The couple is known to have had a son, Anton; a daughter, Berta; and two younger girls. Apparently, the family was well adjusted and lived happily together. Miquel was a devout believer in Christianity and struggled to reconcile his faith with the newly emerging Darwinian concept of evolution. His health was never good: he suffered from a chronic liver disease and exanthematic typhus.

His personal attributes are positively summarized by Frans Stafleu (1966) as follows:

> As a man Miquel comes forward……as a charming personality, full of stories, cheerful, positive, and sometimes almost charismatic in his effect on friends. In the Netherlands of his time [he] may sometimes have felt a lack of opportunity, but the country was privileged to have in him [as] one of the most influential and authoritative botanists of that stimulating and scientifically revolutionary nineteenth century.

Miquel passed away in Utrecht on January 23, 1871, at the age of 59. During his distinguished and wide-ranging career, he had described a staggering total of some 7000 plant species; established himself as an authority on the families Cactaceae, Casuarinaceae, Myrtaceae, Piperaceae, and Polygonaceae; published a total of 440 scholarly works in Latin, German, Dutch, and French; and made significant advances in the study of plant fossils. All in all, he was prodigious in his botanical output across the spectrum of seed plant families and in producing flora treatments. He was as famous in his day, particularly in the field of systematics, as Joseph Hooker, Alphonse de Candolle, and others. Miquel left his estate as a legacy to support botanical studies at the University of Utrecht.

Names of Cycads with Miquel's Authorship
Data from the *World List of Cycads*

Ceratozamia brevifrons Miq. (1847).
Ceratozamia intermedia Miq. (1847) = *C. mexicana* Brong.
Ceratozamia latifolia Miq. (1847).
Ceratozamia longifolia Miq. (1847) = *C. mexicana* Brong.
Ceratozamia robusta Miq. (1847).
Cycas armstrongii Miq. (1868).
Cycas celebica Miq. (1839) = *Cycas rumphii* Miq.
Cycas circinalis var. *angustifolia* Miq. (1840) = *C. circinalis* L.
Cycas circinalis var. *javana* Miq. (1842) ≡ *C. javana* (Miq.) de Laub.
Cycas glauca Hort. *ex* Miq. (1840).
Cycas gracilis Miq. (1863) = *C. media* R.Br.
Cycas madagascariensis Miq. (1840) = *C. thouarsii* R.Br. ex Gaudich.
Cycas revoluta var. *brevifrons* Miq. (1847) = *C. revoluta* Thunb.
Cycas revoluta var. *planifolia* Miq. (1847) = *C. revoluta* Thunb.
Cycas rumphii Miq. (1839).
Cycas rumphii var. *timorensis* Miq. (1840) = *C. glauca* Hort. *ex* Miq.
Cycas siamensis Miq. (1863).
Cycas wallichii Miq. (1842) = *C. circinalis* L.
Dioon angustifolium Miq. (1847).
Dioon edule forma *angustifolium* (Miq.) Miq. (1861) ≡ *Dioon angustifolium* Miq.
Dioon edule forma *imbricatum* (Miq.) Miq. (1861) = *Dioon edule* Lindl.
Dioon edule var. *angustifolium* (Miq.) Miq. (1868) ≡ *Dioon angustifolium* Miq.
Dioon edule var. *imbricatum* (Miq.) Miq. (1868) = *Dioon edule* Lindl.
Dioon imbricatum Miq. (1847) = *Dioon edule* Lindl.
Encephalartos altensteinii var. *angustifolius* Miq. (1842) = *E. altensteinii* Lehm.
Encephalartos altensteinii var. *semidentatus* Miq. (1842) = *E. altensteinii* Lehm.
Encephalartos afer var. *brachyphyllus* (Lehm. & de Vriese) Miq. (1868) = *E. afer* (Thunb.) Lehm.
Encephalartos barteri Carruth. *ex* Miq. (1868).
Encephalartos fraseri Miq. (1863) ≡ *Macrozamia fraseri* Miq.
Encephalartos horridus var. *hallianus* (de Vriese) Miq. (1842) = *E. horridus* (Jacq.) Lehm.
Encephalartos horridus var. *lanuginosus* (Jacq.) Miq. (1838) = *E. longifolius* (Jacq.) Lehm.
Encephalartos horridus var. *latifrons* (Lehm.) Miq. (1838) ≡ *E. latifrons* Lehm.

Encephalartos lanuginosus var. *tridens* Miq. (1842) = *Macrozamia* sp.
Encephalartos lehmannii var. *spinulosus* (Lehm.) Miq. (1868) ≡ *E. spinulosus* Lehm.
Encephalartos longifolius var. *angustifolius* Miq. (1842) = *E. longifolius* (Jacq.) Lehm.
Encephalartos longifolius var. *hookeri* Miq. (1868) = *E. altensteinii* Lehm.
Encephalartos macdonnellii F.Muell. *ex* Miq. (1863) ≡ *Macrozamia macdonnellii* (F.Muell. *ex* Miq.) A.DC.
Encephalartos mauritianus Miq. (1842) = *E. longifolius* (Jacq.) Lehm.
Encephalartos oldfieldii Miq. (1863) = *Macrozamia riedlei* (Gaudich.) C.A.Gardner.
Encephalartos spiralis var. *major* Miq. (1863) = *Macrozamia miquelii* (F.Muell.) A.DC.
Encephalartos spiralis var. *spiralis* Miq. (1863) ≡ *Macrozamia spiralis* (Salisb.) Miq.
Macrozamia fraseri Miq. (1842).
Macrozamia macleayi Miq. (1868).
Macrozamia peroffskyana (Regel) Miq. (1868) ≡ *Lepidozamia peroffskyana* Regel.
Macrozamia spiralis (Salisb.) Miq. (1842).
Zamia angustissima Miq. (1851) = *Z. angustifolia* Jacq.
Zamia calocoma Miq. (1852) ≡ *Microcycas calocoma* (Miq.) A.DC.
Zamia fischeri Miq. (1845).
Zamia kickxii Miq. (1842) = *Z. pygmaea* Sims.
Zamia latifolia Lodd. *ex* Miq. (1849) = *Z. furfuracea* L.*fil.*
Zamia leiboldii Miq. (1847) = *Z. loddigesii* Miq.
Zamia loddigesii Miq. (1843).
Zamia maelenii Miq. (1844) = *Dioon edule* Lindl.
Zamia mexicana Miq. (1861) = *Z. loddigesii* Miq.
Zamia muricata var. *augustifolia* Miq. (1842) = *Z. muricata* Willd.
Zamia muricata var. *obtusifolia* Miq. (1843) = *Z. fufuracea* L.*f.*
Zamia ottonis Miq. (1844) = *Z. pygmaea* Sims.
Zamia potemkinii Hort. *ex* Miq. (1869), *nomen dubium*.
Zamia yatesii Miq. (1851) = *Z. angustifolia* Jacq.
Zamia stricta Miq. (1851).
Zamia verschaffeltii Miq. (1870), *nomen dubium*.

Miquel's Cycad Publications

Miquel, F. A. W. 1838. *Note additionelle sur une autre espèce du même genre. Bulletin des sciences physiques et naturelles en Néerlande* 1: 11–12.

Miquel, F. A. W. 1838. *Observations sur les cycadées de l'herbier royal à Leyde. Bulletin des sciences physiques et naturelles en Néerlande* 1: 82–85.

Miquel, F. A. W. 1838. *Note sur l'Encephalartos horridus Lehm. et sur ses différentes formes. Annales des sciences naturelles botanique* 2(10): 366–369. [*Encephalartos horridus* var. *lanuginosus* (Jacq.) Miq.]

Miquel, F. A. W. 1839. *Note sur la prefoliation des cycadées. Bulletin des sciences physiques et naturelles en Néerlande* 2: 9 and *Annales des sciences naturelles botanique* 2(11): 61–62.

Miquel, F. A. W. 1839. *Mélanges botaniques. Bulletin des sciences physiques et naturelles en Néerlande* 2: 36–48. [*Cycas celebica* Miq., *C. rumphii* Miq.].

Miquel, F. A. W. 1839. *De Encephalartos horrida Lehm. ciusqueformis, scripsi F. A. W. Miquel. Tijdschrift voor natuurlijke geschiedenis en physiologie* 6: 94–106.

Miquel, F. A. W. 1840. *Observations sur les cycadées. Annales des sciences naturelles botanique* 2(14): 60–62.

Miquel, F. A. W. 1840. *De quibus dam cycadeis minus cognitis. Sylloge plantarum novarum* (Part 3): 126–127.

Miquel, F. A. W. 1840. *Commentarii phytographici, quibus varia rei herbariae capita illustrantur.* Part 2. *Sylloge plantarum novarum vel minus cognitarum ex ordinibus Araliacearum, Cactearum, Hypoxidiarum, Cycadearum et Urticacearum.* S. & J. Lutchmans, Leiden. [*Cycas circinalis* var. *angustifolia* Miq., *C. glauca* Hort. *ex* Miq., *C. rumphii* var. *timorensis* Miq.].

Miquel, F. A. W. 1842. Pp.1–82 in *Monographia cycadearum*, R. Nathan, Utrecht. [*Cycas circinalis* var. *javana* Miq., *C. madagascariensis* Miq., *C. revoluta* var. *planifolia* Miq., *C. wallichii* Miq., *Encephalartos altensteinii* var. *angustifolius* Miq., *E. altensteinii* var. *semidentatus* Miq., *E. horridus* var. *hallianus* (de Vriese) Miq., *E. horridus* var. *latifrons* (Lehm.) Miq., *E. lanuginosus* var. *tridens* Miq., *E. longifolius* var. *angustifolius* Miq., *E. longifolius* var. *revolutus* Miq., *E. mauritianus* Miq., *Macrozamia fraseri* Miq., *M. spiralis* (Salisb.) Miq., *Zamia kickxii* Miq., *Z. muricata* var. *angustifolia* Miq.]

Miquel, F. A. W. 1843. *De cycadeis Loddigesianis.Tijdschrift voor natuurlijke geschiedenis en physiologie* 10(1): 68–74. [*Zamia loddigesii* Miq., *Z. muricata* var. *obtusifolia* Miq.].

Miquel, F. A. W. 1843. *Description d'une nouvelle espèce de Zamia. Annales des sciences naturelles botanique* 2(19): 316–317. [*Zamia loddigesii* Miq.]

Miquel, F. A. W. 1843–1844. *Genera et species cycadearum viventium.* Linnaea 17(6): 675–744. *Addenda ad synopsin cycadearum viventium* 18(1): 96–98 of 1844. [*Zamia maelenii* Miq., *Z. ottonis* Miq.]

Miquel, F. A. W. 1844. *Ueber den Baue in eserwachsenen Stammes von Cycas circinalis.* Linnaea 18(1): 125–144.

Miquel, F. A. W. 1845. *Observationes de ovulo et embryonibus cycadearum. Annales des sciences naturelles botanique* 3(3): 193–206.

Miquel, F. A. W. 1845–1846. *Description de plantes nouvelles, rares ou peu connues, introduites dans le jardins de Louis van Houtte.* [*Zamia fischeri* Miq.].

Miquel, F. A. W. 1846. *Recherches sur la structure d'un toncage du Cycas circinalis. Annales des sciences naturelles botanique* 3(5): 11–24.

Miquel, F. A. W. 1846. *Aanteekeningom trenteene nieuw en cycadeënform inAmerika en der zelfsverhoudingtot eeinigefissole typen uit dze planten groep. Het Instituut*: 154–157.

Miquel, F. A. W. 1847–1848. *Collectianea nova ad cycadearum cognitioneum.* Linnaea 19(4): 411–430. *Addendum ad parten primam systematicam.* Linnaea 21: 567–568. [*Zamia leiboldii* Miq.]

Miquel, F. A. W. 1847–1848. *Over eenige nieuwe of zeldzame cycadeen in den Hortus Botanicus te Amsterdam. Tijdschrift voor de wis- en natuurkundige wetenschappen* 1(1): 33–44, 1(2): 103–111, 1(4): 197–209, 297. In German *as Allgemeine Deutsche Garten Blumenzeit* 4: 328–355, 371–375, 518–525. [*Ceratozamia brevifrons* Miq., *C. intermedia* Miq., *C. latifolia* Miq., *C. longifolia* Miq., *C. robusta* Miq., *Cycas inermis* Lour., *C. revoluta* var. *brevifrons* Miq., *Dioon angustifolium* Miq., *D. imbricatum* Miq.].

Miquel, F. A. W. 1849. *Epicrisis systematis cycadearum.Tijdschrift voor de wis- en natuurkundige wetenschappen* 2(4): 280–303, 3: 300. [*Zamia latifolia* Lodd. *ex* Miq.]

Miquel, F.A.W. 1851. *Cycadaceae quaedam Americanae* 9. *Nieuwe verhandelingen van dee erste klasse van het koninklijke Nederlandsche instituut* 3(4): 181–189. [*Zamia angustissima* Miq., *Z. stricta* Miq., *Z. yatesii* Miq.].

Miquel, F. A. W. 1851. *Over de rangschikking der fossiele cycadeae. Tijdschrift voor de wis- en natuurkundige wetenschappen.*4: 205–227.

Miquel, F. A. W. 1852. *Sur une espèce nouvelle de Zamia des Indes occidentales, introduite dans l'establissement L. van Houtte à Gand. Flore des serres, et des jardins de l'Europe* 7(6): 141–142. [*Zamia calocoma* Miq.]

Miquel, F. A. W. 1852. *Cycas rumphii stirps femina.* Linnaea 25: 589–592.

Miquel, F.A.W. 1861. *Prodromus systematis cycadearum*: 1–27. Van der Post, Utrecht. [*Dioon edule* forma *angustifolium* (Miq.) Miq., *D. edule* forma *imbricatum* (Miq.) Miq., *Zamia mexicana* Miq.]

Miquel, F. A. W. 1863. *Over de cycadëen in Nieuwe-Holland.Verslagenen mededeelingen van de afdeeling natuurkunde koninklijke akademie van wetenschappen*15: 363–377. [*Cycas gracilis* Miq., *Encephalartos spiralis* var. *major* Miq., *E. fraseri* (Miq.) Miq., *E. macdonnellii* F.Muell. *ex* Miq., *E. oldfieldii* Miq., *E. spiralis* var. *major* Miq., *E. spiralis* var. *spiralis* Miq.].

Miquel, F. A. W. 1863. *Ueber eine neue* Cycas *aus Siam. Botanische Zeitung* 21: 334. [*Cycas siamensis* Miq.].

Miquel, F. A. W. 1868–69, *Nieuwe bijdragen tot de kennis der cycadeen. Verslagen en mededeelingen van de afdeeling natuurkunde; koninklijke akademie van wetenschappen* (Series 2), 3(1): 1–57, 3(2): 152–165, 196–206, 4: 23–37. [*Zamia potemkinii* Hort. *ex* Miq.].

Miquel, F. A. W. 1868. *Noveaux matériaux pour server à la connaissance des cycadées* 1. *Organes sexuels. Cycas. 2. Encephalartos barteri. 3. Cycadées de la Nouvelle-Hollande extra tropicale. 4. Cycadees del'Afrique. 5 Cycadées de L'Amérique. Archives Néerlandaises des sciences exactes et naturelles* 3(3): 193–254, 3(5): 403–427. [*Cycas armstrongii* Miq., *Dioon edule* var. *angustifolium* (Miq.) Miq., *D. edule* var. *imbricatum* (Miq.) Miq., *Encephalartos barteri* Carruth. *ex* Miq., *Macrozamia macleayi* Miq., *M. peroffskyana* (Regel) Miq.].

Miquel, F. A. W. 1868. On the sexual organs of the Cycadaceae. 1868. English translation from *Noveaux matériaux pour server à la connaissance des cycadées* 1 by W. Thiselton-Dyer (1869). Journal of Botany 7: 64–105.

Miquel, F. A. W. 1870. *Noveaux matériaux pour server à la connaissance des cycadées.* Revision. Classification. Adansonia 9: 352–367.

Miquel, F. A. W. 1870. *Zamia verschaffeltii. Archives Néerlandaises des sciences exactes et naturelles* 5: 82. [*Zamia verschaffeltii* Miq.]

Berthold Carl Seemann (1825–1871)

Berthold Carl Seemann (Fig. 4.28) was born in Hanover, Germany, on February 25, 1825, and was educated at the Lyceum and then at the University of Göttingen. At the age of 19, he moved to England to study botany at the Royal Botanic Gardens Kew. After only four years at Kew and on the recommendation of their Director Sir William Hooker, Seemann was appointed naturalist, along with Thomas Edmondston and John Goodridge, on the HMS *Herald* expedition. Under the command of Captain Henry Kellett, this vessel set off in 1845 on a six-year voyage, the primary purpose being a hydrographical survey of the American west coast and the Pacific (Boulger 1897; Hemsley 1886–1888, 1904). From this voyage, Seemann published a series entitled "*The Botany of the Voyage of HMS Herald*," with the cycads treated in Volume 6 (Fig. 4.29).

While waiting to join the vessel in Colombia in 1847, Seemann investigated the local area. Although there is some confusion in the records, it appears that Seemann discovered the cycad he named *Zamia chigua*, known locally as *helecho*, along the banks of the Río San Juan, Colombia. Three plants of this species were sent to James Yates in Kew in 1847; one of these produced a pollen cone in 1852, and Stevenson and Sabato (1986a) used the illustration (Fig. 4.30) to lectotypify the name.

Fig. 4.28 Berthold Carl Seemann. (Image from Wikipedia)

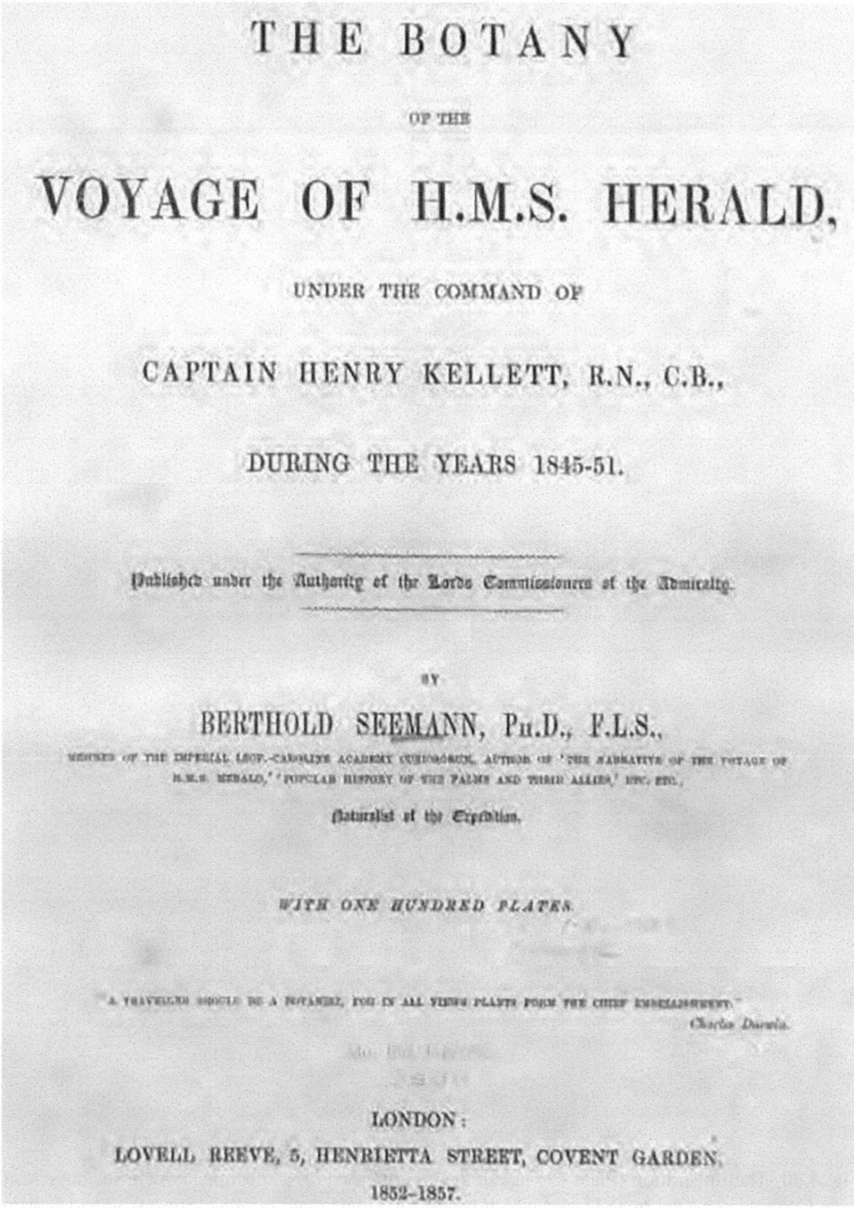

THE BOTANY

OF THE

VOYAGE OF H.M.S. HERALD,

UNDER THE COMMAND OF

CAPTAIN HENRY KELLETT, R.N., C.B.,

DURING THE YEARS 1845-51.

Published under the Authority of the Lords Commissioners of the Admiralty.

BY

BERTHOLD SEEMANN, PH.D., F.L.S.,

MEMBER OF THE IMPERIAL LEOP.-CAROLINE ACADEMY CURIOSORUM, AUTHOR OF 'THE NARRATIVE OF THE VOYAGE OF H.M.S. HERALD,' 'POPULAR HISTORY OF THE PALMS AND THEIR ALLIES,' ETC. ETC.,

Naturalist of the Expedition.

WITH ONE HUNDRED PLATES.

"A TRAVELLER SHOULD BE A BOTANIST, FOR IN ALL VIEWS PLANTS FORM THE CHIEF EMBELLISHMENT."

Charles Darwin.

LONDON:

LOVELL REEVE, 5, HENRIETTA STREET, COVENT GARDEN.

1852–1857.

Fig. 4.29 Frontispiece of Seemann's "Botany of the Voyage of HMS *Herald*" where he describes *Zamia chiqua* in Volume 6

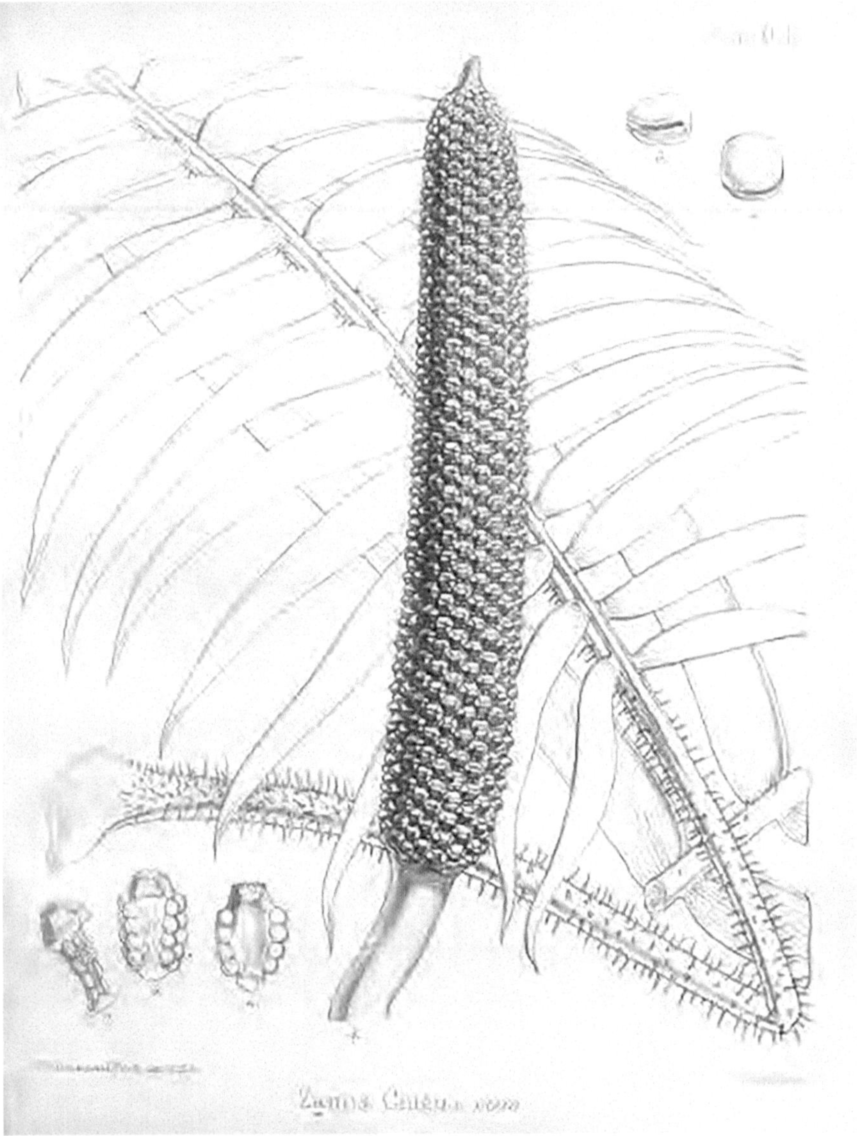

Fig. 4.30 The illustration (Plate 43) of *Zamia chigua* based on a specimen which Seemann sent from Colombia to Kew in 1847 and which produced a male cone in 1852. Stevenson and Sabato (1986) used this illustration to lectotypify the name

There also were populations of *Zamia roezlii* near the mouth of the Río San Juan, and the Indigenous people also call this cycad by the name *chigua* (Norstog, 1986). The two species are quite different in appearance: *Zamia chigua* is a short-trunked plant with crowded leaflets with a smooth surface, while *Z. roezlii* is a taller and salt-tolerant cycad from mangrove habitats with fewer and distinctly grooved leaflets. Typification of these two names by Stevenson and Sabato (1986a) stabilized the nomenclature and the application of *chigua*. These taxa were further clarified and described by Stevenson (2004a).

During a refreshment stop on the northbound voyage of HMS *Herald* at Cabo Corrientes on the Colombian west coast in December 1847, Seemann collected a second arborescent cycad from coastal forests. Later, Gustav Wallis collected more plants for European nurseries from the same locality in 1873. The German botanist Alexander Braun used plants in Berlin to describe the species *Zamia obliqua*, noting the slanted or unequal sides to the leaflets (Braun 1875a, b). Stevenson and Sabato (1986a) identified a two-sheet voucher *Seemann 1569* in the British Museum as the neotype, this being consistent with a lithograph of a coning plant at Kew illustrated by Walter Fitch as Plate 7542 in *Curtis's Botanical Magazine* of 1897 (Fig. 4.31).

The *Herald* explored most of the west coast of America. Seemann was able to make excursions ashore in Peru, Ecuador, and Mexico. In addition, the vessel was ordered in 1848 to the Bering Strait to search for Sir John Franklin's ill-fated expedition through the Northwest Passage. During that time, Seemann was able to collect plants and artifacts relating to the Inuit people. The ship then returned home via Hong Kong, Singapore, the Cape of Good Hope (where Seemann met other visiting botanists with whom he climbed Table Mountain), Saint Helena, and Ascension, arriving back in England in June 1851. Seemann's (1853) report of the expedition was published in the two-volume "*Narrative of the voyage of HMS Herald during the years 1845–1851*" and the multi-volume work "Botany of the Voyage of HMS *Herald.*" He was awarded a doctorate by the University of Göttingen for his endeavors.

In 1859, Hooker sent Seeman to Fiji for a two-year period to survey the islands for agricultural development and to document the native flora. Seemann reported his work in "Viti: An Account of a Government Mission to the Vitian or Fijian Islands in the Years 1860–1861" and also produced *Flora Vitiensis*: a description of the plants of the Viti or Fiji Islands with an account of their history and properties, a monumental work published in 10 parts between 1865 and 1873.

Seemann was aware of the native cycad of Fiji, referring to it as *Cycas circinalis* and by its Fijian name *roro* and writing in 1862 of it

> ...a tree of thirty feet high…by no means a common plant on the islands, having been encountered only at Viti Levu and Ovalau in isolated specimens; and as the pith-like substance contained in the trunk was reserved for the sole use of the chiefs, and forbidden to the lower casts, no inducement existed on the part of those debarred from it to extend it by cultivation as is done on the Tongan Islands.

Recognizing Seemann's work, Alexander Braun (1876) revised the name of the cycad from Fiji and other parts of the western Pacific to *Cycas seemannii* (Fig. 4.32).

Fig. 4.31 A lithograph of *Zamia obliqua*, Plate 7542, from Hooker in *Curtis' Botanical Magazine* of 1897, based on plants grown at Kew imported from the Cabo Corrientes locality in Colombia where Seemann originally found the species

In a recent visit to Fiji, Roy Osborne has noticed that the native cycads are becoming more abundant, probably as they are no longer harvested for their starch value, and that there is quite vigorous natural regeneration from seed crops.

During and in between these various excursions abroad, Seemann edited and wrote for the journal *Bonplandia* and the *Journal of Botany British and Foreign*, as

Fig. 4.32 *Cycas seemannii* near Sigota on Viti Levu (Fiji) in 2016. Vigorous regeneration is occurring in the population as the plants, seedlings, and seeds are no longer being harvested for their starch value. (Photo: Roy Osborne)

well as published numerous other botanical articles and books in both English and German. Berthold's brother, Wilhelm Seemann, also a botanist, was co-editor of *Bonplandia*. Berthold also became known as an authority on the family Araliaceae. In addition to his botanical endeavors, he composed music and, in the 1860s, wrote three plays that were successfully produced in Germany.

In 1864, Seemann was appointed by a firm in Amsterdam to report on the natural resources of Venezuela, after which he managed a sugar estate in Panama and the Javali gold mine in Chontales, Nicaragua. Tragically, while working at the mine, he succumbed to malaria and died on October 10, 1871, aged only 46.

Seemann was a Fellow of the Linnean Society and the Royal Geographical Society. His herbarium vouchers are filed mainly at Kew and at the British Museum. As well as the recognition of his name in Braun's *Cycas seemannii*, the genera *Seemannia* (Gesneriaceae), *Seemannantha* (Fabaceae), and *Seemannaralia* (Araliaceae) are named for him.

Name of the Cycad with Seemann's Authorship
Data from the *World List of Cycads*

Zamia chigua Seem. (1854).

Seemann's Cycad Publications

Seemann, B. C. 1854. Cycadeae, Flora of the Isthmus of Panama. In: Botany of the Voyage of HMS *Herald*, Vol. 6: 201–203, Plate 43. [*Zamia chigua* Seem.]

Seemann, B. C. 1862. Viti: An Account of a Government Mission to the Vitian or Fijian Islands in the Years 1860–1861. [The use of cycad starch is described on pp. 288–289.]

Seemann, B. C. 1865–1873. Flora Vitiensis: a description of the plants of the Viti or Fiji Islands with an account of their history and properties. This was published in 10 parts between 1865 and 1873. L. Reeve & Co., London.

Willem Hendrik de Vriese (1806–1862)

Willem Hendrik de Vriese (Fig. 4.33) was born in Oosterhout in the southern part of the Netherlands, on August 11, 1806. He left Oosterhout at an early age, studying at the Leiden Latin School and later marrying Charlotta Bosse daughter of the school Rector. He continued studies in medicine at the University of Leiden where his interest in botany earned him a Gold Medal. He was awarded a doctorate in 1831. For the next two years, de Vriese practiced medicine in Rotterdam while also giving botany classes at the local medical college. He then moved northwards, being appointed Associate Professor of Botany at the *Athenaeum Illustré* in Amsterdam in 1834, taking the full Professorship role in 1841, and, also, being involved with the Amsterdam Botanical Garden (*Hortus Botanicus*). In 1845, he returned to Leiden as

Fig. 4.33 An oil-on-canvas portrait of Willem Hendrik de Vriese by celebrated Dutch artist Jan Hendrik Neuman. The work, painted 29 years after de Vriese's death, is displayed in the Senate Room of Amsterdam Municipal University

Professor of Botany, also succeeding his mentor Caspar Reinwardt as Head of the Leiden Botanical Garden (*Hortus Botanicus Leiden*).

In October 1857, de Vriese was commissioned to carry out botanical surveys in the Dutch East Indies, with particular emphasis on the quality and viability of the native agricultural crops such as coffee, cacao, camphor, nutmeg, vanilla, indigo, rubber, and gutta-percha (Hooker 1857). Accompanied by his wife Charlotta, he made brief visits to Singapore and Sri Lanka (then Ceylon) and spent four years studying the agriculture and flora in areas of what is now Indonesia (e.g., Java, Moluccas, Timor, Sumatra, Sulawesi, and Borneo).

During his time in Amsterdam, de Vriese wrote the first section of *Hortus Spaarn-Bergensis* (1839), a description of the exotic plant collection assembled by the wealthy Dutch banker and politician, Adriaan van der Hoop. In 1841, he produced a biography (de Vriese 1841) on the German physician, plant collector, anatomist, and cytologist, Franz Meyen (1804–1840). After taking his position in Leiden, de Vriese published more prolifically. He produced a paper on the palms of Suriname, then a Dutch colony in South America (1848). His interests also included an appreciation of tropical orchids, as seen in his scientific text accompanying the beautifully illustrated publication on the orchids of the East Indies (Pahud et al. 1854). In 1855, he compiled the multi-authored *Tuinbouw-Flora* a three-volume work on new and important horticultural plants of Holland and its colonies. This project was carried out under Royal auspices. He summarized and extended the works of his mentor, Caspar Reinwardt, including the latter's botanical treatise on the Dutch East Indian flora, *Plantae Indiae Batavae Orientis* (de Vriese 1856). He wrote several papers on the economic botany of cinchona, vanilla, camphor, and other tropical products.

The cycad-related systematic works of de Vriese are described separately below. Apart from those important publications, he wrote monographs on the genus *Rafflesia*, the families Goodeniaceae and Lobeliaceae, and the fern family Marattiaceae, the latter co-authored with Pieter Harting. He also named several angiosperms, including *Cyrtandra decurrens* (Gesneriaceae); *Dampiera diversifolia*, five species of *Goodenia* and two species of *Scaevola* (all Goodeniaceae); *Lobelia rhombifolia* and *Wahlenbergia preisii* (both Campanulaceae); and *Saurauia microphylla* (Actinidiaceae). Remarkably, ten of these 12 taxa are Australian endemics—although de Vriese had never visited that country. The publications in which de Vriese was involved as editor, co-author, or principal author extend in range and number far beyond the few examples given in this text.

The science of botany was much respected and underwent something of an explosion of interest in Europe in the nineteenth century. Holland was a convenient meeting point geographically, and de Vriese would certainly have hosted many of the workers in this volume. Johann Lehmann (1792–1860) visited de Vriese in Amsterdam in 1836, and their interaction catalyzed the cycad publications we summarize below; Lehmann had only recently named the genus *Encephalartos* in his 1834 *Pugillus*. We also know that de Vriese had Robert Brown stay at his Leiden home on at least one occasion. Perhaps most importantly, de Vriese had a profound influence as a mentor to Friedrich Miquel (1811–1871) whose 1842 *Monographia cycadearum* subsequently became the most important cycad work of the nineteenth century. Furthermore, we can reasonably speculate that he interacted with Berthold Seeman (1825–1871), Alphonse de Candolle (1806–1893), and other contemporary European botanists and that cycad biology would have been on the agenda of many meetings and much correspondence. We know too that he visited and was much impressed by the exotic plants on display at the Loddiges nursery at Hackney near

London, and it is probable that he met with both William Hooker (1785–1865) and his son, Joseph Hooker (1817–1911), at Kew.

The first major cycad taxonomy work by de Vriese was his 1837 monograph of *Encephalartos*, in which he reviews Lehmann's *E. elongatus* Lehm., *E. latifrons* Lehm., E. *nanus* Lehm., and *E. spinulosus* Lehm. and then describes two new taxa: *E. brachyphyllus* Lehm. and de Vriese and *E. van-hallii* de Vriese. Of the six taxa in the treatment, only *E. latifrons* Lehm. remains a currently accepted name. *E. brachyphyllus* Lehm. and de Vriese (Fig. 4.34) is a synonym for *E. afer* (Thunb.) Lehm., and *E. van-hallii* de Vriese is a synonym for *E. horridus* (Jacq.) Lehm. The text, in Latin, is illustrated with line drawings.

The next cycad publication concerning systematics by de Vriese was his 10-page work on *Encephalartos lehmannii* Lehm. published in 1843. Named initially, but not validly, by Danish plant collector Christian Ecklon (1833) as *Zamia lehmanniana*, this species was later validly published by Lehmann (1834) as *E. lehmannii* Lehm. This recognition of species was confirmed again by Lehmann when he saw the living specimen in the Amsterdam *Hortus* in 1836. An interesting aside is the

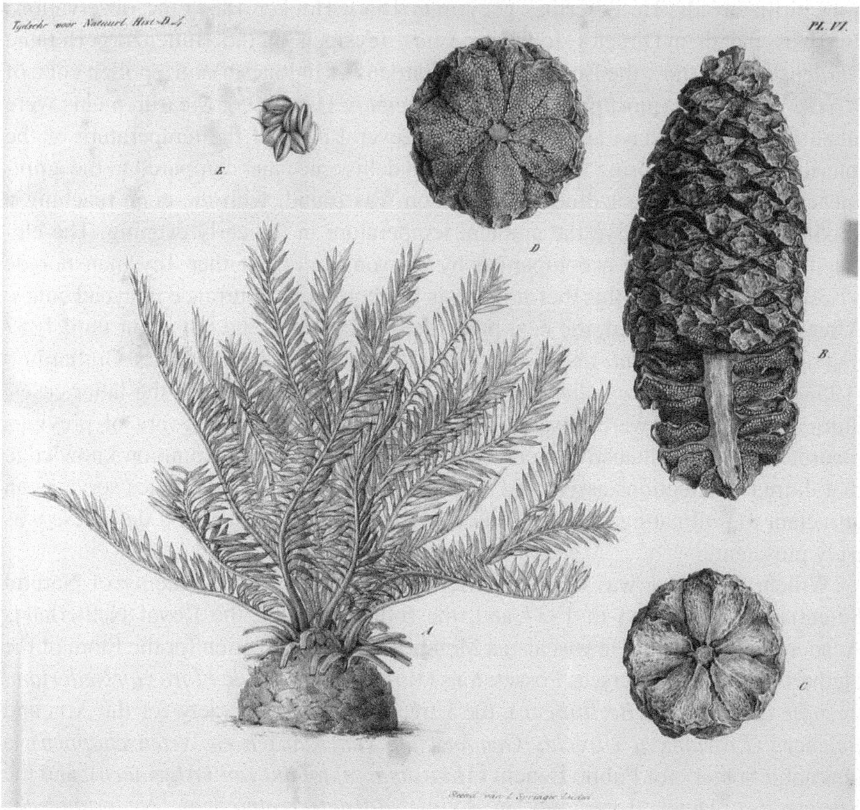

Fig. 4.34 The line drawings of *Encephalartos brachyphyllus* Lehm. and de Vr. (now *E. afer* (Thunb.) Lehm.). Plate VI from de Vriese (1837)

publication of *E. lehmanni* Ecklon (Anonymous. 1836) Allgemeine Gartenzeitung. This 1843 paper, unillustrated and in Latin, was dedicated by de Vriese (somewhat effusively) to Friedreich Miquel in acknowledgment of the latter's 1842 monograph. Miquel responded in the same journal with a review of nine species of *Encephalartos*, two of *Cycas*, and six of *Zamia* (Miquel 1843).

The third important cycad publication by de Vriese was his 1847 description, in French, of four cycad taxa: *Encephalartos altensteinii* Lehm. (as seen in the botanic gardens in Amsterdam, Leiden, and Hamburg and in Adriaan van der Hoop's private garden at Spaarn-Bergen near Harlem), *Zamia muricata* Willd. (also, from the Spaarn-Bergen collection), *Cycas circinalis* Linn. (seen in the Bogor Gardens in Java and elsewhere; now known to be incorrectly named), and *C. rumphii* Miq. (again, from the van der Hoop collection). The work is exquisitely illustrated with colored lithographs by Pieter Trap and Elisabeth Koning (Figs. 4.35, 4.36, 4.37, and 4.38). The same plants that de Vriese studied are still extant in the botanic gardens in Amsterdam and Leiden (Fig. 4.39).

Apart from these three major cycad-related publications, de Vriese was the first botanist to report on cycad cone thermogenesis, a phenomenon known previously only in the aroids. He describes (Teysman 1850; Hooker 1851) the observations made by the then Director Johannes Elias Teysman of the Buitenzorg Botanic Gardens in Java (now the Bogor Botanic Gardens of Indonesia), on a pollen cone of "*Cycas circinalis*" (most probably *Cycas rumphii,* Hill 1995). Measurements were taken morning, midday, and evening over several days of the temperature of the interior of the cone during the time of pollen dehiscence and compared to the ambient temperature. A clear diurnal fluctuation was found, with the cone reaching a maximum of 13 °F above the ambient temperature in the early evening. The elevated temperature was accompanied by a strong odor. Neither Teysman nor de Vriese were aware that this thermogenesis is a common occurrence in cycad cones. After this original record, the next published observations did not occur until 1857 by Poissan in *Dioon edule* and the next after over a century later by Jacot-Guillardmo (1958) in *Encephalartos altensteinii* and *E. lehmannii.* In both of the latter cases, diurnal fluctuations were observed, and both authors were unaware of previous records. After the exhaustive work of Tang (1987), it became common knowledge that diurnal fluctuations associated with the emission of cone volatiles serve as an attractant to pollinating insects. Thus, the research of Teysman and de Vriese was truly prescient.

Willem de Vriese was elected a Member of the German Academy of Natural Scientists (*Leopoldina*) in 1837 and, the following year, the Royal Netherlands Academy of Sciences. He was also a Member of the Association for the Flora of the Netherlands and its Overseas Possessions (*Vereeniging voor de Flora van Nederland en zijne Overzeesche Bezittingen*), the Utrecht Provincial Society for the Arts and Sciences (*Provinciaal Utrechts Genootschap van Kunsten en Wetenschappen*), a charitable society for Public Benefit (*Maatschappij tot nut van't Algemeen*), and the Society for Dutch Literature in Leiden (*Maatschappij van Nederlandsche Letterkunde van Leiden*). In 1842, he was awarded foreign membership of the Royal Horticultural Society in London, and in 1855, he became a Knight of the Order of the Netherlands Lion.

Fig. 4.35 The lithograph of *Zamia muricata* Willd. with a seed cone, from the private collection of Adriaan van der Hoop at Spaarn-Bergen, near Harlem. This fine work is one of many drawings by the Dutch artist Elisabeth Koning (1816–1887). Unnumbered plate from de Vriese (1847)

His name was commemorated when British botanist John Lindley (1843) named the bromeliad genus *Vriesea* (orthographically incorrect but a conserved generic name) and the segregate genus *Neovriesia* (Britton 1923).

Unfortunately, de Vriese's health deteriorated during his time in the East Indian tropics; he returned to the Netherlands in a weakened state and passed away in Leiden on January 23, 1862, at the age of 55. Possibly also suffering from the same

Fig. 4.36 A lithograph of *Cycas rumphii* Miq. bearing a pollen cone, again from the van der Hoop collection and painted by Elisabeth Koning. Unnumbered plate from de Vriese (1847)

Fig. 4.37 A lithograph of a *Cycas* identified at the time as *C. circinalis* L. However, the well-defined spongy flotation layer seen in the seed section places this plant firmly in the *C. rumphii* group (Hill 1995)). Unnumbered plate from de Vriese (1847)

Fig. 4.38 An illustration of pollen (left) and seed (right) cones of *Encephalartos altensteinii* Lehm. prepared by Dutch lithographer and printer Pieter Trap. Unnumbered plate from de Vriese (1847)

Fig. 4.39 Roy Osborne with Dutch horticulturist Jan Maarten Visser admiring the prize specimen of *Encephalartos altensteinii* at the Amsterdam Botanic Garden (*Hortus Botanicus Amsterdam*). Originally owned by King Willem II (1792–1849), this plant, a pollen cone–bearing plant, now about 350 years old, was bought by the Gardens in 1850 for 150 guilders (Hinterding 1995). This important specimen, which has produced male cones over several years, was well known to both de Vriese and Miquel. A different view of the same plant is seen with Oliver Sacks in Fig. 7.60. (Photo: Angela Osborne, May 2023)

malady, his wife had died in 1861, and very sadly, he lost his only son at about the same time. Two months after his death, de Vriese's library and herbarium collection were sold off by public auction.

The de Vrieses had four children: Marie Anna (b. 1835), Willem Hendrik junior (b. 1839), François Antionette (b. 1841), and Charlotta Dorothea (b. 1847).

Józef Warszewicz (1812–1866)

Józef Warszewicz (Fig. 4.40), Ritter von Rawicz (Julius von Warscewicz), was born on September 8, 1812, in Vilnius (Wilno), Lithuania, son of a poor Polish family. He worked in the botanical garden of Vilnius University but became embroiled in the failed "November uprising" against Russia and was forced to flee to Germany. From 1840–1844, he worked as a gardener at the Berlin Botanical Garden (Gaerdt 1868; Köhler 2014).

In 1844, the influential German explorer, Alexander von Humboldt, convinced Louis van Houtte to recruit Warszewicz for a plant-collecting operation based in Guatemala. It was there that Warszewicz came across a variegated leaf cycad (Fig. 4.41) at an unrecorded locality in Guatemala. His brief mention of the cycad, printed in the *Allgemeine Gartenzeitung* of 1845, was designated the lectotype for *Zamia variegata* by Stevenson and Sabato (1986a) but that process was retroactively excluded in The International Code of Botanical Nomenclature. Thus, a neotype specimen from a population at Lacandona in Chiapas, Mexico, is now accepted and more informative than a written description (Nicolalde-Morejón et al. 2009).

After about one year of working for Van Houtte, Warszewicz decided to become independent. He commenced collecting large quantities of plants, especially orchids, and extended his repertoire to include cultural artifacts, hummingbirds, herpetological specimens, and other items sought by European institutions and private collectors. He explored extensively through Guatemala, Nicaragua, Costa Rica, and Panama, living in crude accommodation, subsisting on maize and fruit, and shipping specimens back to Europe whenever the opportunity arose (Savage 1970). There is an excellent summary of the plant collections of Warszewicz to horticulture and to the study and descriptions of new world orchids (Seemann 1854).

In 1848, Warszewicz visited the Chiriquí region of Panama, collecting *Zamia lindleyi* from a population in the Fortuna Valley in the western highlands and *Z. skinneri* along the Guarúmo River toward the Atlantic coast, sending live plants and seeds to England and Berlin (Calonje et al. 2012). In a slight diversion from the Boquete-Róbalo trail, he climbed the 3478-meter Volcán Chiriquí (now Volcán Barú). After further fieldwork in South America, a yellow fever infection forced Warszewicz to return to Europe in 1850.

In Germany, he worked for a period with his German orchidologist colleague, Heinrich Reichenbach. He also provided notes and sketches (Fig. 4.42) of two Panamanian cycads to Albert Dietrich, who used them to publish the names *Zamia lindleyi* and *Z. skinneri* (Dietrich 1851).

Fig. 4.40 A bust of Józef Warszewicz, erected in 1880, in the Botanical Garden of Jagiellonian University, Krakow

Warszewicz then returned to Central America in early 1851. In Panama, he collected *Zamia pseudoparasitica* specimens and more plants from the previous cycad localities for British plant collector James Yates, but apparently, only the *Z. skinneri* plants survived the passage back to Europe. He then extended his explorations to the western parts of South America but was robbed of all his possessions after arriving

Fig. 4.41 Foliage of a specimen of Warszewicz's *Zamia variegata* from Guatemala. Also known in synonymy as Thiselton-Dyer's *Z. picta*, this is the only naturally occurring cycad with variegated leaves. (Photo: Roy Osborne)

in Guayaquil, Ecuador. Despite that, he continued traveling through Colombia, Brazil, Bolivia, and Peru, collecting new plant species, animals, and ethnographic artifacts. Unfortunately, he was affected by a second bout of yellow fever and had to return to Europe in 1853, taking yet more plants and animals back home on that final voyage.

After settling down in Krakow, Poland, Warszewicz re-established many of his tropical American plant specimens in the greenhouses of Jagiellonian University

Fig. 4.42 Sketches made in Panama by Józef Warszewicz of *Zamia lindleyi* (left) and *Z. skinneri* (right) as used in Dietrich's publication of the two species in 1851

Botanic Garden, where he served as Curator until his death on December 29, 1866. He also brought a herbarium and a collection of tropical animals to Krakow. One of Krakow's streets is named for him, and his statue, erected in 1880, stands in the university gardens. His name is commemorated in the orchid genera *Warczewiczella* and *Warscaea*, the South American tree genus *Warszewiczia* (Rubiaceae), and some 22 specific epithets. Most of his herbarium vouchers were bequeathed to the Berlin Botanic Garden but were lost during World War II.

Names of Cycads with Warszewicz's Authorship
Data from the *World List of Cycads*

Zamia lindleyi Warsz. (1851).
Zamia skinneri Warsz. (1851).
Zamia variegata Warsz. (1845).

Warszewicz's Cycad Publications

Warszewicz, J. 1845. *Correpondenz: Nachricht.* Allgemeine Gartenzeitung 13: 252–253. [*Zamia variegata* Warsz.].

Dietrich, A. 1851. *Bericht über zwei neue Zamien.* Allgemeine Gartenzeitung 19:145–146, illustrations by Warszewicz, p. 418. [*Zamia lindleyi* Warsz., *Z. skinneri* Warsz.].

Carl Ludwig Willdenow (1765–1812)

Carl (Karl) Ludwig Willdenow (Fig. 4.43) was born in Berlin on August 22, 1765, and was taught the elements of botany by his uncle Johann Gleditsch, a pioneer in the study of plant sexuality and reproduction and for whom the legume genus *Gleditsia* is named. Willdenow went on to study pharmacy in the medieval town of Langensalza and then medicine at the University of Halle, during which time he published his *Florae Berolinensis Prodromus* (1787), a work that brought him into contact with Alexander von Humboldt. The two became lifelong friends and working colleagues.

Fig. 4.43 A copper plate print of Carl Ludwig Willdenow. Artist unknown, 1812. (From Willdenow 1822)

The young Willdenow earned his income by working in his father's Berlin pharmacy *Apotheke under den Linden* for eight years but meanwhile published the first edition of his *Grundriss der Kräuterkunde zu Vorlesungen* (1792), translated into English as "Principles of Botany," a basic textbook intended to replace Linnaeus' obsolete *Philosophia Botanica.* In 1798, he was appointed Professor of Natural History at the Berlin Medical-Surgical College; then in 1801, he took over the Directorship of the Berlin Botanical Garden; and in addition, became the Professor of Botany at the newly founded University of Berlin. In 1822, Willdenow published his last general botany textbook entitled: *Anleitung zum Selbststudium der Botanik.*

Willdenow's publications are many and varied. One of the most significant was his fourth edition update of Linnaeus' *Species Plantarum* in 1806 where he described three novel cycads among the 16 named taxa at that time. Apart from *Zamia muricata*—as discussed below—he included two other novel cycads. One was *Zamia tenuis* "from the islands of Bahamas" but without further information of any locality; that name is now taken as a synonym for *Z. integrifolia*. Similarly, he mentioned a *Zamia tridentata*, again without locality data, but which appears to represent an Australian *Macrozamia*. The only other work of Willdenow with cycad data is seen in his editorship of Juan de Loureiro's *Flora Cochinchinensis* of 1793.

Plants from the South American collections of Alexander von Humboldt, Aimé Bonpland, and others were re-established in the Berlin Botanical Garden, and vouchers were distributed between Willdenow's personal herbarium now in Berlin (Stafleu, 1972) and the herbarium at the Paris Museum of Natural History. The only cycad known from the Humboldt and Bonpland's collections is a species they found near Puerto Cabello, Venezuela, which Willdenow described in 1806 as *Zamia muricata* (Figs. 4.44, 4.45, and 4.46), the epithet referring to the characteristic small teeth that occur along the distal portion of the leaflets. The holotype (Fig. 4.44) is still present in the Willdenow herbarium collection now part of the Berlin-Dahlem (B) herbarium (Stafleu 1972), under the reference *Humb. and Bonpl.18536*, with an isotype in the herbarium at Leiden (L) (Stevenson and Sabato 1986a). An excellent illustration of *Zamia muricata* by Elisabeth Koning was printed in de Vriese (1847; Fig. 4.45), and a rather comprehensive description, together with general notes on cycad biology, was provided by Gustav Karsten (1857; Fig. 4.46).

Apart from his work in systematics, Willdenow was a pioneer in the field of phytogeography. His work, together with that of Alexander von Humboldt, led to the novel notions that plant distribution patterns changed over time, the barriers to dispersal were not absolute, climate influenced the numbers and character of plant species, new species arose, and previously existing ones could become extinct. These fundamental concepts subsequently influenced many of the cycad biologists discussed in this volume. Willdenow also mentored Christian Sprengel, a forerunner in studies of floral biology and plant pollination.

Willdenow died on July 10, 1812, at age 46. His name was remembered by Carl Thunberg in the genus *Willdenowia* (Restionaceae), and the journal of the same name was published by the Berlin Botanic Gardens.

Fig. 4.44 Humboldt and Bonpland's holotype of *Zamia muricata* that they collected near Puerto Cabello in Venezuela and described by Willdenow in 1806. The voucher, *Humb. and Bonpl. 18,536*, is now filed with other Willdenow material in the Berlin-Dahlem herbarium (B)

Fig. 4.45 A painting of *Zamia muricata* by artist Elisabeth Koning and published in de Vriese (1847)

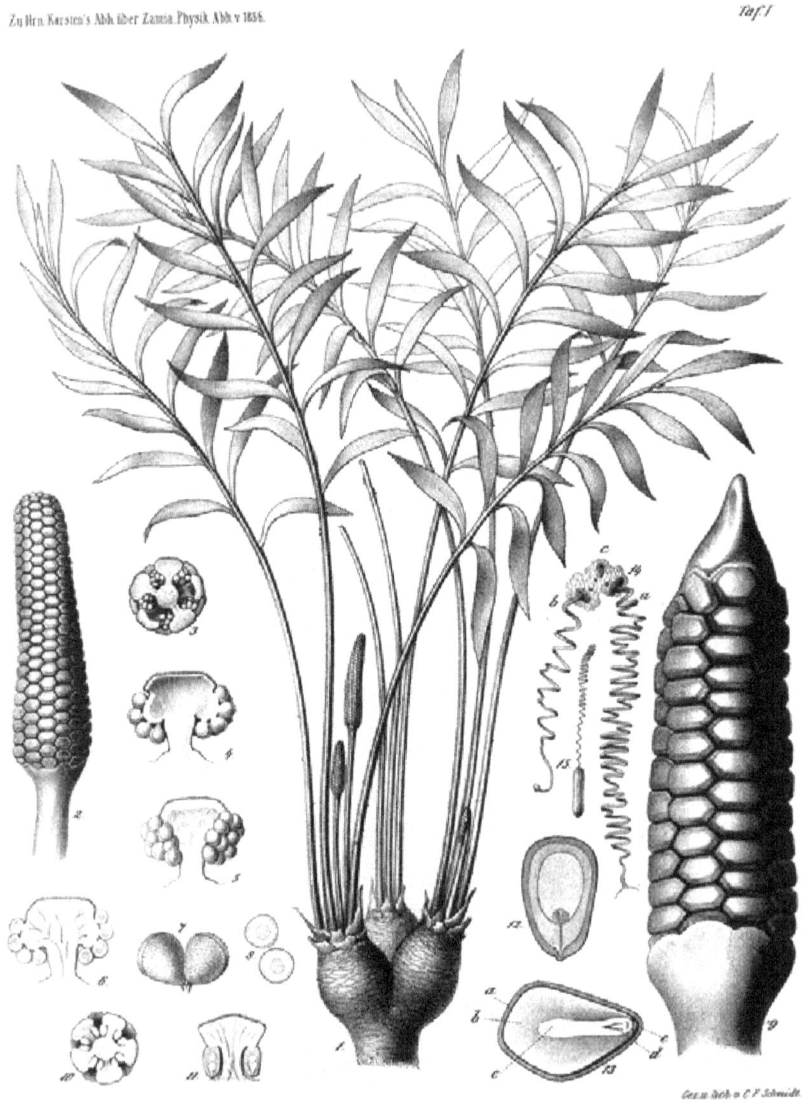

Fig. 4.46 Line drawings of *Zamia muricata* in Gustav Karsten's descriptive text of 1857

Names of the Cycads with Willdenow's Authorship
Data from the *World List of Cycads.*

Zamia muricata Willd. 1806.
Zamia tenuis Willd. 1806 = *Z. integrifolia* L.*f.*
Zamia tridentata Willd. 1806 = *Macrozamia* sp.

Willdenow's Cycad Publication

Willdenow, C. L. 1806 in Linnaeus' *Species Plantarum*, Fourth Edition. *Dioecia polyandra* 4: 845–849. [*Zamia muricata* Willd., *Z. tenuis* Willd., *Z. tridentata* Willd.]

Chapter 5
The Late Nineteenth Century

Alexander Karl Heinrich Braun (1805–1877)

Alexander Karl (Carl) Heinrich Braun (Fig. 5.1) was born on May 10, 1805, in Regensburg, Bavaria. He studied botany in Heidelberg, Paris, and Munich. He took a post as a botany lecturer at the Karlsruhe Polytechnic from 1833 to 1846. He then became Professor of Botany at the Universities of Freiburg and Giessen but at a time when there was serious revolutionary unrest in those cities. The Prussian army restored order in late 1849, and Braun was involved in efforts to release prisoners in Rastatt. Despite these difficult times, he published his most important volume *Betrachtungen über die Erscheinung der Verjüngung in der Natur, inbesondere in der Lebens-und Bildungsgeschichte der Planze* ("The phenomenon of rejuvenescence in nature especially in the life and development of plants") in parts over the period 1849–1850. From 1851 to 1877, he relocated to the University of Berlin as a Professor, where he also assumed duties as the Director of the Berlin Botanical Garden, joint positions that he occupied until his death (Anonymous 1877; Gregersen 2021).

Throughout his career, Braun was a prodigious writer. Much of his research was in plant morphology; he made important contributions to cell theory and was a pioneer in the study of phyllotaxis. He was a proponent of vitalism, a popular nineteenth-century theory that claimed a regulative force existed within living matter to maintain functionality. He also ventured into taxonomy as evidenced by his several monographs and flora treatments of the algal family Characeae, monographs of the lycophytes *Isoetes* and *Selaginella*, and revisions of some monocotyledon genera such as the aquatic genus *Najas*.

Braun's writings on cycads were incidental to his main themes and arose from observations of plants and plant vouchers collected for the Berlin Botanical Garden (Braun 1849–1850). An example of particular interest to cycad enthusiasts is the history of the discovery of *Encephalartos hildebrandtii* (Fig. 5.2), an account of

R. Osborne, D. Wm. Stevenson, *The Cycad Botanists*, Memoirs of The New York Botanical Garden 123, https://doi.org/10.1007/978-3-031-79082-9_5

Fig. 5.1 Alexander Karl Heinrich Braun. An engraving by Weger, ca. 1875. From *Archiv für Kunst und Geschichte*, Berlin

which is given by Otto Stapf (1914) in Kew Bulletin. Stapf tells that Sir John Kirk, British Consul General to Zanzibar, wrote to Joseph Hooker in Kew telling him that he had found a new *Encephalartos* plant on the coast near Dar-es-Salaam, Zanzibar, in 1868. Over the next 16 years, Kirk sent various leaf, cone, seed, and plant samples to Kew. Some of the seeds survived. At least one plant was in cultivation of Kew at least until Stapf's publication in 1914.

In the interim, the German traveler and collector, Johannes Hildebrandt, working along the entire north-eastern coast of Africa and funded on a modest budget from Berlin, was sending numerous natural history specimens back to Berlin. Included among these was a small consignment of cycads that he had found near Mombasa in Kenya (Beentje 1998). Alexander Braun, seeing these plants in the Berlin Botanic Garden and working with the Garden's Curator Carl Bouché, was quick to describe *Encephalartos hildebrandtii* in 1874, thus gaining priority over any possible description from Kew when it became clear that the species was widespread along the east African coast. After some debate about the name and intraspecific relationships of this taxon, Kew accepted Braun's name and published illustrations of *Encephalartos hildebrandtii* in *Curtis's Botanical Magazine* of 1915.

Fig. 5.2 *Encephalartos hildebrandtii*, an East African cycad discovered separately by Sir John Kirk and Johannes Hildebrandt, sent variously to Kew and Berlin, named by Alexander Braun in 1874, and illustrated in the two plates above (plant and pollen cone details, Plate 8582; plant and seed cone details, Plate 8593) from *Curtis's Botanical Magazine* of 1915. Color lithographs by Walter Fitch, reproduced by Vincent Brookes, Day and Son, London

Another four cycad species were described by Braun and are mentioned elsewhere in this volume. In our summary of Berthold Seemann's seafaring expeditions, we tell of Seemann's discovery of a cycad at Cabo Corrientes on the Colombian west coast in December 1847. Gustav Wallis collected material of this cycad and two other species in 1873. This material and Wallis' field notes allowed Braun (1875a) to describe *Zamia montana*, *Z. obliqua*, and *Z. wallisii*. Braun (1875b) also published a rather complete description of *Lepidozamia*. Finally, in the account of Seemann's further voyages of exploration is a report of the native cycad that he found on the island of Fiji, the plant then referred to as *Cycas circinalis* but subsequently named *Cycas seemannii* by Braun in 1876.

Alexander Braun passed away in Berlin on March 29, 1877. His name is remembered in the moss genus *Braunia* and the fern *Polystichum braunii*. His son, Johannes Braun, was a plant collector who worked in Cameroon and Madagascar.

Names of Cycads with Braun's Authorship
Data from the *World List of Cycads*.

Cycas seemannii A. Braun (1876).
Encephalartos hildebrandtii A. Braun & C.D.Bouché (1874).
Zamia montana A. Braun (1875).
Zamia obliqua A. Braun (1875).
Zamia wallisii A. Braun (1875).

Braun's Cycad Publications

Braun, A. & C. D. Bouché. 1874. *Plantae novae a cl. Hildebrandt ex Africa orientali allatae, a. 1874 in horto Berolinensi cultae. Index seminum in Horto Botanico Berolinensi collectorum* 18 [*Encephalartos hildebrandtii* A. Braun & C.D.Bouché].

Braun, A. 1875a. *Die Frage nach der Gymnospermie der Cycadeen erläutert durch der Stellung dieser Familie im Stufengang des Gewächsreichs. Monatsberichte der Königlichen Preussische Akademie des Wissenschaften zu Berlin* 1879: 241–267, 289–377. [*Zamia montana* A. Braun, *Z. obliqua* A. Braun, *Z. wallisii* A.Braun].

Braun, A. 1875b. *Ueber Lepidozamia. Sitzungsberichte der Gesellschaft Naturforschen der Freunde zu Berlin* 1875: 29–37.

Braun, A. 1876. *Ueber zwei von Hildebrandt angeführte Cycadeen nebst Bermerkungen über einige Cycadeen. Sitzungsberichte der Gesellschaft Naturforschen der Freunde zu Berlin* 1876: 113–125. [*Cycas seemannii* A.Braun].

William Bull (1828–1902)

William Bull (Fig. 5.3) was born in Winchester, Hampshire, England, on June 1, 1828. Little is known of his family, childhood, or education, but it seems that his first employment was as "a traveller" for two nursery businesses, Messrs. William Rollisson and Sons of Tooting and Messrs. E. G. Henderson of St. John's Wood. In 1861, he purchased the nursery of John Weeks and Co. in King's Road, Chelsea, and soon commenced sourcing, promoting, and selling rare plants that were unavailable elsewhere. He described a large number of novel plant species in his Retail Lists and in *The Gardeners' Chronicle*, as well as exhibiting specimens regularly at the Royal Horticultural Society shows and international events. He was well known in English scientific circles, being a Fellow of the Linnean Society of London, the Royal Gardening Society, the Royal Zoological Society, the Royal Botanical Society, and other learned institutions. He was apparently delighted when awarded the Victoria Medal of Honour by the Council of Royal Horticultural Society (Anonymous 1902).

Bull's rare plant inventories were both impressive and diverse, focusing on a large range of orchids, bulbs, ferns, palms, and flowering and foliage ornamentals. He sourced and introduced unusual plants from North, Central, and South America and the islands of the Western Pacific, Asia, Australia, and Tropical and Southern Africa. In later issues of his Retail Lists (Fig. 5.4), Bull rather grandly boasts that his establishment "has been favoured by the distinguished patronage of …" proceeding to list a lengthy entourage of British and foreign royalty and aristocracy.

Among his suppliers were Edward Shuttleworth and John Carder who, as mentioned in our section on Gustav Wallis, were commissioned in 1882 to collect a consignment of the *Zamia montana* plants that Wallis had located in the cloud forests of Antioquia, Colombia. Bull was generous in sharing some of these cycads and several other rare plant imports with botanists at Kew.

The cycads that Bull listed for sale in the Retail List over the period 1861–1914 comprised over 50 different species in eight genera. Many of the names he used do

Fig. 5.3 William Bull, English botanist, nurseryman, and plant collector. (Image from The Gardeners' Chronicle series 3, 31: 381. 1902)

not qualify as being validly published, and many of those names have never been used elsewhere. His lists include such oddities as *Encephalartos denticulatus*, *E. glaucus*, *E. majesticus*, *E. multiflexus*, *Macrozamia eburnea*, *M. elegantissima*, and *M. plumosa*, names which we have difficulty relating to any known taxa.

Of the various cycads that Bull introduced to horticulture is *Zamia prasina* which he sourced from British Honduras (now Belize), a cycad better known to collectors under its previous name Z. *polymorpha* (Calonje and Meerman 2009). The Retail List 176 of 1881 reads:

Another of Bull's important cycad introductions was *Bowenia serrulata*. This he described as "a most distinct and remarkable plant which gained first prize at the International Horticultural Exhibitions held at Carlisle in 1877 and in Ghent in 1878."

1879.

A RETAIL LIST

NEW

BEAUTIFUL & RARE

PLANTS

WILLIAM BULL,

F.L.S., F.R.G.S., F.Z.S., F.R.H.S., M.A.I., F.R.B.S. & M.S.A.,

NEW PLANT MERCHANT,

KING'S ROAD, CHELSEA,

LONDON, S.W.

WILLIAM BULL, F.L.S.,

BEAUTIFUL PLANTS,

PERFECTLY NEW TO THIS COUNTRY,

MADAGASCAR, PERU, JAVA, PHILIPPINE ISLANDS, NEW CALEDONIA, ST. CATHERINE'S, RIO NEGRO, NICARAGUA, UNITED STATES OF COLOMBIA, COSTA-RICA, PARA, NATAL, JAPAN, ASCENSION, SOUTH SEA AND FIJI ISLANDS, &c., &c.

AN INSPECTION OF WHICH IS INVITED.

THE PRINCE CONSORT'S GOLD MEDAL,

THE LINDLEY MEDAL,

MANY GOLD MEDALS,

UPWARDS OF ONE THOUSAND PRIZES

AND AWARDS

WILLIAM BULL, F.L.S.,

FOR NEW AND RARE PLANTS ALONE,

Mr. W. BULL mentions this as conveying some idea of the importance of his collection.

Fig. 5.4 Two frontispiece pages to William Bull's Retail List 154 of 1879 in which *Ceratozamia fuscoviridis* is described and offered for sale by "price on application"

Bull died in Chelsea, London, on June 1, 1902, at the age of 74. The obituary written in *The Gardens' Chronicle* that year concludes that "he was a most remarkable man and his death will cause deep regret among a large circle of friends and correspondents."

Zamia prasina
A handsome and distinct-looking Cycad, introduced from [British] Honduras. The stems are in the young plants roundish or roundish-ovate, clothed with the imbricating scales left by the falling of the leaves. The leaves are equally pinnate, the dark dull-green leaf stalks terete, with an indistinct furrow in front, and furnished with a few scattered white prickles. The leaf-blade is oblong-ovate, pinnate, with some sixteen or seventeen pairs of oblanceolate pinnae, entire, and tapered at the base, where they are set on by a distinctly swollen articulation, denticulate towards the apex, the upper surface of a bright grass-green colour, whence the name. 1½ guineas.

Names of Cycads with Bull's Authorship

Data from the *World List of Cycads*

Bowenia spectabilis var. *serrulata* W. Bull (1878) ≡ *B.serrulata* (W. Bull) Chamb.

Ceratozamia fuscoviridis W. Bull (1879).

Encephalartos mackenii W. Bull (1870), *nomen dubium.*

Encephalartos majesticus W. Bull (1870), *nomen dubium.*

Encephalartos regalis W. Bull (1889) = *E. altensteinii* Lehm.

Encephalartos villosus var. *amplicatus* W. Bull (1873), *nomen dubium.*

Macrozamia corallipes var. *gyrata* W. Bull (1873), *nomen dubium.*

Macrozamia cylindrica W. Bull (1874), *nomen dubium.*

Macrozamia eburnea W. Bull (1873), *nomen dubium.*

Macrozamia elegantissima W. Bull (1873), *nomen dubium.*

Macrozamia mackenzii W. Bull (1877) = *M. miquelii* (F. Muell.) A.DC.

Macrozamia plumosa W. Bull (1874), *nomen dubium.*

Macrozamia spiralis var. *eburnea* W. Bull (1873), *nomen dubium.*

Macrozamia spiralis var. *princeps* W. Bull (1873), *nomen dubium.*

Stangeria schizodon W. Bull (1872) = *S. eriopus* (Kunze) Baill.

Zamia amplifolia W. Bull *ex* Mast. (1878).

Zamia miquelii W. Bull (1876), *nomen dubium.*

Zamia prasina W. Bull (1881).

Zamia tenuifolia W. Bull (1872), *nomen dubium.*

Bull's Cycad Publications

Bull, W. 1870. New plants: *Encephalartos mackenii.* Retail List No. 54: 116. [*Encephalartos mackenii* W. Bull].

Bull, W. 1872. New plants: *Stangeria schizodon, Zamia tenuifolia.* Retail List No. 72: 8. [*Stangeria schizodon* W. Bull, *Zamia tenuifolia* W. Bull].

Bull, W. 1873. New plants: *Macrozamia eburnea, M. elegantissima.* Gardeners' Chronicle 30(11): 358. [*Macrozamia eburnea* W. Bull, *M. elegantissima* W. Bull].

Bull, W. 1873. New plants: *Encephalartos majesticus, E. villosus* var. *amplicatus.* Retail List No. 83: 8, 187. [*Encephalartos majesticus* W. Bull, *E. villosus* var. *amplicatus* W. Bull, *Macrozamia corallipes* var. *gyrata* W. Bull, *M. spiralis* var. *eburnea* W. Bull, *M. spiralis* var. *princeps* W. Bull].

Bull, W.1874. New plants: *Macrozamia plumosa.* Gardeners' Chronicle n.s. 1(17): 532. [*Macrozamia plumosa* W. Bull].

Bull, W. 1876. New plants: *Zamia miquelii.* Retail List No. 121: 141. [*Zamia miquelii* W. Bull].

Bull, W. 1877. New plants: *Macrozamia mackenzii.* Retail List No. 129: 7. [*Macrozamia mackenziei* W. Bull].

Bull, W. 1878. New plants: *Bowenia spectabilis* var. *serrulata.* Retail List No. 143: 4, Plate 5. [*Bowenia spectabilis* var. *serrulata*].

Bull, W. 1878. New plants: *Zamia amplifolia* Hort. Bull. Gardeners' Chronicle 10(261): 810. [*Zamia amplifolia* W. Bull *ex* Mast.]

Bull, W. 1879. New plants: *Ceratozamia fuscoviridis.* Retail List No. 154: 4. [*Ceratozamia fuscoviridis* W. Bull].

Bull, W. 1889. New plants: *Encephalartos regalis.* Retail List No. 250: 8, Plate 4. [*Encephalartos regalis* W. Bull].

Alphonse Louis Pierre Pyramus de Candolle (1806–1893)

In our section on Eduard von Regel (1815–1892), we mentioned that there were five generations of that family involved in horticultural and botanical endeavors. We now turn to another European family, the Swiss de Candolle clan, where four generations were botanists of considerable note (Briquet 1940): Augustin Pyramus (1778–1841), Alphonse Louis Pierre Pyramus (1806–1893), Anne Casimir Pyramus (1836–1918), and Richard Émile Augustin (1868–1920). Of these, it is Alphonse (Fig. 5.5) who was influential in cycad biology, and it is his botanical work that we focus on in the text that follows.

He was born in Paris on October 28, 1806, the son of Augustin Pyramus and Fanny née Forras. His boyhood was spent in Montpellier, southern France, until his father relocated to take up the Professorship of Natural History at the Academy of Geneva (later the University of Geneva) in 1816. Alphonse studied law at the university, graduating in 1829. He then assisted his father in the Natural History department, succeeded him as Chair in 1835, and also assumed the Directorship of the Botanical Garden of Geneva.

After 15 years, de Candolle stepped down from the academic and administrative responsibilities to concentrate on his self-funded research and writing, working from the ancestral home high (Fig. 5.6) on a hill in Geneva's Cour de St. Pierre precinct. De Candolle's personal library and herbarium became a sanctuary where researchers from all over the world came to discuss their botanical projects with the Genevan master.

The de Candolle succession assembled one of the most important private herbarium collections known, estimated as about 400,000 specimens and containing thousands of types. In 1921, one year after the premature death of Richard Augustin, this important herbarium was donated to the Conservatory and Botanic Gardens of Geneva. This institution ranks among the six richest herbaria worldwide; the de Candolle specimens are filed in the general collection (G) or in the specific subcollection used in the preparation of the *Prodromus* (G-DC).

Various important botanical works were published by the de Candolle family, but it was their *Prodromus systematis naturalis regni vegetabilis* (1824–1873; Fig. 5.7) that is best known; this mammoth seventeen-volume publication was commenced by Augustin, continued and extended by Alphonse, with the final volume completed by Casimir. It was in the penultimate volume of the *Prodromus* (Vol. 16, 1868: Fig. 5.8) that Alphonse dealt with the Cycadaceae. In 26 pages, he reviewed all known species in nine genera but excluded *Lepidozamia* as a *genus dubium*. The new cycads that he described or revised were *Dioon strobilaceum* (now *D. edule*); three species of *Macrozamia*, *Microcycas calocoma*; and six species of *Zamia*. One of the interesting cases is the typification of *Zamia furfuracea* var. *trewii* where the only material cited by de Candolle in the protologue is the illustration *Palmifoliajructu clavato polypireno* [*polyspermo*] in C. J. Trew, *Plantae Selectae*. Table 26. 1752 (Stevenson and Sabato 1986a) that was done from a plant in Lord Peter King's Garden. This, therefore, must serve as the type, and because the plate is mounted as

Fig. 5.5 Alphonse Louis Pierre Pyramus (or Pyrame) de Candolle. (From Wikipedia)

a specimen at Geneva, it can serve as the holotype (Fig. 5.9). Stevenson and Sabato (1986a) discovered the original specimen of 1694 at the British Museum of Natural History from Lord Peter King's (as Lord Petro on the label) garden (Fig. 5.10) and from which the plate was based. They seriously doubted that de Candolle ever saw this specimen and designated the original specimen as the type of the type

Fig. 5.6 The home of the de Candolle family, with associated plaque, in Cour de Saint-Pierre in the historical part of Geneva. This is where Alphonse de Candolle prepared his *Prodromus* of 1868 dealing with the Cycadaceae, and it was the repository for the de Candolle herbarium for almost 105 years. (Photo: Fred Stauffer)

Fig. 5.7 The de Candolle seventeen-volume *Prodromus systematis naturalis regni vegetabilis (1824–1873)* commenced by Augustin, continued and extended by Alphonse, with the final volume completed by Casimir. (Photo courtesy of the *Conservatoire et Jardin Botaniques* (CJB, Geneva))

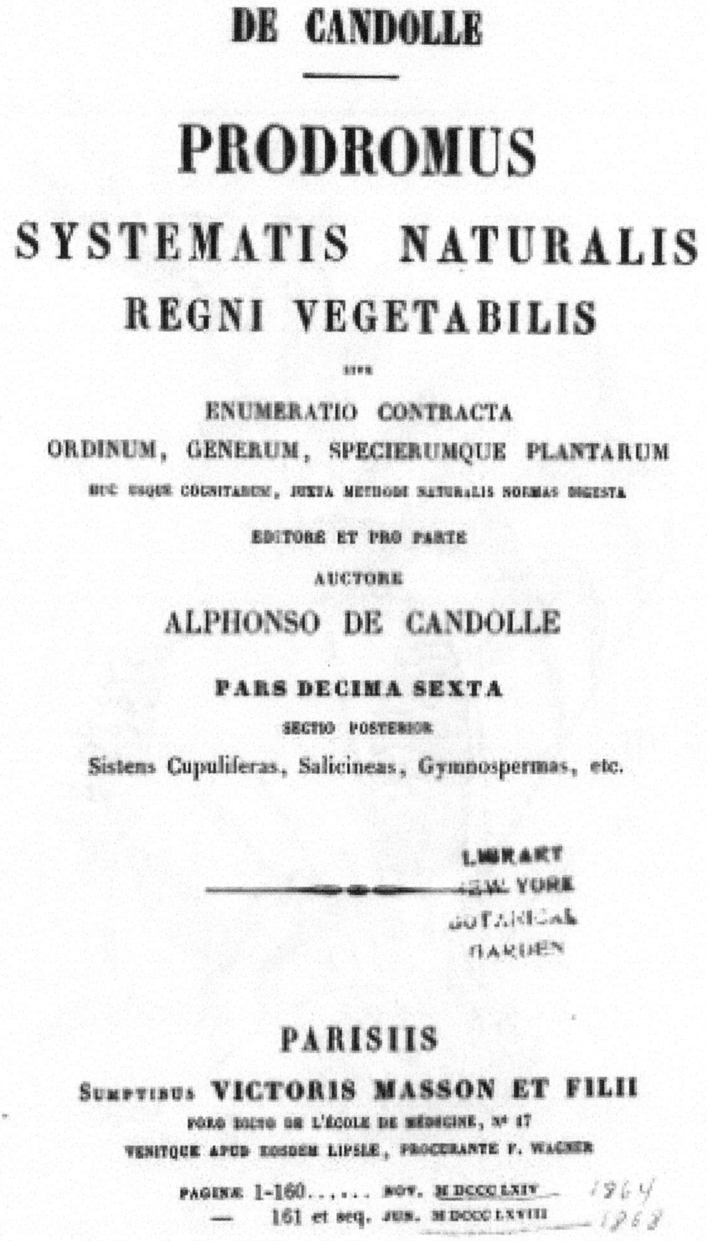

DE CANDOLLE

PRODROMUS

SYSTEMATIS NATURALIS

REGNI VEGETABILIS

SIVE

ENUMERATIO CONTRACTA

ORDINUM, GENERUM, SPECIERUMQUE PLANTARUM

HUC USQUE COGNITARUM, JUXTA METHODI NATURALIS NORMAS DIGESTA

EDITORE ET PRO PARTE

AUCTORE

ALPHONSO DE CANDOLLE

PARS DECIMA SEXTA

SECTIO POSTERIOR

Sistens Cupuliferas, Salicineas, Gymnospermas, etc.

PARISIIS

SUMPTIBUS VICTORIS MASSON ET FILII

FORO DICTO DE L'ÉCOLE DE MÉDECINE, Nº 17

VENITQUE APUD EOSDEM LIPSIÆ, PROCURANTE F. WAGNER

PAGINÆ 1-160...... NOV. MDCCCLXIV 1864

— 161 et seq. JUN. MDCCCLXVIII 1868

Fig. 5.8 The title page of Volume 16, Part 2, of Alphonse de Candolle's *Prodromus systematis naturalis regni vegetabilis*, in which he deals with the Cycadaceae in his section *Gymnospermaes*

Fig. 5.9 The illustration in the Geneva Herbarium of Alphonse de Candolle's *Zamia furfuracea* var. *trewii*. The mezzotint portrait by Georg Ehret, originally printed in *Plantae Selectae* (1752), is the type for the variety, which has since been included in the broader concept of *Z. furfuracea*. The actual plate taken from the book is mounted on a herbarium sheet

Fig. 5.10 The typotype of *Zamia furfuracea var. trewii* at the British Museum of Natural History

(typotype) because the holotype must be material seen by the author of a description at the time the description was prepared.

Alphonse's other major botanical volumes included his *Géographie botanique raisonnée* (1855), *La phytographie* (1880), and *Origine des plantes cultivées* (1883, also published in English as the "Origin of Cultivated Plants" in 1886), all of which contributed greatly to establishing the basis of modern biogeography and in part to ethnobotany. Influenced by his training in law, he was also instrumental in formulating *Lois der la nomenclature botanique*, the first set of Laws of Botanical Nomenclature, as adopted by the International Botanical Congress in Paris in 1867, with Alphonse being the President at that time. In addition, he produced monographs on the families Apocynaceae, Begoniaceae, Campanulaceae, Myristicaceae, Myrsinaceae, Santalaceae, and Styracaceae. He also published various papers on plant morphology and physiology and a remarkable historical summary *Histoire des sciences et des savants depuis deux siècles* (1873), "which displayed the objectivity of a naturalist with the clarity of a jurist" (Pilet n.d.).

Apart from these significant contributions to science, Alphonse de Candolle was also active in politics and community affairs; among other achievements in that area, he was responsible for the introduction of postage stamps to Switzerland.

Alphonse de Candolle, like Jacquin, Thunberg, and Robert Brown, was a Foreign Member of the Royal Swedish Academy of Sciences and the Royal Netherlands Academy of Arts and Sciences. He was a Fellow of the Royal Society of London. He was awarded the Gold Medal of the Linnean Society of London in 1889. He passed away at the age of 87 in Geneva on April 4, 1893, and was survived by his son, Casimir (1836–1918), and grandson, Richard (1868–1920). In an obituary, de Candolle's American friend, William Farlow (1893), wrote fondly that:

> Alphonse de Candolle was tall in stature, with a prominent nose, and small, rather deep-set eyes. His appearance was strikingly dignified, for his manners were polished and courteous, and he had the happy faculty of making all, no matter how different their ages or condition, feel perfectly at ease in his presence.

William Thiselton-Dyer (1893) wrote from his office at Kew similar to de Candolle:

> Distinguished in appearance, his manners though reserved, were always exquisitely urbane. He was singularly kind to all who were disposed to engage in biological work; and would spare no pains to help and even aid, with his own accumulated material, those who were willing to undertake research.

Names of Cycads with Alphonse de Candolle's Authorship
Data from the *World List of Cycads.*

Dioon strobilaceum Lem. *ex* A.DC. (1868) = *Dioon edule* Lindl.
Macrozamia macdonnellii (F. Muell. *ex* Miq.) A.DC. (1868).
Macrozamia miquelii (F. Muell.) A.DC. (1868).
Macrozamia oldfieldii (Miq.) A.DC. (1868) = *M. riedlei* (Gaudich.) C.A.Gardner.

Microcycas calocoma (Miq.) A.DC. (1868).
Zamia boliviana (Brong.) A.DC. (1868).
Zamia floridana A.DC. (1868) = *Z. integrifolia* L.*fil.*
Zamia furfuracea var. *trewii* A.DC. (1868) = *Z. fufuracea* L.*fil.*
Zamia multifoliolata A.DC. (1868) = *Z. angustifolia* Jacq.
Zamia spartea A.DC. (1868).
Zamia pygmaea var. *wrightii* A.DC. (1868) = *Z. pygmaea* Sims.

Alphonse de Candolle's Cycad Publication

De Candolle, A. P.1868. Cycadaceae, in *Prodromus systematis naturalis regni vegetabilis* 16(2): 522–548.

August Wilhelm Eichler (1839–1887) and Robert Knuds Friedrich Pilger (1876–1953)

August Wilhelm Eichler (Fig. 5.11) was born on April 22, 1839, in Neukirchen, Hessen, Germany, attending school at Eschwege and then studying mathematics and natural sciences at the University of Marburg. His first role was an assistant to Karl von Martius at Munich, where he was appointed teacher in botany in 1865, taking over editorial responsibilities for the *Flora Brasiliensis* in1868 when Martius passed away. He worked as a Professor of Botany at the Technical High School in Graz, also supervising the local botanical garden. A year later, he was appointed Professor of Botany at Kiel. Finally, in 1877, he succeeded Alexander Braun as Professor of Botany and Director of the Berlin Botanic Garden.

His graduate dissertation in 1861, on the subject of leaf emergence and development, set the tone for numerous further publications over the next 25 years. Considered one of the most prominent systematic and morphological botanists of his time, he was the proponent of the "Eichler System," an early phylogenetic classification where the plant kingdom is separated into the Cryptogamae (non-flowering plants) and Phanerogamae (flowering plants), the latter divided into gymnosperms and angiosperms, the latter in turn separated into monocotyledons and dicotyledons. The Eichler system became the foundation for Adolf Engler's widely accepted basis of plant classification. Equally important is his two-volume work on the organization of seed pant reproductive axes, which begins with cycads and other gymnosperm cones followed by angiosperm families all in the sequence of the Eichler classification system. The *Blütendiagramme* (Eichler 1875–1878) has remained the standard for such description to this day.

In 1863, the treatment of the Cycadaceae in *Flora Basiliensis* by Eichler included the description and illustration of *Zamia poeppigiana* (Fig. 5.12), a cycad that the German plant collector Eduard Poeppig had found in the Maynas Alto region of

Fig. 5.11 August Wilhelm Eichler (1839–1887), considered one of the most prominent systematic and morphological botanists of his time. (From Verhandlungen des Botanischen Vereins der Provinz Brandenburg 29: 1. 1888)

Peru in 1830. Eichler also mentioned a *Zamia parasitica*, but the validity of that name was never confirmed. Some twenty years later, Eichler told of another Mexican cycad, *Dioon spinulosum*, sample material of which had reached both William Thiselton-Dyer in Kew and Eichler in Berlin. Although the origin of the plants was not clear and no coning material was seen, Eichler (1883a), with Thiselton-Dyer's help, published a description in *Gartenzeitung* (Fig. 5.13). In the same year, Eichler (1883b) published a short paper on *Lepidozamia peroffskyana*.

Eichler's (1887a) treatment of the Cycadaceae in *Die natürlichen Pflanzenfamilien* (Fig. 5.14) commences with a discussion of morphological and anatomical features illustrated with 14 figures, most of which are reproductions from earlier authors' works, and followed by a taxonomic treatment of 83 species in nine genera. Eichler (1887b) also covered fossils in his cycad treatment in *Die natürlichen* Pflanzenfamilien. A complete bibliography of Eichler's publications may be found in Balfour (1887).

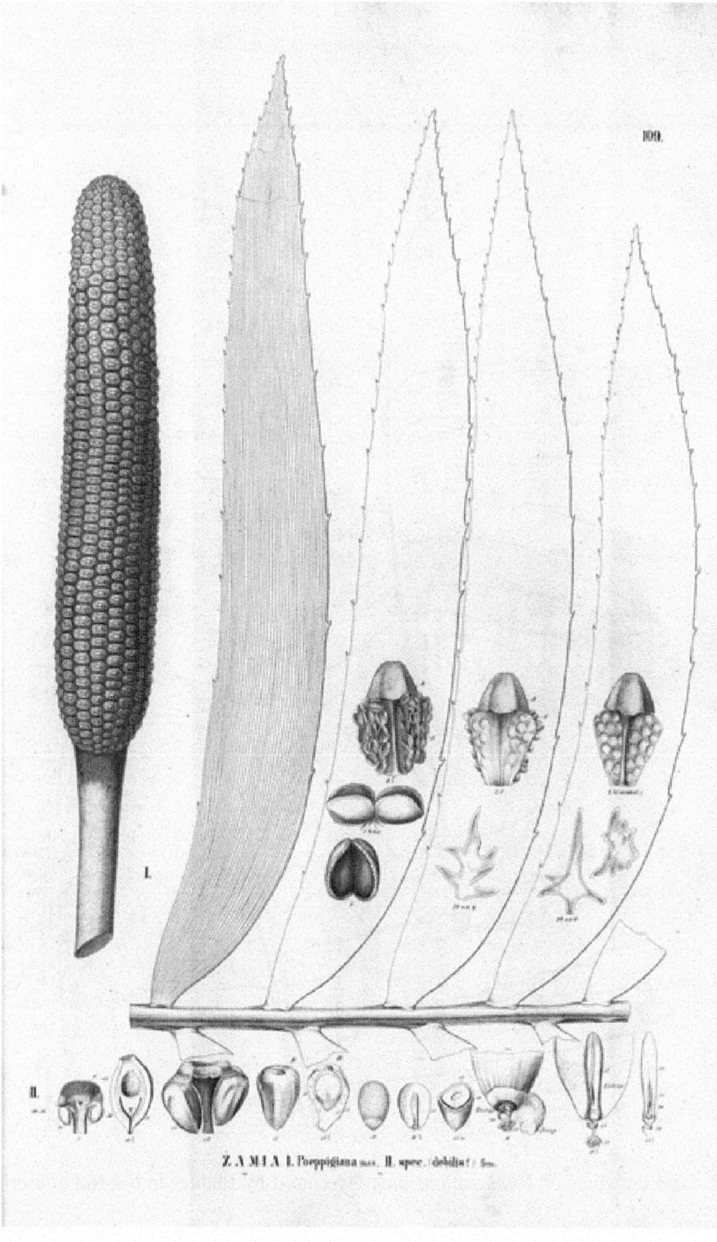

Fig. 5.12 Plate 109, the illustration of *Zamia poeppigiana* that Martius and Eichler used in the species description in *Flora Brasiliensis* in 1863

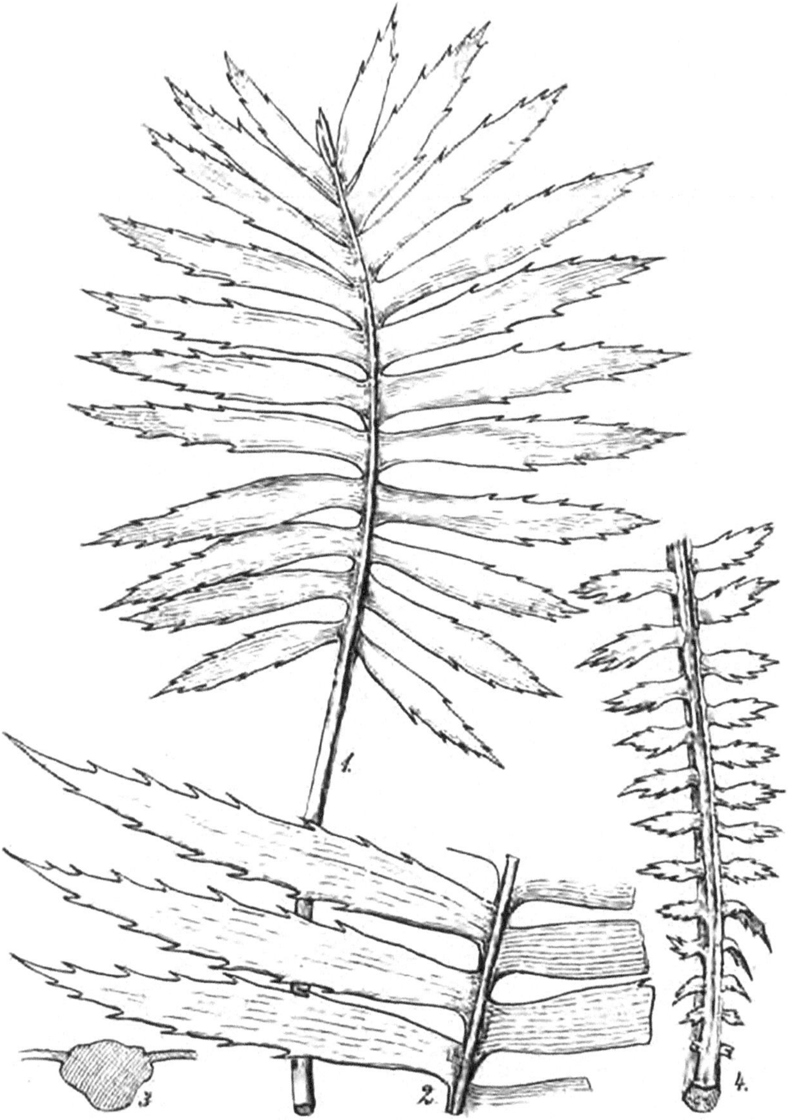

Fig. 5.13 Leaf drawings of *Dioon spinulosum* Dyer used by Eichler in his 1883 description of the species

August Eichler contracted leukemia in 1878 and died in Berlin on March 2, 1887, at age 47 (Anonymous 1888). He had served as Secretary to the International Botanical Congress in Paris in 1867; he was a Member of the Berlin Academy of Sciences, the French Academy of Sciences, the Royal Academy of Belgium, the Academy of Sciences of Munch, and the Linnaean Society of London. His name is

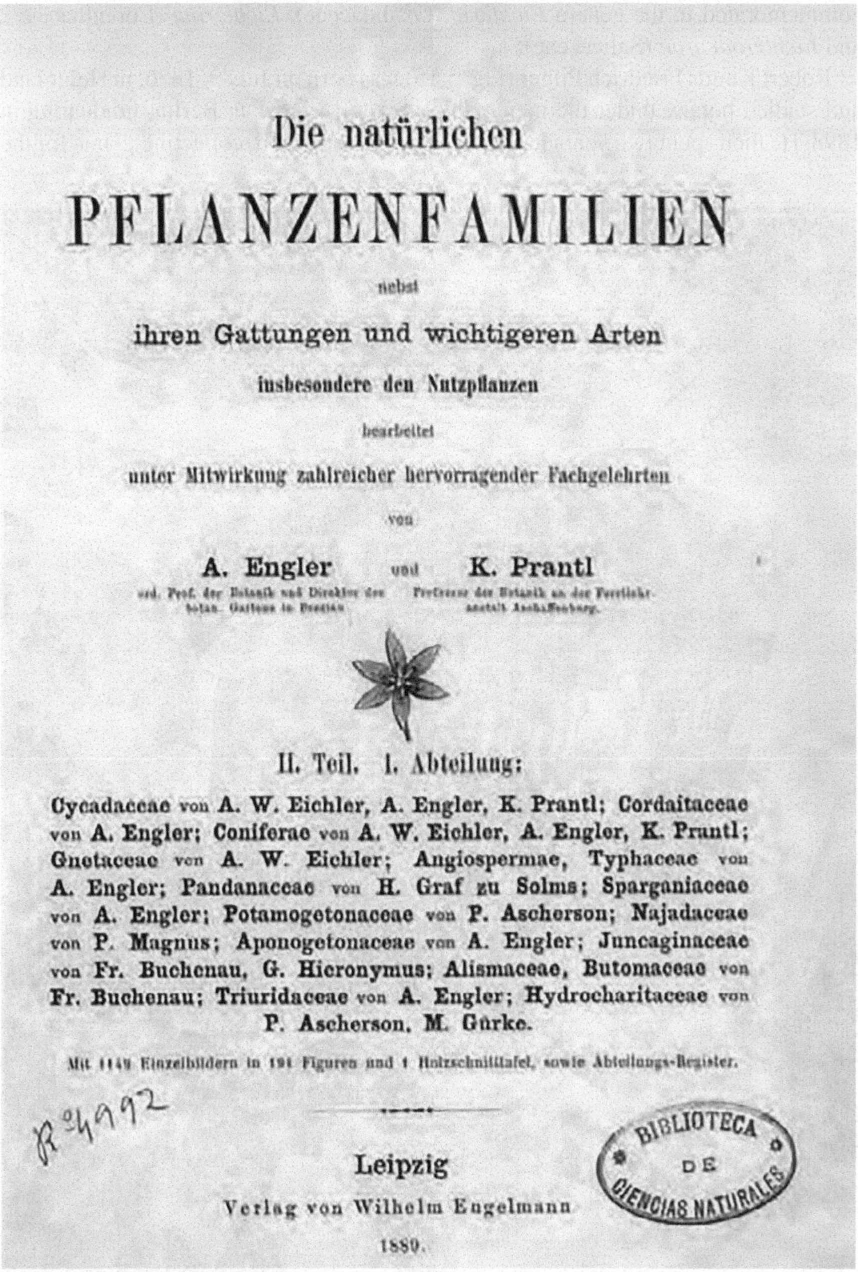

Die natürlichen

PFLANZENFAMILIEN

nebst

ihren Gattungen und wichtigeren Arten

insbesondere den Nutzpflanzen

bearbeitet

unter Mitwirkung zahlreicher hervorragender Fachgelehrten

von

A. Engler und K. Prantl

ord. Prof. der Botanik und Direktor des botan. Gartens in Breslau

Professor der Botanik an der Forstlehranstalt Aschaffenburg.

II. Teil. 1. Abteilung:

Cycadaceae von A. W. Eichler, A. Engler, K. Prantl; Cordaitaceae von A. Engler; Coniferae von A. W. Eichler, A. Engler, K. Prantl; Gnetaceae von A. W. Eichler; Angiospermae, Typhaceae von A. Engler; Pandanaceae von H. Graf zu Solms; Sparganiaceae von A. Engler; Potamogetonaceae von P. Ascherson; Najadaceae von P. Magnus; Aponogetonaceae von A. Engler; Juncaginaceae von Fr. Buchenau, G. Hieronymus; Alismaceae, Butomaceae von Fr. Buchenau; Triuridaceae von A. Engler; Hydrocharitaceae von P. Ascherson, M. Gürke.

Mit 1149 Einzelbildern in 191 Figuren und 1 Holzschnitttafel, sowie Abteilungs-Register.

Leipzig

Verlag von Wilhelm Engelmann

1889.

Fig. 5.14 The title page of Volume 2 of Engler and Prantl's *Die natürlichen Pflanzenfamilien* of 1889 in which August Eichler published his monograph on the Cycadaceae

commemorated in the genera *Eichleria* (Oxalidaceae), *Eichlerina* (Loranthaceae), and *Eichlerodenron* (Salicaceae).

Robert Knuds Friedrich Pilger (Fig. 5.15) was born on July 3, 1876, in Helgoland and studied botany under the mentorship of Adolf Engler in Berlin, graduating in 1898. He then spent two years in the Mato Grosso of Brazil, collecting plants for the

Fig. 5.15 Robert Pilger's grave in the cemetery of Kaiser Wilhelm Memorial Church in Berlin. https://www.findagrave.com/memorial/187972227/robert-knud-friedrich-pilger

Berlin Botanical Garden of which he ultimately became Professor and Director. His scientific contributions were varied and extensive but with a strong focus on gymnosperms (Pilger 1926a) and grasses. Pilger's update of the Cycadaceae in *Die natürlichen Pflanzenfamilien* (1926b) is similar in presentation and content to that of Eichler but more detailed, with 36 figures and covering 88 species. Pilger also revised Eichler's rank of tribe to subfamily.

Robert Pilger passed away in Berlin on September 1, 1953, at age 76. His name is remembered in the algal genus *Pilgeria*, the fungus *Pilgeriella*, the grass *Pilgerochloa*, and the cypress *Pilgerodendron*.

Names of Cycads with Eichler's Authorship

Data from the *World List of Cycads*

Dioon spinulosum Dyer *ex* Eichler (1883).
Zamia parasitica Poepp. *ex* Eichler (1863), *nomen dubium*.
Zamia poeppigiana Mart. & Eichler (1863).

Eichler & Pilger's Cycad Publications

Eichler, A.W. 1863. Cycadaceae. Pp 410–416, Plates 108–109 in C.F.P. de Martius (ed.), *Flora Brasiliensis,* Frid. Fleischer, Leipzig. [*Zamia parasitica* Poepp. ex Eichler, *Z. poeppigiana* Mart. & Eichler].

Eichler, A.W. 1875. Gymnospermae, *Blüthendiagramme* Vol. 1. Eichler, A.W. 1879. *Samen von Ceratozamia mexicana. Sitzungsberichte der Gesellschaft Naturforschender Freunde zu Berlin.*

Eichler, A.W. 1883. *Lepidozamia peroffskyana Rgl. Gartenzeitung* 2(1): 38–42.

Eichler, A.W. 1883. *Ein neues Dioon Gartenzeitung* 2: 411–413, 438–439, Plate 80. *[D. spinulosum Dyer].*

Eichler, A.W. 1887. Cycadaceae, Sago- oder Farnpalmen. Pp. 6–23 in: A. Engler & K. Prantl (eds), *Die natürliche Pflanzenfamilien nebst ihren Gattungen und wichtigeren Arten.* Vol 2. Wilhelm Engelmann, Leipzig.

Eichler, A.W. 1887. Fossile Gattungen der Cycadaceae. Pp. 24–26 in: A. Engler & K. Prantl (eds), *Die natürliche Pflanzenfamilien nebst ihren Gattungen und wichtigeren Arten.* Vol. 2. Wilhelm Engelmann, Leipzig.

Pilger, R.K.F. 1926. Cycadaceae. Pp. 44–82 in: A. Engler & K. Prantl (eds), *Die natürliche Pflanzenfamilien nebst ihren Gattungen und wichtigeren Arteninsbesondere den Nutzpflanzen.* Vol. 13. Wilhelm Engelmann, Leipzig.

William Jackson Hooker (1785–1865) and Joseph Dalton Hooker (1817–1911)

William Jackson Hooker (Fig. 5.16) was born in Norwich, England, on July 6, 1785, son of a merchant's clerk and descendant of the sixteenth-century historian John Hooker and the theologian, Richard Hooker. Soon after completing his schooling at Norwich Grammar and Stanton Hall, he inherited the considerable estate of his godfather, which allowed him to travel to Iceland, France, Switzerland, and Italy

Fig. 5.16 William Jackson Hooker (1785–1856). (A lithograph by Thomas Herbert Maguire, from the Wellcome Collection, London)

and to meet many of the leading European botanists while also securing the patronage of James Smith and Joseph Banks. In 1815, he married Maria Turner, daughter of botanist and antiquarian, Dawson Turner, and the couple produced five children.

In 1820, William Hooker was appointed Regius Professor of Botany at Glasgow University, where he also developed their botanic garden while spending 21 years in teaching and research. In 1841, he succeeded William Aiton as the first full-time Director of the Royal Botanic Gardens at Kew. Under his enthusiastic and capable leadership, the Gardens were expanded and enhanced to include a 270-acre arboretum, a museum for economic botany, and the iconic Palm House, and Kew became recognized as the world's leading botanical institution (Allan 1967; Desmond 1975; FitzGerald 2009).

During his career, William Hooker published more than 20 major works and numerous periodicals, mainly in cryptogamic botany, floristic studies, and economic botany. He both contributed to and edited many issues of *Curtis's Botanical Magazine*, the *Journal of Botany*, and Kew's *Botanical Miscellany*. His many distinctions included memberships in the Linnean Society, the British Royal Society, the Royal Swedish Academy of Sciences, the *Académie Française*, the American Academy of Arts and Sciences, and the American Philosophical Society. He was awarded an honorary doctorate from the University of Glasgow and was made a Knight of the Royal Guelphic Order, a Knight Bachelor, and a Companion of the French *Légion d'Honneur*. His name is remembered in the moss genus *Hookeria* and at least 17 *hookeri* epithets. Sir William Jackson Hooker died on August 12, 1865, at the age of 80 of a septic paralysis of the throat and was buried in St Anne's Churchyard in Kew Green.

Joseph Dalton Hooker (Fig. 5.17), the younger son of William and Maria, was born on June 30, 1817, in Halesworth, Suffolk. Unlike his father, Joseph had the advantage of a formal education, graduating with a medical degree from the University of Glasgow in 1839. At the age of 22, he was appointed surgeon-botanist on the 4-year expedition of HMS *Erebus* to Antarctica. This was followed by a 3-year expedition to India and the Himalayas and somewhat shorter voyages to Palestine, Morocco, and the U.S.A., each of which resulted in the discovery of new plant species and the publication of important papers. He was twice married and fathered nine children. In 1855, he was appointed Assistant Director of the Royal Botanic Gardens at Kew, ten years later succeeding his father as Director and holding that position for 20 years. He continued his father's work with acclaim but had to overcome somewhat strained relations with the British Museum as both institutions vied for supremacy in their herbarium collections.

Joseph Hooker was the author of numerous papers and monographs; his main works included the seven-volume *Flora of British India* and the *Genera Plantarum*, a huge work co-authored with George Bentham (1862–1863) covering approximately 97,000 species of seed-bearing plants in more than 7500 genera. He also continued his father's work by editing the various Kew journals and producing volumes 11–17 of the *Icones Plantarum*. He became a close friend and supporter of Charles Darwin and was greatly involved with the various meetings addressing Darwin's theories and debating the priority of Darwin's work over that of Alfred Wallace.

Fig. 5.17 Sir Joseph Dalton Hooker (1817–1911). (A lithograph by Thomas Herbert Maguire, from the Wellcome Collection, London)

Among Joseph Hooker's many distinctions are his presidencies of the British Association and the Royal Society, his Fellowship of the Linnean Society, and his memberships of the Royal Swedish Academy of Sciences, the Royal Netherlands Academy of Arts and Sciences, and the American Philosophical Society. He was awarded the Royal Society's Royal Medal, Copley Medal and Darwin-Wallace Medal, the Gold Medal of the Royal Geographical Society, and a Commonwealth Order of Merit. He was made a Companion of the Order of the Bath and Knight Grand Commander of the Order of the Star of India (Turrill 1963).

The plant species *Banksia hookeriana*, *Grevillea hookeriana*, *Iris hookeriana*, *Polygonatum hookeri*, and *Sarcococca hookeriana* are named for Joseph Hooker, as well as the Antarctic snail *Notodiscus hookeri*, the New Zealand sea-lion *Phocarctos hookeri*, the town of Hooker Oak in California, and Hooker Island in Franz Josef Land. He passed away on December 10, 1911, after a short illness and was buried alongside his father in St Anne's Church in Kew Green.

The Hookers' contribution to cycad biology is most important through their roles at Kew, principally the stewardship of the cycad vouchers in the herbarium and the development of a living cycad collection in the glasshouses. They were involved directly in the taxonomy of only five cycad taxa. In 1859, William Hooker published a beautiful illustration and informative text on *Stangeria* as Plate 5121 in Curtis's Botanical Magazine. Later, William Hooker (1863a) was puzzled by the South African plant that had been growing in Kew under the German botanist Carl Willdenow's label of *Zamia tridentata,* but which Hooker considered it to be "a variety of *Encephalartos horridus*" using the colloquial name "three-toothed *Encephalartos*" and publishing the name *Encephalartos horridus* var. *trispinosus* (Fig. 5.18) as a note with lithographic Plate 5371 in *Curtis's Botanical Magazine*. It was over 100 years later that Robert Dyer (1965) elevated the variety to a specific rank as *Encephalartos trispinosus*. In the same issue of *Curtis's Botanical Magazine* (Hooker 1863b), the Australian species *Bowenia spectabilis* was illustrated and documented. This cycad had been first noted by explorer-botanist Allan Cunningham in 1819, but it was only in 1863 that Walter Hill, then Curator of Brisbane Botanic Gardens, sent plant and cone material to Kew. Recognizing the uniqueness of this interesting plant, the acquisition allowed the Hookers to name a new genus in honor of the Governor of Queensland at the time, Sir George Bowen. That was completed with notes and as Plate 5398 (Fig. 5.19) in the *Botanical Magazine*.

The fourth cycad, featured as Plate 5943 (Fig. 5.20) in *Curtis's Botanical Magazine* of 1872, was another Australian species imported from New South Wales with male and female plants grown to maturity by the English nurseryman and plant collector, William Bull. Joseph Hooker (1872) named this plant *Macrozamia corallipes*, a name later placed in synonymy with *M. spiralis*.

The fifth cycad was a plate and text contributed by J. D. Hooker (1890) on *Zamia wallisii* A. Braun as Plate 7103 in Curtis's Botanical Magazine (Fig. 5.21). The pollen cones are illustrated, but the ovulate cones are still unknown, and thus, those still remain enigmatic.

Fig. 5.18 William Hooker's *Encephalartos horridus* var. *trispinosus*, an illustration by Walter Fitch and lithographers Vincent Brooks, Day and Son. Plate 5371 in *Curtis's Botanical Magazine* 89 of 1863

Fig. 5.19 An illustration by Walter Fitch and lithographers Vincent Brooks, Day and Son of William and Joseph Hooker's *Bowenia spectabilis*. Plate 5398 in *Curtis's Botanical Magazine* 89 of 1863

Fig. 5.20 The illustration by Walter Fitch and lithographers Vincent Brooks, Day and Son of Joseph Hooker's *Macrozamia corallipes* [= *M. spiralis*]. Plate 5943 in *Curtis's Botanical Magazine* 98 of 1872

Fig. 5.21 The illustration by Walter Fitch and lithographers Vincent Brooks, Day and Son of Joseph Hooker's *Zamia wallisii Plate* 7103 in *Curtis's Botanical Magazine* 116 of 1890

Names of Cycads with Hooker and Hooker Fil.'s Authorship
Data from the *World List of Cycads.*

Bowenia spectabilis Hook. *ex* Hook.*fil*. (1863).
Encephalartos horridus var. *trispinosus* Hook.(1863) ≡ *E. trispinosus* (Hook.) R.A.Dyer.
Macrozamia corallipes Hook.*fil*. (1872) = *M. spiralis* (Salisb.) Miq.

Hooker and Hooker Fil.'s Cycad Publications

Hooker, W. J. 1859. *Stangeria eriopus*. Curtis's Botanical Magazine 85: Plate 5121 and text.
Hooker, W. J. 1863. *Encephalartos horridus* var. *trispinosus*, three-toothed *Encephalartos*. Curtis's Botanical Magazine 89: Plate 5371 and text. [*Encephalartos horridus* var. *trispinosus* Hook.].
Hooker, W. J. 1863. *Bowenia spectabilis,* Australian *Bowenia*. Curtis's Botanical Magazine 89: Plate 5398 and text. [*Bowenia spectabilis* Hook. *ex* Hook.*fil*.].
Hooker, J. D. 1872. *Macrozamia corallipes*, native of New South Wales. Curtis's Botanical Magazine 98: Plate 5943 and text. [*Macrozamia corallipes* Hook.*fil*.].
Hooker, J.D. 1890. *Zamia wallisii*. Plate 7103 in Curtis's Botanical Magazine 116.

Charles Antoine Lemaire (1800–1871)

Charles Antoine Lemaire (Fig. 5.22) was born in Paris on November 1, 1800, the son of Antoine Charles Lemaire and Marie Jeanne Davio. During his early education, he was considered an outstanding scholar by his teachers. He entered the University of Paris where he became a well-respected senior graduate in the Department of Classical Literature.

However, a chance contact by Lemaire with the horticultural staff at the Paris Museum of Natural History sparked an interest in botany and horticulture which led to a major career change. He first worked as an assistant at a large Paris nursery where he was responsible for developing a large collection of cacti. Lemaire greatly admired the intrinsic geometric elegance of these plants, and the Cactaceae became a dominating feature for the rest of his life. A second aspect of his career change was a passion for professional writing, illustrating, and publishing. Between 1839 and 1854, he edited six volumes of *L'Horticulteur universel* in addition to the *Revue horticole*, the *Herbier général de l'amateur*, and four volumes of *Jardin fleuriste*, contributing greatly to their content and working closely with botanical illustrators such as Jean-Christophe Heyland (1792–1866).

In addition to the editorial work, Lemaire also wrote numerous papers on different aspects of cacti and other succulents: the main works being his *Cactearum aliquot novarum* (1838), *Cactearum genera nova speciesque novae* (1839), and *Iconographie descriptive des Cactées* (1841–1847). One of the well-known genera he named is *Schlumbergera,* which includes the popular "Christmas Cactus." During this period, Lemaire struck up a relationship with the nurseryman, explorer, and

Fig. 5.22 Charles Antoine Lemaire (1800–1871). (From Wikipedia)

publisher Louis Van Houtte in Ghent, Belgium. Arising from that was Lemaire's description of the Mexican cycad, *Dioon aculeatum* (later *D. angustfolium*), published in Van Houtte's catalog of 1846.

In 1854, Van Houtte invited Lemaire to relocate to Belgium to co-edit the first ten volumes of *Flore des serres et des jardins del'Europe*. He also then produced the first 16 volumes of *L'Illustration horticole* for another Belgian horticulturist and businessman, Ambroise Verschaffelt. During this period in Ghent, Lemaire published descriptions of another three cycads. In 1863, he suggested the name *Dioon strobilosum* but questioned whether it was a new species (that name was later corrected and published by A. de Candolle (1868) as *D. strobilaceum,* now *D. edule* (see Stevenson and Sabato 1986b)). In 1867, he revised two South African cycad names from *Zamia ghellinckii* and *Z. villosa* to *Encephalartos ghellinckii* (Fig. 5.25) and *E. villosus,* respectively; his decisions were based on plants sent to Verschaffelt in February 1867 by Mark McKen, then Curator of the Durban Botanic Gardens. Several of Lemaire's cycad-related contributions *L'Illustration horticole* are accompanied by excellent plates (Figs. 5.23, 5.24, and 5.25) prepared by the well-known Belgian lithographers, Paul and Louis Stroobant.

Fig. 5.23 A sepia lithograph by Paul Stroobant, supposedly representing *Dioon edule* in habitat, as published by Lemaire in *L'illustration Horticole* (1855). The figures in the illustration do not appear to be ethnic Mexicans

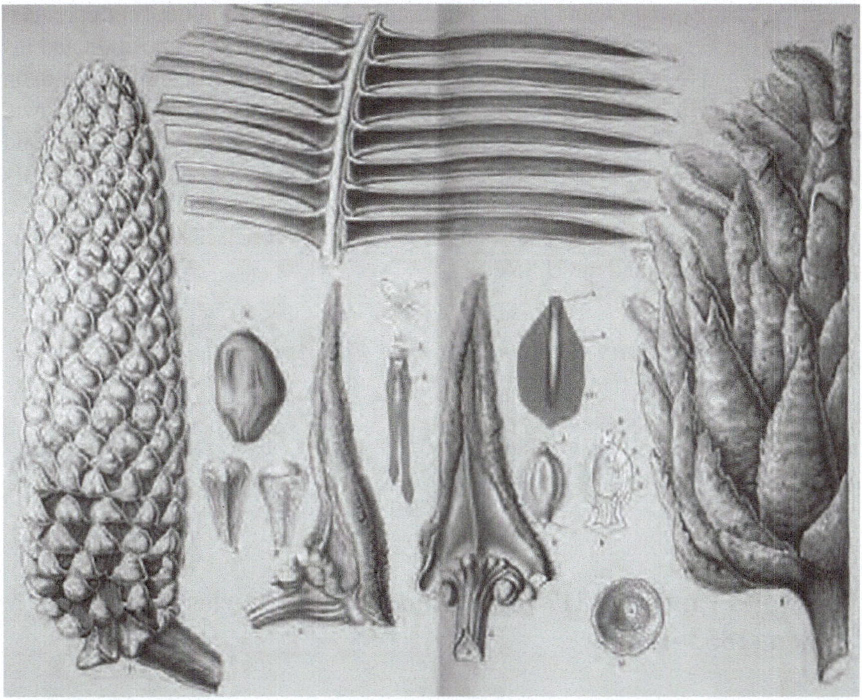

Fig. 5.24 *Dioon edule*, leaf and cone detail, from *L'illustration horticole* 2 of 1855

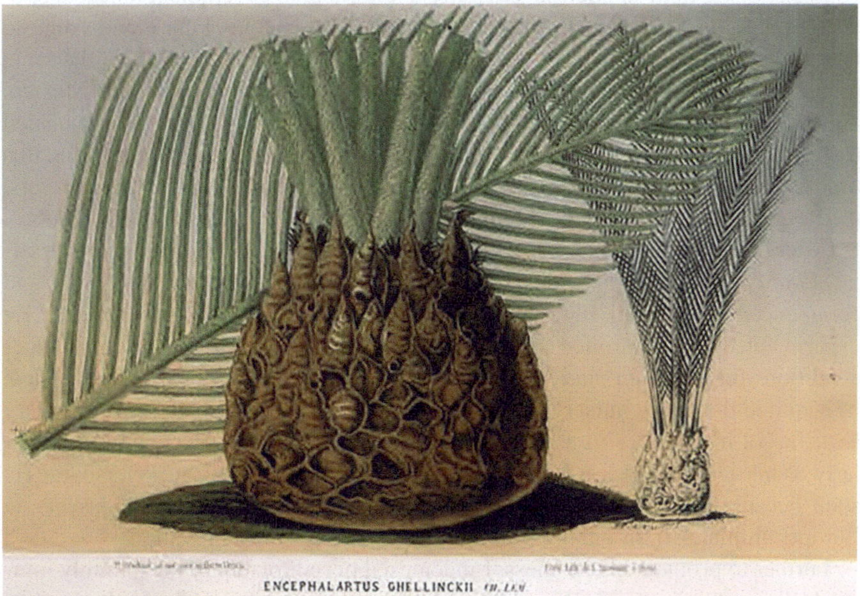

Fig. 5.25 A lithograph by Paul Stroobant of *Encephalartos ghellinckii* growing in the garden of Belgian horticulturist Ambroise Verschaffelt that Charles Lemaire (1868) used as Planche 567 in *L'Illustration horticole* 1868 to supplement his description of 1867

Unfortunately, when Jean Linden, yet another Belgian horticultural entrepreneur, took over the Verschaffelt establishment in 1869, Lemaire lost his position and his income and had to return to Paris where he lived almost in poverty until his death on June 22, 1871. His herbarium vouchers were dispersed by sale to persons unknown. Édouard André, who succeeded Lemaire as editor of *L'Illustration horticole*, said rather poignantly of him that "posterity will esteem Monsieur Lemaire more highly than did his contemporaries." His surname was later commemorated in *Lemairea*, a genus of Asian shrubs (Rubiaceae), and *Lemaireocereus* (Cactaceae), a cactus genus occurring in Mexico and the U.S.A.

Names of Cycads with Hooker and Hooker Fil.'s Authorship
Data from the *World List of Cycads*.

Bowenia spectabilis Hook. *ex* Hook.*fil*. (1863).
Encephalartos horridus var. *trispinosus* Hook.(1863) ≡ *E. trispinosus* (Hook.) R.A.Dyer.
Macrozamia corallipes Hook.*fil*. (1872) = *M. spiralis* (Salisb.) Miq.

Jean Jules Linden (1817–1898) and Lucien Charles Antoine Linden (1853–1940)

Jean Jules Linden (Fig. 5.26) was born in Luxembourg on January 12, 1817. He studied at the Royal Athénée in Luxembourg and then the Free University of Brussels. In 1835, he applied successfully to the Belgian government to lead a two-year expedition to Brazil, together with naturalists Nicolas Funck and Auguste Ghiesbreght. That was the start of a lifelong passion for orchids. In 1837, the same group spent four years collecting plants and live animals in Cuba and Mexico, although Linden suffered an acute attack of yellow fever in Campeche. Accompanied by his stepbrother, Louis-Joseph Schlim, Linden undertook a third expedition, this time to Venezuela and Colombia, from 1841 to 1844. As a reward for his endeavors, the Belgian government granted him a Venezuelan estate, which he sold in 1866.

After returning to Belgium, Linden launched the first of several rare plant nurseries, *Etablissiment d'introduction de plantes*, in partnership with Nicolas Funck, and commenced business in 1846 using the stock he had collected in Venezuela. He experimented with controlled environments to determine the temperatures and other conditions for optimum plant growth and development. As his business expanded, he acquired the Ghent nursery of Ambroise Verschaffelt and also opened subsidiary branches of his *Horticulture Internationale* in Brussels and Paris, exhibiting regularly at horticultural events in London, Paris, Amsterdam, and St. Petersburg. He soon became known as the most important nineteenth-century orchid-grower in Europe, ultimately introducing over a thousand species into cultivation.

Further to promote his business, Linden used his editorship of the monthly journal *L'Illustration horticole* (Fig. 5.27), which he took over from Charles Lemaire in 1870, to describe novel orchids and other rare and unusual plant discoveries. In that

Fig. 5.26 Jean Jules Linden. (Portrait from the archives of the University of Ghent)

magazine and other plant publications such as his *Journal des orchidées* and *La Semaine horticole*, *Pescatorea*, and *Hortus Lindeneanus*, he employed the leading botanical artists of the time. Foremost, among these was the prolific watercolor artist and engraver, Pieter De Pannemaeker, whose legacy of botanical art has become greatly sought after in its own right.

In order to supply the burgeoning demand for orchids and other rare plants in Europe in the mid-1800s, Linden employed a number of professional plant collectors. In our pages on the German plant hunter, Gustav Wallis, we describe a dispute between Linden and Wallis that led to the termination of their contract. Another collector working in South America was Benedikt Roezl, who included two cycad species in a major plant consignment that he sent to Linden. Our section on Roezl describes the events that led to the discovery and naming of *Zamia roezlii* in 1873 and of *Z. lindenii* (Fig. 5.28) in 1875. Linden also had a hand in Regel's naming of *Zamia manicata* in 1876.

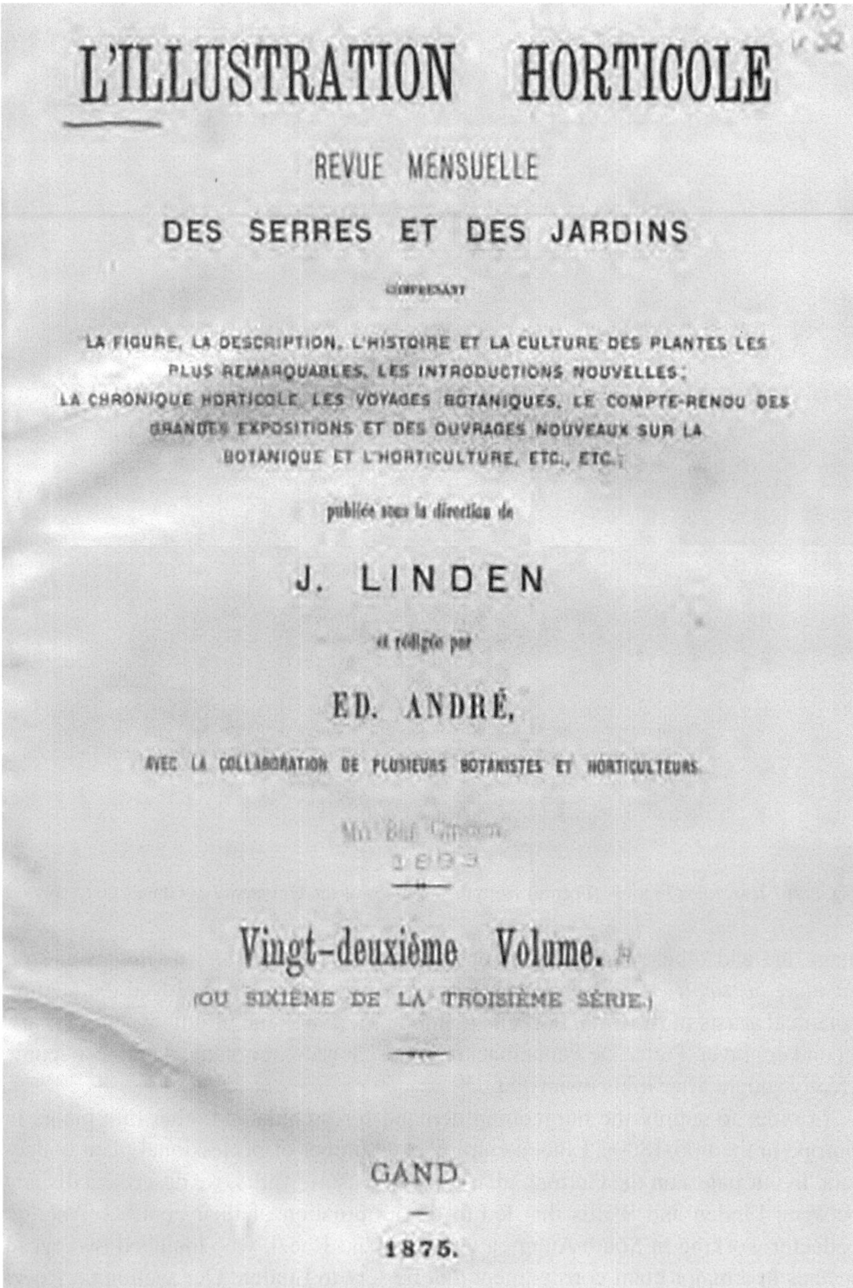
L'ILLUSTRATION HORTICOLE

REVUE MENSUELLE

DES SERRES ET DES JARDINS

COMPRENANT

LA FIGURE, LA DESCRIPTION, L'HISTOIRE ET LA CULTURE DES PLANTES LES PLUS REMARQUABLES, LES INTRODUCTIONS NOUVELLES ;
LA CHRONIQUE HORTICOLE, LES VOYAGES BOTANIQUES, LE COMPTE-RENDU DES GRANDES EXPOSITIONS ET DES OUVRAGES NOUVEAUX SUR LA BOTANIQUE ET L'HORTICULTURE, ETC., ETC. ;

publiée sous la direction de

J. LINDEN

et rédigée par

ED. ANDRÉ,

AVEC LA COLLABORATION DE PLUSIEURS BOTANISTES ET HORTICULTEURS

Vingt-deuxième Volume.

(OU SIXIÈME DE LA TROISIÈME SÉRIE.)

GAND.

1875.

Fig. 5.27 The title page of *L'illustration horticole* 22 of 1875 in which Édouard André named and described *Zamia lindenii*

Fig. 5.28 *Zamia lindenii*, a cycad from Ecuador and Peru, discovered by Benedict Roezl and named by Édouard André (1875) for Jean Linden, as depicted by Ghent botanical artist Pieter De Pannemaeker in *L'illustration horticole* 22 of 1875. Stevenson and Sabato (1986a) chose this plate to lectotypify *Zamia lindenii*

Fig. 5.29 Pieter De Pannemaeker's lithograph of *Cycas tonkinensis*, published as *Zamia tonkinensis* in *L'Illustration horticole* (Linden & Rodigas,1885). Given the specific epithet, this taxon may represent a cycad from the north of Vietnam, but we have been unable to match it to any known species. It therefore remains a *nomen dubium*, a scientific name of unknown or doubtful application

Among these various business activities, Linden also assumed the scientific directorship of the Royal Belgian Society of Zoology and Horticulture, a position that also involved running the Brussels Botanical Garden at Parc Léopold.

Linden married Anna Reuter in October 1845, and on December 12, 1853, the couple had a son, Lucien Charles Antoine, who joined his father's business and with whom Linden published 17 volumes of *Iconographie des Orchidées*, also known as *Lindenia*. Lucien, in collaboration with Émile Rodigas, also continued production of *L'Illustration horticole* from 1882 until 1896. Lucien was responsible for coining a number of cycad names that we have not been able to relate to any known species; these *nomina dubia* include *Cycas bellefonti*, *Cycas/Zamia tonkinesis* (Fig. 5.29), and *Zamia noeffiana*.

Jean Linden passed away in Ixelles, Belgium, on January 12, 1898. He was a patron of King Léopold II and Queen Marie-Henriette. He was awarded the titles Knight and Commander of the Order of Léopold and Commander of the Order of Franz-Joseph of Austria and was appointed Consul of Colombia and Consul of the Grand-duchy of Luxembourg. Apart from being recognized in *Zamia lindenii*, his name is commemorated in *Gloxinella lindeniana* (Gesneriaceae), *Iresine lindenii* (Amaranthaceae), and two orchids, *Phalaenopsis lindenii* and *Polyrrhiza lindenii*. Jean's son Lucien died on October 17, 1940.

Names of Cycads with Linden and Linden Fil.'s Authorship
Data from the *World List of Cycads*

Zamia roezlii Linden (1873).
Cycas bellefontii L.Linden & Rodigas (1886), *nomen dubium.*
Zamia tonkinesis L. Linden and Rodigas (1885) ≡ *Cycas tonkinesis* (L.Linden & Rodigas) Rodigas, *nomen dubium.*
Zamia manicata Linden *ex* Regel (1876).
Zamia noeffiana L.Linden *ex* Wittm. (1897), *nomen dubium.*

Linden and Linden Fil.'s Cycad Publications

Linden, J. 1873. *Zamia roezlei. Catalogue des plantes nouvelles* 90: 10. [*Z. roezlii* Linden].
Linden, L. 1897. *Zamia noeffiana. Gartenflora* 46: 358 [*Z. noeffiana* L.Linden *ex* Wittm., *nomen dubium*].
Linden, L. & E. Rodigas. 1885. *Zamia tonkinesis. L'Illustration horticole* 32: 27, Plate 547. [*Z. tonkinesis* L.Linden & Rodigas].
Linden, L. & E. Rodigas. 1886. *Cycas bellefontii. L'Illustration horticole* 33: 27, Plate 586. [*C. bellefontii* L.Linden & Rodigas, *C. tonkinensis* (L.Linden & Rodigas) Rodigas].

Charles Moore (1820–1905)

Charles Moore (Fig. 5.30) was born in Dundee, Scotland, on May 10, 1820, son of Charles and Helen Muir (née Rattray). After being schooled in Dundee, he emigrated with his family to Ireland where the family name was changed to Moore. After completing training at Trinity College in Dublin, Charles worked as a botanist for the Ordnance Survey of Ireland, while his brother David became the Director of the Glasnevin Botanic Garden. After further training at Kew and Regent's Park and on recommendations by Professors Lindley and Henslow to the British Colonial Office, Moore was appointed in 1848 to the post of Government Botanist and Director of the Botanic Gardens in Sydney, Australia.

Being a "new" immigrant, Moore was not made particularly welcome by the rather staid and somewhat backward Garden's administration at that time. Despite some adversity, he worked hard to improve the facility, striking a compromise between design for scientific value and design for recreation. He addressed issues of poor soil, inadequate water supplies, and shortage of funds. The Palm Grove (Fig. 5.31), a central feature of the Gardens enjoyed by visitors today, was developed on reclaimed land as an extension of the original area and is testimony to Moore's skill and foresight. He labeled plants according to their natural order, scientific name and authority, English name, and country of origin—a system still followed at present.

Moore undertook collecting expeditions to the New Hebrides (now Vanuatu), Solomon Islands, New Caledonia, New Zealand, and Lord Howe Island. He shared

Fig. 5.30 Charles Moore (1820–1905). (Portrait from the State Library of New South Wales)

samples of living and dried material with Ferdinand von Mueller in Melbourne and with George Bentham at Kew. But most of his herbarium material was a result of his numerous field trips to areas along the Australian east coast. This is when he investigated the various *Macrozamia* populations that led to his 1884 "Notes on the genus" detailing five species, *Macrozamia cylindrica*, *M. fawcettii*, *M. flexuosa*, *M. heteromera*, and *M. secunda*, and two varieties, *M. heteromera* var. *glauca* and *M. heteromera* var. *tenuifolia*, both of which were subsequently elevated to species level.

Moore retired as Director of the Sydney Royal Botanic Gardens in May 1896, at the age of 76, and returned to Europe. He was succeeded at the Gardens by Joseph Maiden (1908). He had been an active Member of the Royal Society of New South Wales for 49 years and held the positions of Councillor, Vice President, and

Fig. 5.31 Part of the Palm Grove, a much-appreciated feature of the Royal Botanic Gardens, Sydney. This extension to the gardens was developed on reclaimed land by Charles Moore in the late 1800s and is testimony to Moore's skill and foresight. (Photo: James Clugston)

President. He was a Commissioner for the Philadelphia and Melbourne Exhibitions; a Member of the Hyde Park Improvement Committee; a Trustee for Hyde, Phillip, and Cook Parks; a Trustee of the National Park and Australian Museum; Chairman of the Vine Diseases Board; Fellow of the Linnean Society of London and the Royal Horticultural Society; and an Associate of the Royal Botanic Society in London (King 1974).

Moore's name is acknowledged in the epithets of some 30 plant taxa; 19 of these were in descriptions by Ferdinand von Mueller, including that of the Queensland cycad *Macrozamia moorei* (Fig. 5.32). Predeceased by his wife, Elizabeth Bennett (née Edwards), he died childless on April 30, 1905, in Paddington, London, and was buried beside his wife in the Anglican section o Rookwood Cemetery.

Fig. 5.32 A hand-colored glass slide of *Macrozamia moorei* in Queensland, prepared by Charles Chamberlain after his visit to Springsure in November 1911. This cycad was one of 19 plant taxa named for Moore by Ferdinand von Mueller

Cycad Names with Charles Moore's Authorship
Data from the *World List of Cycads*

Macrozamia cylindrica C. Moore (1884) = *M. macleayi* Miq.
Macrozamia denisonii C. Moore & F. Muell. (1858) = *Lepidozamia peroffskyana* Regel.
Macrozamia fawcettii C. Moore (1884).
Macrozamia flexuosa C. Moore (1884).
Macrozamia heteromera C. Moore (1884).
Macrozamia heteromera var. *glauca* C. Moore (1884) = *M. glaucophylla* D.L.Jones.
Macrozamia heteromera var. *tenuifolia* C. Moore (1884) = *M. stenomera* L.A.S.Johnson.
Macrozamia secunda C. Moore (1884).

Charles Moore's Cycad Publications

Moore, C. 1884. A census of the plants of New South Wales. Government Printer, Sydney.
Moore, C. 1884. Notes on the genus *Macrozamia*. Journal and Proceedings of the Royal Society of New South Wales 17: 115–122. [*Macrozamia cylindrica* C. Moore, *M. fawcettii* C. Moore, *M. flexuosa* C. Moore, *M. heteromera* C. Moore, *M. heteromera* var. *glauca* C. Moore, *M. heteromera* var. *tenuifolia* C. Moore, *M. secunda* C.Moore].
Moore, C. & E. Betche. 1893. Handbook of the Flora of New South Wales. Government Printer, Sydney.

Thomas Moore (1821–1887)

Thomas Moore (Fig. 5.33) was born in Stoke, Surrey, England, on May 21, 1821. Trained as a gardener, he first worked as a nurseryman and then assisted in the landscaping of Regent's Park in London. In 1848, under the mentorship of John Lindley, then Head of Botany at University College, London, he was appointed Curator of the Apothecaries' Garden at Chelsea, later named the Chelsea Physic Garden, a post he was to retain for almost 40 years. It is a curious coincidence that, also in 1848, Lindley was arranging for Charles Moore (no relation of Thomas) to take up the Directorship of the Royal Botanic Gardens in Sydney, Australia.

Moore soon developed a special interest in ferns, plants that became increasingly popular during the Victorian era and a period that horticulturists sometimes refer to as a time of "pteridomania." Moore wrote several texts on the group, including "Popular History of British Ferns" (1851), "Ferns of Great Britain and Ireland" (1855), and British ferns and their allies as well as *Icones filicum* (1857–1862). The first three of these went through several later editions. He also edited a number of horticultural magazines and was involved with the administration of several horticultural societies. He was also a co-author of a series entitled "The Orchid Album" that is comprised of colored figures and descriptions of new, rare, and beautiful orchidaceous plants (Bery 2021). He was elected a Fellow of the Linnean Society in 1851.

After three or four years of declining health, Thomas Moore passed away at the Chelsea Botanical Garden on January 1, 1887, and was buried in Brompton Cemetery.

The reason for including Thomas Moore in this compendium of cycad biologists relates to the somewhat chequered nomenclatural history of the South African species we now call *Stangeria eriopus.* Initially thought to be the fern, *Lomaria coriacea*, this taxon was named *Lomaria eriopus* in 1839 by Gustav Kunze, a German biologist with a particular interest in ferns. In 1851, Dr. William Stanger, Natal Provincial Surveyor-General with a passion for natural history, sent live material of the plant to England. When it produced its characteristic cones, Thomas Moore

Fig. 5.33 Thomas Moore (1821–1887). (Image from the Streatham Society)

realized that this was not a fern at all, but an entirely different entity that he called *Stangeria paradoxa* (Fig. 5.34) in 1853. The name was technically invalid as it did not include Kunze's original epithet. Subsequent names were W. Bull's *Stangeria schizodon* (1872) and Regel's *Stangeria katzeri* (1874), but it was left to the French botanist and physician, Henri Baillon, to correct the name to *Stangeria eriopus* (Fig. 5.35) in 1892 (Vorster, 1985; Grobbelaar, 2002). *Stangeria* is the single species in the genus, and the originally proposed family Stangeriaceae has since been merged into the Zamiaceae on the basis of molecular evidence. A further aspect of interest is that tissue from the swollen caudices of this cycad was (and still is) used regularly as a particularly effective purgative by the Zulu, Pondo, and Xhosa tribespeople in South Africa. Osborne et al. (1994) analyzed material being sold at the Durban *muthi* (traditional medicine) markets and found that it contained a remarkably high concentration of sodium sulfate (Glauber's salt) that explains its efficacy as an emetic.

Fig. 5.34 Plate 5121 from William Hooker's entry for *Stangeria paradoxa* (= *Stangeria eriopus*) in Curtis' Botanical Magazine 85 of 1859

Fig. 5.35 Copy of a 1997 watercolor painting of *Stangeria eriopus* by John Donaldson. Reproduced by permission of the artist

Names of the Cycad with Thomas Moore's Authorship
Data from the *World List of Cycads*

Stangeria paradoxa T. Moore (1853) = *S. eriopus* (Kunze) Baill.

Thomas Moore's Cycad Publication
Moore, T. 1853. *Stangeria paradoxa*. Hooker's Journal of Botany and Kew Garden, Misc. 2: 228. [*Stangeria paradoxa* T. Moore].

Ferdinand Jacob Heinrich von Mueller (1825–1896)

Ferdinand von Mueller (Fig. 5.36) was born (as Ferdinand Müller) in Rostock, northern Germany, on June 30, 1825. Following the early loss of his father, he was raised by his mother; educated at Tönning, Schleswig; and indentured to a pharmacist at the age of 14. After obtaining a qualification in pharmacy, he went on to study botany at Kiel University, graduating with a Ph.D. at the age of 22 (Mueller 1853). That same year, he published his first botanical paper, a work on the flora of Schleswig. Because both Müller and his sister Bertha suffered from poor health, Ludwig Preiss (another great German botanist who had just returned from Australia) suggested that a move to the warmer climate of that country would be beneficial. Thus, in 1847, Müller and his two sisters, Clara and Bertha, made the bold move to the colony of South Australia, arriving in Adelaide on December 15 that year.

Müller's first employment in Australia was as a pharmacist in central Adelaide, but his passion was in "the great outdoors," and he explored South Australia from Mount Gambier to the Flinders Ranges and Lake Torrens. He immersed himself in his new homeland, changed his surname from Müller to Mueller, became naturalized as a British subject, and even attempted a brief farming venture. During the early 1850s, he commenced a series of publications in German and English on the flora of South Australia.

After 4 years in South Australia, Mueller moved to Victoria intending to establish a pharmaceutical business either in Melbourne or on the goldfields. However, after a meeting with colonial Lieutenant-Governor Charles La Trobe, he accepted the newly created position of Victorian Government Botanist. He was extraordinarily active in exploring the relatively unknown parts of the territory, including high-altitude alpine areas, the Buffalo Ranges, the upper reaches of the Goulburn River, lengthy stretches of the coast, and more. Traveling yet further afield, he joined expeditions to more remote areas, including the challenging Gregory Expedition of 1855–1856 through Queensland, across much of the Northern Territory, and into the Great Sandy Desert of Western Australia.

Fig. 5.36 Baron Sir Ferdinand Jacob Heinrich von Mueller, KCMG. Photo ca. 1880–1890 by Johnstone, O'Shannessy and Co. kindly supplied by the National Library of Norway

Mueller collected nearly 800 plants new to science and established the National Herbarium of Victoria, which still holds many of his vouchers and types and also boasts the country's most comprehensive botanical library. His publications were as phenomenal in number as they were Catholic in diversity. He was responsible in part for planning several major expeditions, including those to Western Australia, Papua New Guinea, and the Pacific islands, Antarctica. He was also involved with arrangements for the ill-fated Burke and Wills Expedition of 1860–1861 but became disenchanted when his advice was ignored (Mueller 1858).

In 1857, Mueller was appointed the first Director of Melbourne's Royal Botanic Gardens and its herbarium (Fig. 5.37). However, there was subsequent criticism that his approach was too scientific and educational and not sufficiently aesthetic, and differences of opinion led to his dismissal from that position in 1873.

Mueller collaborated widely on the international front, providing botanical and zoological material to numerous botanical gardens and museums. He supplied vouchers to the British Museum and to herbaria in Australia, Kew, Paris, Vienna, and elsewhere. Botanists remember his name in a remarkable and diverse list of taxa: the genera *Austromuellera* and *Sirmuellera* (Proteaceae), *Muelleranthus* (Fabaceae), *Muellerina* (Loranthaceae), *Muellerothamnus* (Monimiaceae), *Muelleriella* and *Muelleriopsis* (diatoms), *Muellerena* (a seaweed), *Muellerella* (a lichen), the species *Allocasuarina muelleriana* (Casuarinaceae), *Callitris muelleri* (Cupressaceae), *Eucalyptus muelleriana* (Myrtaceae), *Persoonia muelleri* (Proteaceae), *Terminalia ferdinandiana* (Combretaceae), and *Verticordia muelleriana* (Myrtaceae)—to which we can add the Australian lizard *Lerista muelleri* and a number of fish and invertebrates.

Mueller was friendly with Charles Darwin, although his orthodox religious views clouded his acceptance of Darwin's theories. His international contributions were recognized by decorations from France, Spain, Denmark, and Portugal, the conferring in 1867 of the title Baron (*Freiherr*) by King Karl of Württemberg, a Doctor of Medicine degree from the University of Rostock, a Fellowship of the Royal Society of London, and many knighthoods including a KCMG in 1879.

Cycads were not Mueller's main plants of interest; that accolade would go to the eucalypts, which he studied in much greater detail. However, he was instrumental in describing *Cycas papuana* F. Muell. and *C. scratchleyana* F. Muell. from Papua New Guinea and three species of *Cycas* from Queensland including *C. normanbyana* (Fig. 5.38) and *C. cairnsiana* (Fig. 5.39). He also wrote brief notes about many other Australian cycad taxa under the name *Encephalartos*, later to be assigned to *Macrozamia*. Mueller reported the taxon now known as *Lepidozamia peroffskyana* Regel under several genus names and the epithet *denisonii*; it was only when he collaborated with Regel in Saint Petersburg that the genus, and both species of *Lepidozamia* were correctly assigned (Osborne 2018).

Mueller was a key Member of the Victorian Institute for the Advancement of Science, this later being the Philosophical Institute of Victoria and then the Royal Society of Victoria. He was awarded a Clarke Medal by the Royal Society of New South Wales. A number of Australian mountain ranges, peaks, rivers, creeks, and lakes are named after him, and further afield, his name is remembered in Mount

Fig. 5.37 The facade of the National Herbarium of Victoria (MEL), a 1988 extension to the 1934 building. The bust outside is that of Mueller and inside is the appropriately named Mueller Hall. MEL holds almost 1.5 million specimens, approximately 26,000 of which were collected by Mueller himself. (Photo from the State Botanical Collection, Royal Botanic Gardens Victoria)

Fig. 5.38 A line drawing from *Fragmenta Phytographiae Australiae*, Volume 66, Part 8 (1874), where Mueller has notes on *Cycas normanbyana* F. Muell. from the mouth of the Burdekin River and *Cycas media* R.Br. from near Port Douglas. These two cycads are now known as *Cycas media* subsp. *media* and *Cycas media* subsp. *banksii*, respectively

Mueller and Mueller Glacier (South Island, New Zealand; Fig. 5.40) and yet another Mueller Glacier in Spitzbergen, northern Norway. His name is also commemorated in the journal "Muelleria," a publication of the National Herbarium of Victoria, and in the issue of two postage stamps, a tuppence ha'penny red of 1948 (Fig. 5.41a),

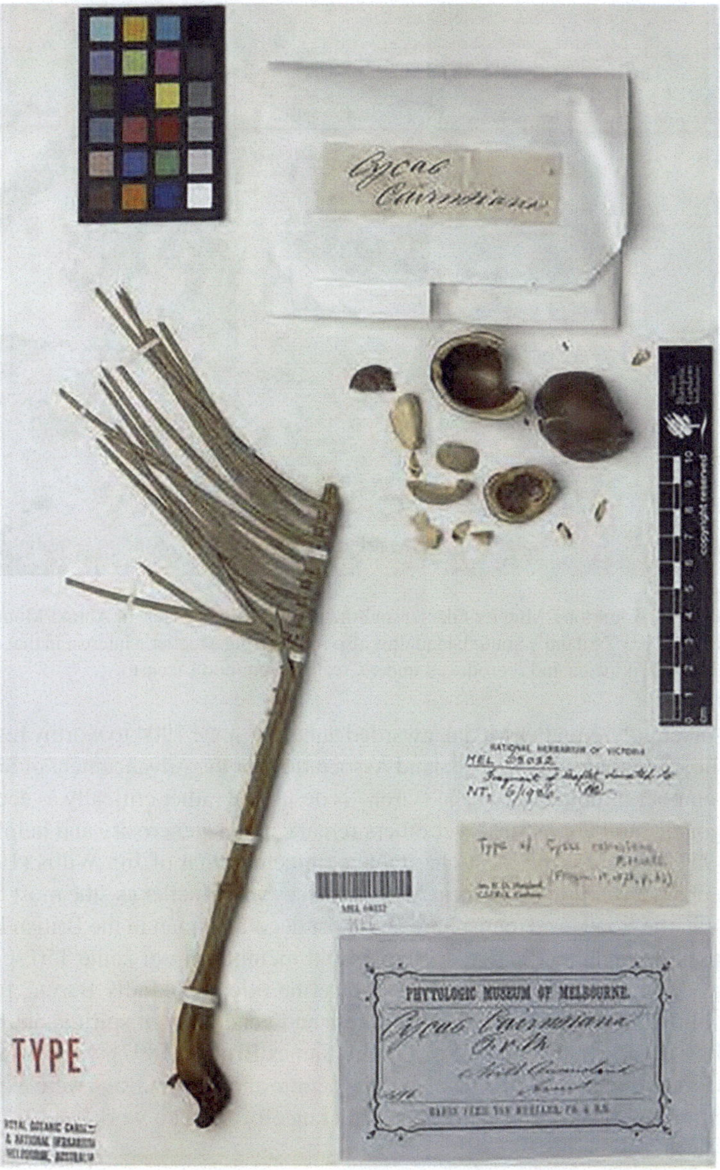

Fig. 5.39 The holotype of *Cycas cairnsiana* at the National Herbarium of Victoria (MEL), prepared from material collected in 1876 by William Armit in North Queensland

and a joint Australian/German $1.20/100 pfennings pair of 1996 (Fig. 5.41b), released 100 years after his death.

Mueller died at the age of 71 in Melbourne on October 10, 1896, and was buried at the St. Kilda Cemetery. Although he was twice engaged, he never married and was survived only by his sister Clara. Funds were subsequently raised for the

Fig. 5.40 The 13 km long Mueller Glacier and the hikers' Mueller Hut, in Aoraki/Mount Cook National Park, New Zealand's South Island, this also recognizing Mueller's interest in that country. (Sourced from Wikipedia and reproduced under Creative Common License)

establishment of a Mueller Medal, awarded annually since 1904 to worthy researchers by the Australian and New Zealand Association for the Advancement of Science.

In a number of obituaries, "The Baron" is described rather critically as eccentric, vain, egoistic, and unpractical, yet others remark on his generosity and helpfulness (Best 1952). Perhaps, the most charitable comment is that of Jim Willis (1949) of the National Herbarium of Victoria, who describes von Mueller as "the most famous scientist in the Southern Hemisphere and most decorated man in the British Empire [with his] 20 knighthoods, five doctorates and membership of some 150 scientific societies throughout the world." We embrace the title "the kindly Baron" from an obituary in which von Mueller's friendliness and generosity of spirit is abundantly clear (Best 1952). In the Journal of Botany, James Britten (1897) writes somewhat sadly of von Mueller that "the whole of his income and all his means were expended either in private generosity or in the furtherance of science work, and he died a poor man."

Readers wanting further details of von Mueller's achievements and publications should consult the reference works by Morris (1974), Churchill et al. (1978), and Home (1997).

Fig. 5.41 Mueller is commemorated in two postage stamps; (**a**). a tuppence ha'penny red in 1948 and (**b**). a joint Australian/German $1.20/100 pfenning issue in 1996, 100 years after the Baron's death

Names of Cycads with Mueller's Authorship
Data from the *World List of Cycads*

Cycas cairnsiana F. Muell. (1876).
Cycas kennedyana F. Muell. (1882) = *C. media* R.Br.
Cycas normanbyana F. Muell. (1874) = *C. media* R.Br.
Cycas papuana F. Muell. (1876).
Cycas scratchleyana F. Muell. (1885).
Encephalartos denisonii (C. Moore & F. Muell.) F. Muell. (1859) = *Lepidozamia peroffskyana* Regel.
Encephalartos dyeri F. Muell. (1885) = *Macrozamia dyeri* (F. Muell.) C.A. Gardner.
Encephalartos miquelii F. Muell. (1862) = *Macrozamia miquelii* (F. Muell.) A. DC.

Encephalartos moorei (F. Muell.) F. Muell. (1881) = *Macrozamia moorei* F. Muell.

Encephalartos pauli-guilielmi (W. Hill & F. Muell.) F. Muell. (1859) = *Macrozamia pauli-guilielmi* W. Hill & F. Muell.

Encephalartos preissii (Lehm.) F. Muell. (1858) = *Macrozamia riedlei* (Gaudich.) C.A.Gardner.

Encephalartos spiralis var. *diplomera* F. Muell. (1866) = *Macrozamia diplomera* (F. Muell.) L.A.S.Johnson.

Macrozamia denisonii C. Moore & F. Muell. (1858) = *Lepidozamia peroffskyana* Regel.

Mueller's Cycad Publications

Mueller, F.J.H. 1852. *Diagnoses et descriptiones plantarum novarum, quas in Nova Hollandia australi praecipue in regionibus interioribus detexit et investigavit.* Linnaea 25: 589–593.

Mueller, F.J.H. 1855. Notes on South Australian botany, definitions of rare or hitherto undescribed Australian plants. Goodhugh & Trembath, Melbourne.

Mueller, F.J.H. 1858–1882. *Fragmenta phytographiæ Australiæ* 1, 3, 5, 8, 10 & 11. [*Cycas cairnsiana* F. Muell., *C. normanbyana* F. Muell., *Encephalartos miquelii* F. Muell., *E. moorei* (F. Muell.) F. Muell., *E. spiralis* var. *diplomera* F. Muell., *Macrozamia denisonii* C. Moore & F. Muell., *M. pauli-guilielmi* W. Hill & F. Muell.].

Mueller, F.J.H. 1858–1859. Notes on some rare and medicinal plants of Australia. Quarterly Journal & Transactions of the Pharmaceutical Society of Victoria 2: 44–46, 90–91. [*Encephalartos denisonii* (C.Moore & F. Muell.) F. Muell., *E. pauli-guilielmi* (W.Hill & F. Muell.) F. Muell., *E. preissii* (Lehm.) F. Muell.].

Mueller, F.J.H. 1859. Australian Cycadeae. Gardeners' Chronicle: 992–993.

Mueller, F.J.H. 1863–1878. *Flora Australiensis*, a description of the plants of the Australian territory. G. Bentham assisted by F. von Mueller. Seven volumes. L. Reeve & Co, London.

Mueller, F.J.H. 1874. Additions to the lists of the principal timber trees and other select plants readily available for Victorian industrial culture. Stillwell & Knight, Melbourne.

Mueller, F.J.H. 1874. On transplanting aged cycads. Gardeners' Chronicle n.s.1: 18.

Mueller, F.J.H. 1876. Descriptive notes on Papuan plants 1(4): 71–72, 2(6): 15. [*Cycas papuana* F. Muell.].

Mueller, F.J.H. 1881. Plants of north-western Australia. Government Printer, Perth.

Mueller, F.J.H. 1881–1885. Australasian Chemist and Druggist 4: 80, 84-85; 5: 80–81; 8: 12. [*Cycas kennedyana* F. Muell., *Encephalartos douglasiil* F. Muell., *E. dyeri* F. Muell., *Macrozamia moorei* F. Muell.].

Mueller, F.J.H. 1882–1890. Systematic census of Australian plants with chronologic, literary and geographic annotations. Several supplements and editions. M'Carron Bird & Co., Melbourne.

Mueller, F.J.H. 1885. Succinct notes on some plants from New Guinea. Victorian Naturalist 2(2): 18–20. [*Cycas scratchleyana* F. Muell.].

Mueller, F.J.H. 1885. Description of a new cycadeous plant from south-western Australia. *Botanisches Centralblatt* 23: 225–226. [*Encephalartos dyeri* F. Muell.].

Mueller, F.J.H. 1871–1895. Select extra-tropical plants readily eligible for industrial culture or naturalization. Eight editions. Government Printer, Melbourne.

Eduard August von Regel (1815–1892)

There are not many families that can claim to have horticultural and/or botanical careers spanning five generations, but the Regel family was one. (See also our section on Alphonse Louis de Candolle and the four generations of the de Candolle family.) This text deals with something of the life and works of Eduard August von Regel (Fig. 5.42), but we note also that Eduard's grandfather, Döring, and his father, Ludwig, were fruit growers and enthusiastic gardeners. Eduard's oldest Swiss-born son, Johann Albert (1845–1908), became a Russian physician and botanist, while his grandson, Constantin Andreas (1890–1970), followed as a Russian and Lithuanian horticulturalist and botanist.

Eduard August von Regel (Edward von Regel, Edward de Regel, Édouard von Regel or Eduard-Avgust Ludwigovich Regel) was born on August 13, 1815, in

Fig. 5.42 Eduard August von Regel. Based on an engraving by an unknown artist. From Fischer von Waldheim (1913–1915) in Acta Horti Petropolitani 3: 128–132. 1875

Gotha, Thuringia, Germany, one of 11 children of Ludwig, a teacher, preacher, and gardener. As a young boy, he spent much time in the family gardens and gained a basic knowledge of practical horticulture. Eduard had elementary schooling at the Gymnasium in Gotha and was indentured at the age of 15 to the Limonaia Royal Garden in Gotha, after which he worked as a volunteer in the Göttingen Botanical Garden. He then found employment at the botanic gardens in Bonn (1837–1839) and Berlin (1839–1842). At the age of 27, Eduard moved to Switzerland, eventually to take up the position of Head Gardener at the Zürich University Botanical Garden. In 1852, he founded *Gartenflora*, a magazine that he edited until 1884 and for which he wrote an enormous number of short articles over many years. He was also a student and a teacher at Zürich University where he obtained both formal and honorary doctorates.

In 1855, at the age of 40, Regel relocated to Saint Petersburg, Russia, where he joined the staff of Imperial Botanical Garden as Scientific Director, then later as First Botanist. Founded in 1714 by the order of Peter the Great, this establishment was originally an "Apothecary's Garden" and focused on plants of medicinal value. Under Regel's enthusiastic influence, the garden expanded to include numerous rare and exotic plants, and 25 greenhouses spread over an area of 19 hectares. In 1872–1874, Regel designed the Admiralty Garden on the bank of the River Neva and supervised its creation; this garden remains a popular venue for locals and tourists alike. In 1875, he was appointed Director of the Saint Peterburg Garden, a post that he held until his death. The garden is now part of part of the Komarov Botanical Institute (Fig. 5.43), a leading international botanical institution and herbarium.

Fig. 5.43 The impressive entrance gates to the Komarov Botanical Institute and Saint Petersburg Botanical Garden. (Photo: Roy Osborne)

During his tenure at Saint Petersburg, Regel was involved with describing over 3000 plant species, publishing over 3100 scientific papers, raising funds for plant-collecting expeditions, and corresponding widely with plant collectors internationally to acquire both herbarium vouchers and living specimens. His own vouchers and types were deposited at herbaria in Saint Petersburg, Berlin (mostly destroyed during WW II), Geneva, Göttingen, Hamburg, Kew, Moscow, Munich, and Paris in Europe and in Philadelphia in the U.S.A. Regel served under three Russian Tsars, Nicholas I, Alexander II, and Alexander III. He appears to have struck up a particular acquaintance with Tsar Nicholas, naming *Strelitzia nicolai* for him. It seems that he was also commissioned to teach elements of botany to the sons of either Nicholas I (Knapp 1892) or Alexander II (Rodionenko 1986).

Regel was the Founder and Vice President of the Russian Gardening Society, an Associate Member of the Saint Petersburg Academy of Sciences, elected a Foreign Member of the Linnean Society of London in 1890, and received many awards and medals, including the prestigious Prussian Red Eagle medal. The genera *Regelia* (Myrtaceae), *Euregelia* (Campanulaceae), *Eduardoregelia* (Liliaceae), and *Neoregelia* (Bromeliaceae) and the species *Cestrum regeli* (Solanaceae) are named for him.

Among the many plants that captured his attention, Regel had a particular fascination with cycads. His name is associated with 45 cycad species, seven of which are currently valid names. Most of the cycad papers were published in *Gartenflora* over the period 1857–1880, but Regel's major cycad treatise was his *Cycadearum, generum specierumque revisio* of 1876, generally referenced as *Acta Horti Petropolitani*, Volume 4 (Fig. 5.44), a large comprehensive work in which he reviewed 67 species of cycads in 11 genera. Like most European nineteenth-century botanists, Regel published his major works in at least two languages, usually Latin and German, and usually in the same year.

Like many of his contemporaries, Regel did not see cycads in habitat but relied on information, notes, live plants, seeds, and herbarium specimens from travelers. As one of many examples, Regel's collaborations with Australian Government botanist Ferdinand von Mueller led him to describe the genus and both species of the Queensland genus *Lepidozamia*.

The living cycad collections in the glasshouses at Saint Petersburg presently comprise over 100 specimens in nine genera (Osborne 2018). These include six large healthy specimens of *Lepidozamia peroffskyana* (Fig. 5.45), several of which have produced cones. The original plants grown by Regel from seeds in the late 1800s were destroyed during the Siege of Leningrad in WW II, but new Australian seeds had been obtained about 45 years ago, and it is these plants that are now among the garden's prized exhibits (Osborne 2018). A related matter is that Regel took an offset of the large *Encephalartos altensteinii* plant—brought from South Africa to Hamburg in 1832 by the Danish plant collector Christian Ecklon—from Hamburg to Zürich in 1855; both the parent plant and its offspring are still living in their respective botanic gardens.

Regel passed away on April 15, 1892, in Saint Petersburg at the age of 77 after a gradual decline in health. He was buried in the Smolensky Lutheran Cemetery on

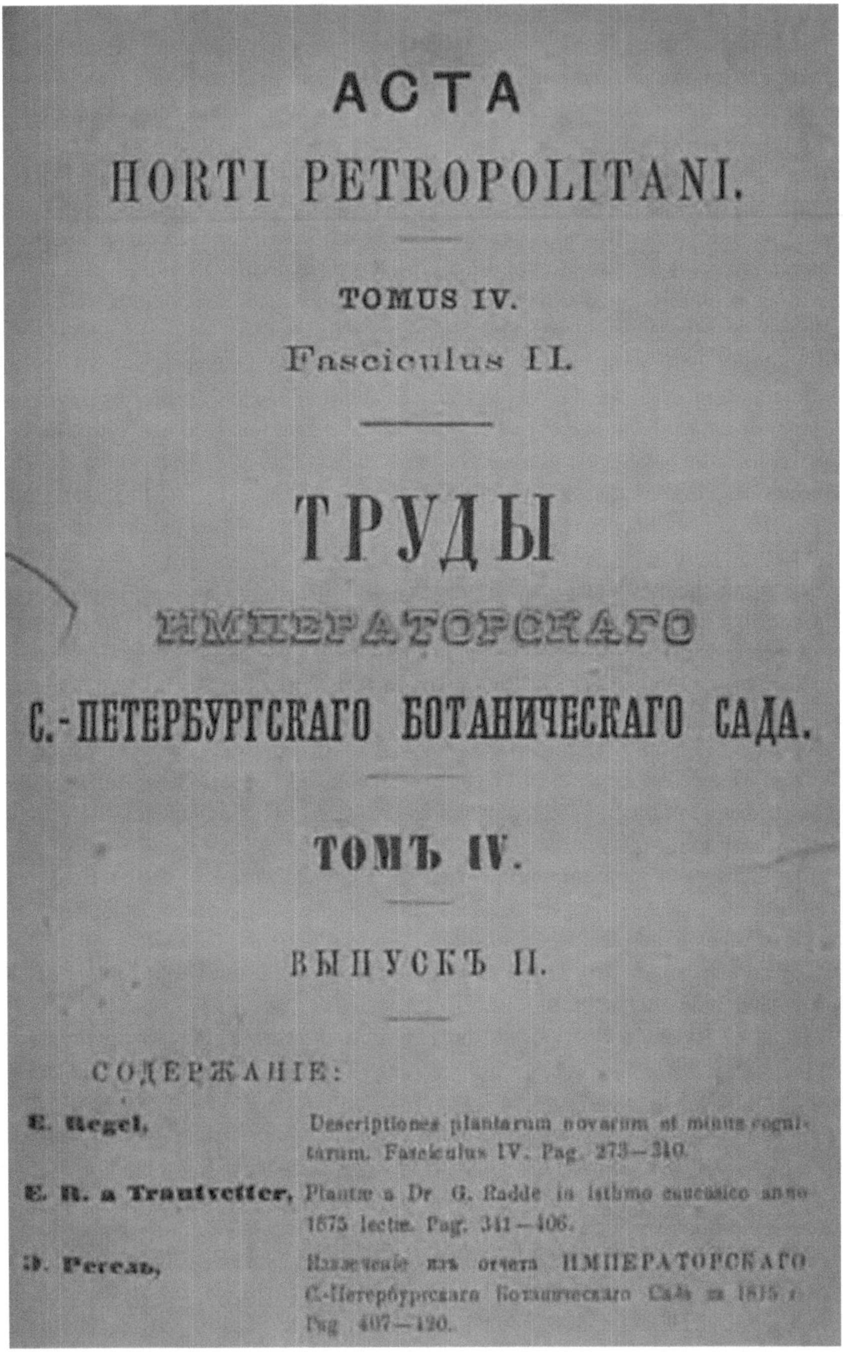

ACTA
HORTI PETROPOLITANI.

TOMUS IV.
Fasciculus II.

ТРУДЫ
ИМПЕРАТОРСКАГО
С.-ПЕТЕРБУРГСКАГО БОТАНИЧЕСКАГО САДА.

ТОМЪ IV.

ВЫПУСКЪ II.

СОДЕРЖАНІЕ:

E. Regel,	Descriptiones plantarum novarum et minus cognitarum. Fasciculus IV. Pag. 273—310.
E. R. a Trautvetter,	Plantæ a Dr. G. Radde in isthmo caucasico anno 1875 lectæ. Pag. 341—406.
Э. Регель,	Извлеченіе изъ отчета ИМПЕРАТОРСКАГО С.-Петербургскаго Ботаническаго Сада за 1875 г. Pag. 407—420.

Fig. 5.44 The cover page for Vol. 4 of *Acta Horti Petropolitani* containing Regel's *Cycadearum generum specierumque revisio* (1876), a mammoth treatise of 48 pages covering 67 species of cycads in 11 genera

Fig. 5.45 Roy Osborne (right) and Dr. Kirill Tkachenko inspecting *Lepidozamia peroffskyana* in the Komarov Botanical Institute glasshouses, Saint Petersburg, September 2018. (Photo: Angela Osborne)

Dekabristov Island, almost within sight of his Admiralty Garden. Regel's personality has been mentioned both kindly and adversely. In one obituary, Herder states him as having "an unsatisfiable ambition and a ruthless egoism"; others were less harsh, mentioning his "strictest performance of duty and iron diligence" (Dammer 1892). Fellow Russian botanist Josef Knapp (1892) wrote of Regel:

> *Auch die botanisches Gärten und Museen haben vollens Ursache um ihn za trauern, thielte er doch redlichst mit denselben die in Saint Petersburg aufgespeichertern botanischen Schätze. Friede seiner Asche und Ehre seinem Andenken*! [Our translation: Also the botanical gardens and museums have ample reason to mourn him, as he gladly shared with them all the botanical treasures amassed in Saint Petersburg. Peace to his ashes and honour to his memory.]

Whatever one's point of view, this man was clearly a person of remarkable determination, perseverance, productivity, and influence.

Names of Cycads with Regel's Authorship
Data from the *World List of Cycads*

Aulacophyllum Regel, various species (1876) = *Zamia* spp.
Ceratozamia katzeriana Regel (1876) = *Zamia katzeriana* (Regel) E.Rettig.
Ceratozamia kuesteriana Regel (1857).
Cycas gracilis var. *glauca* Regel (1876) = *C. media* R.Br.
Cycas gracilis var. *viridis* Regel (1876) = *C. media* R.Br.
Cycas riuminiana Porte ex Regel (1863).
Encephalartos altensteinii var. *distans* Regel (1874) = *E. altensteinii* Lehm.
Encephalartos altensteinii var. *grandis* Regel (1875) = *E. altensteinii* Lehm.
Encephalartos altensteinii var. *macrophyllus* Regel (1875) = *E. altensteinii* Lehm.
Encephalartos altensteinii var. *parvifolius* Regel (1874) = *E. altensteinii* Lehm.
Encephalartos altensteinii var. *paucidentatus* Regel (1874) = *E. paucidentatus* Stapf & Burtt Davy.
Encephalartos altensteinii var. *spinosior* Regel (1875) = *E. altensteinii* Lehm.
Encephalartos afer var. *integrifolius* Regel (1874) = *E. afer* (Thunb.) Lehm.
Encephalartos afer var. *unidentatus* Regel (1874) = *E. afer* (Thunb.) Lehm.
Encephalartos cycadifolius var. *friderici-guilielmi* (Lehm.) Regel (1874) = *E. friderici-gulielmi* Lehm.
Encephalartos cycadifolius var. *glaber* Regel (1874) = *E. cycadifolius* (Jacq.) Lehm.
Encephalartos longifolius var. *latifolius* Regel (1874) = *E. longifolius* (Jacq.) Lehm.
Encephalartos verschaffeltii Regel (1874) = *E. afer* (Thunb.) Lehm.
Lepidozamia denisonii (C. Moore & F. Muell.) Regel (1875) = *Lepidozamia peroffskyana* Regel.
Lepidozamia hopei Regel (1876).

Lepidozamia peroffskyana Regel (1857).
Macrozamia tridentata (Willd.) Regel (1876) = *Macrozamia* sp.
Stangeria katzeri Regel (1874) = *S. eriopus* (Kunze) Baill.
Zamia angustifolia var. *angustissima* (Miq.) Regel (1878) = *Z. angustifolia* Jacq.
Zamia angustifolia var. *floridana* (A.DC.) Regel (1878) = *Z. integrifolia* L. *fil.*
Zamia angustifolia var. *stricta* (Miq.) Regel (1878) = *Z. stricta* Miq.
Zamia angustifolia var. *yatesii* (Miq.) Regel (1878) = *Z. angustifolia* Jacq.
Zamia baraquiniana Hort. ex Regel (1876) = unknown.
Zamia leiboldii var. *angustifolia* Regel (1876) = *Z. loddigesii* Miq.
Zamia leiboldii var. *latifolia* Regel (1876) = *Z. loddigesii* Miq.
Zamia lindenii Regel ex André (1875).
Zamia loddigesii var. *angustifolia* Regel (1857) = *Z. loddigesii* Miq.
Zamia loddigesii var. *obtusifolia* Regel (1857) = *Z. loddigesii* Miq.
Zamia manicata Linden ex Regel (1876).

Regel's Cycad Publications

Regel, E.A. 1857. *Zwei neue Cycadeen. Bulletin de la société impériale des naturalistes de Moscou* 30(1): 161–191. [*Ceratozamia kuesteriana* Regel, *Lepidozamia peroffskyana* Regel, *Zamia loddigesii* var. *angustifolia* Regel, *Z. loddigesii* var. *obtusifolia* Regel].

Regel, E.A. 1857. *Die Cycadeen des Botanishen Gartens in Petersburg*. Gartenflora 6: 5–16.

Regel, E.A. 1863. *Cycas riuminiana* Porte. Gartenflora 12: 16–17. [*Cycas riuminiana* Porte ex Regel].

Regel, E.A. 1874. *Descriptiones plantarum novarum et minus cognitarum in regionibus Turkestanicis a Cl. P. et O. Fedschenko, Korolkow, Kuschakewicz et Krause collectis:* Cum *adnotationibus ad plantas vivas in horto imperiali botanico Petropolitano cultas. Trudy imperatorskago S.-Petersburgskago botaniceskago sada* 3(1): 97–168. [*Encephalartos altensteinii* var. *distans* Regel, *E. altensteinii* var. *parvifolius* Regel, *E. altensteinii* var. *paucidentatus* Regel, *E. afer* var. *integrifolius* Regel, *E. afer* var. *unidentatus* Regel, *E. cycadifolius* var. *friderici-guilielmi* (Lehm.) Regel, *E. cycadifolius* var. *glaber* Regel, *E. longifolius* var. *latifolius* Regel, *E. verschaffeltii* Regel].

Regel, E.A. 1874. *Stangeria Katzeri* Rgl. Gartenflora 23: 163–165. [*Stangeria katzeri* Regel].

Regel, E.A. 1875. *Die Cycadeen unserer Garten*. Gartenflora 24: 35–43. [*Encephalartos altensteinii* var. *grandis* Regel, *E. altensteinii* var. *macrophyllus* Regel, *E. altensteinii* var. *spinosior* Regel, *Lepidozamia denisonii* (C.Moore & F. Muell.) Regel].

Regel, E.A. 1876. *Cycadearum generum specierumque revisio. Acta Horti Petropolitani* 4(4): 275-320. [*Ceratozamia katzeriana* Regel, *Cycas gracilis* var. *glauca* Regel, *C. gracilis* var. *viridis* Regel, *Zamia baraquiniana* Hort. ex Regel, *Z. leiboldii* var. *angustifolia* Regel, *Z. leiboldii* var. *latifolia* Regel, *Z. manicata* Linden ex Regel].

Regel, E.A. 1876. *Die Cycadeen, deren Gattungen und Arten*. Gartenflora 25: 3-6, 45–51, 140–144, 202, 205, 227–230, 259–262, 370–373. [*Aulacophyllum* Regel (various species), *Ceratozamia kuesteriana* Regel, *Lepidozamia hopei* Regel, *Macrozamia tridentata* (Willd.) Regel].

Regel, E.A. 1878. *Neue und empfehlenswerthe Zierpflanzen*. Gartenflora 27: 13, 114–124. [*Zamia angustifolia* var. *angustissima* (Miq.) Regel, *Z. angustifolia* var. *floridana* (A.DC.) Regel, *Z. angustifolia* var. *stricta* (Miq.) Regel, *Z. angustifolia* var. *yatesii* (Miq.) Regel].

Benedikt Roezl (1824–1885)

Benedikt (Benedict) Roezl was born on August 13, 1824, in Horoměřice, Bohemia (now part of the Czech Republic). After training as a gardener, he moved to Belgium and joined the establishment of Louis van Houtte.

However, he was lured by the New World and traveled to the U.S.A. and then Mexico, where he dabbled in various business and farming activities and also designed parks for Mexico City at the invitation of President Lerdo de Tejada.

A major change in his life was the result of a visit to Cuba where he was demonstrating a machine that he had constructed to extract fiber from agave leaves. Unfortunately, the apparatus ran out of control, and Roezl's hand was amputated and replaced by an iron hook. At that stage, he decided to travel through South America collecting rare plants, particularly orchids, for nurserymen such as Henry Sander in England and Jean Linden in Belgium. He has been considered the most famous, but also perhaps the most notorious, orchid hunter of his time as his collecting avarice was such that vast numbers of plants were indiscriminately stripped from habitat to cater to the European orchid frenzy of Victorian times. His iron hook prosthesis made him quite popular among the local Indians who kept on bringing him plants.

Roezl collected specimens of at least two cycads. This first was at an unspecified locality "on the Pacific Coast of South America." The cycad specimens, together with many other hundreds of plants, were sent to Linden's nursery for onward selling. Linden, acknowledging his supplier, called the cycad "Zamia roezlei" (= *Z. roezlii*) in his 1873 plant catalog/price list, with the comment that the plant was a "*grande et magnifique espèce à frondes longissimes*." In *L'Illustration horticole* of the same year, Linden printed a two-page Illustration (Fig. 5.46) and notes on the cycad prepared by Édouard André. In the initial absence of any herbarium vouchers from the original collection, Stevenson and Sabato (1986) selected this double plate as the neotype for the *Zamia roezlii*. However, Lindström (2009) located a Kew specimen of leaf material (Fig. 5.47a) that had been collected by Roezl in April 1872, and this is now the designated *Zamia roezlii* holotype on the World List of Cycads. *Zamia roezlii* is indeed a magnificent plant when seen in habitat (Fig. 5.48).

Roezl collected material of at least one other cycad population, possibly on the coastal plain or foothills of the Cordillera Occidental in Ecuador, again shipping plants to Linden who once more had illustrations and notes cycad prepared by Édouard André in 1875. In this case, it was Regel who described the species under the name *Zamia lindenii*. Stevenson and Sabato (1986a) designated André's Plate 195 as the lectotype for the species name, while Lindström (2009) identified a Kew voucher, prepared in 1880 from a plant in Linden's nursery and originally supplied by Roezl, as an epitype.

Following his collecting expeditions in South America, Roezl returned to Prague in 1875. During the course of his career, Roezl had described several hundred plants; remarkably 133 of them in the single genus, *Pinus*. He was the Founder of the Czech botanical magazine *Flora*. Toward the end of his life, he was given the Order of St. Stanislaus by Tsar Alexander III. He passed away in Prague on October 14, 1885.

Fig. 5.46 *Zamia roezlii*, the cycad from Colombia and Ecuador named by Jules Linden for Benedict Roezl. The artwork, by botanical artist Pieter de Pannemaeker, was published in *L'Illustration horticole* 20 of 1873, and Stevenson and Sabato (1986a) chose the composite plate to neotypify the name

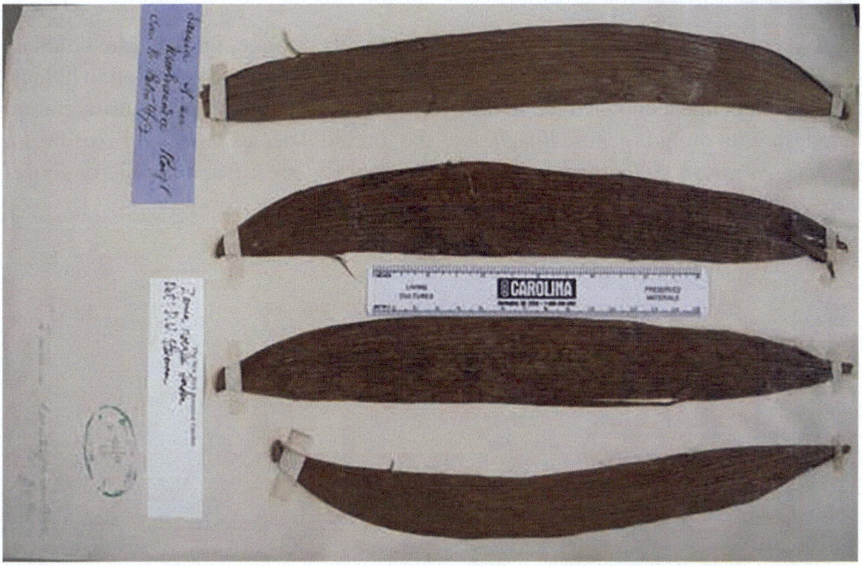

Fig. 5.47 A Kew herbarium specimen of *Zamia roezlii* leaflets. Although the annotations are not entirely clear, it seems that the material was collected in April 1872 by Roezl while in the company of another Belgian plant collector, Charles Patin. This voucher is identified as the holotype for the species on the World List of Cycads

Fig. 5.48 *Zamia roezlii* in dense coastal forest in the Valle de Cauca, Colombia. (Photo: Michael Calonje)

A statute of Roezl in Charles Square (Fig. 5.49), Prague, in which he holds an orchid and has a native American guide alongside, marks his contribution to botany. Apart from being acknowledged in the cycad *Zamia roezlii*, his name is also remembered in the orchid genus *Roezliella*; the orchid species *Miltoniopsis roezlii*, *Pescatorea roezlii*, *Selenipedium roezlii*, and *Sobralia roezlii*; and the Sierra gooseberry, *Ribes roezlii*.

Fig. 5.49 The statue in Charles Square, Prague, in which Benedikt Roezl holds an orchid and has a native American guide alongside

As an addendum to this section, Dennis Stevenson tells of this incident when he and Knut Norstog were investigating *Zamia roezlii* during a field trip in South America. It demonstrates to some extent the influence that the famous explorer Alexander von Humboldt (1769–1859) had on local villagers that he encountered during his expedition.

We were in Parque Nacional Natural Uramba Bahía Málaga or near there on the coast in Colombia. We were staying in a small village, enjoying ourselves and spending an evening with local people one of whom was the equivalent of the village mayor. He turned to Knut and asked: "What do you know about Humboldt?", which completely flabbergasted us. It turned out that Humboldt had stayed there with ancestors in his family, some 150 years previously – but it to him it was like yesterday. He was quite proud of that aspect of their village history.

William Turner Thiselton-Dyer (1843–1928)

William Turner Thiselton-Dyer (Fig. 5.50), referred to in most botanical texts simply as Dyer, was born in Westminster, London, on July 28, 1843, son of physician William George Thiselton-Dyer and botanist Catherine née Firminger. He was educated at King's College School and King's College London before entering Christ Church, Oxford, to study mathematics and natural science. His first position was as Professor of Natural Science at the Royal Agricultural College in Cirencester in Gloucestershire and, soon thereafter, as Professor of Botany at the Royal College of Science in Dublin. In 1872, on the recommendation of Sir Joseph Hooker, he assumed teaching responsibilities at the Royal Horticultural Society in London before joining Hooker in Kew as Assistant Director of the Royal Botanic Gardens in 1875 and then succeeding Hooker as Director 10 years later.

During his 30-year tenure at Kew, Thiselton-Dyer spent considerable time working on projects for the British colonies, introducing Brazilian rubber plants to Ceylon and Malaysia and West Indian cacao to Ceylon (now Sri Lanka). In 1877, he was given charge of the Jodrell Laboratory, which was to become one of the finest of such research facilities in Europe. He was noted for designing a new rock garden for Kew's alpine plants and employing the first women gardeners on the staff. In terms of personality, he was described as decisive, direct, methodical, and well organized, but more than a little stubborn, insensitive, and autocratic (Geison n.d.).

Thiselton-Dyer's published works include an English adaptation of Samuel Johnson's popular American text "How Plants Grow" (with Arthur Church, 1869), "The Flora of Middlesex" (with Henry Trimen, 1869), and an English edition of Sachs' "Textbook of Botany" (1875). He served periods as editor of *Icones*

Fig. 5.50 Sir William Thiselton-Dyer at his office in Kew. (Image from Kew Bulletin No. 3 of 1929)

Plantarum, *Index Kewensis*, and Curtis's Botanical Magazine. In 1887, he founded the Kew Bulletin, destined to become a leading botanical journal. Although he did not travel to see any cycads in their habitat, he used data supplied by others to describe seven New World and three Asian cycad taxa. For example, in 1885, he named *Cycas beddomei* (Fig. 5.51) for the British military officer and naturalist, Colonel Richard Beddome, Conservation Officer of the Madras Forestry Department, India. His main work was the Cycadaceae monograph in William Hemsley's *Biologia Centrali-Americana*, "Contributions to the Knowledge of the Flora and Fauna of Mexico and Central America" (1884). He also provided much of the cycad data that was used in the publications of the Cycadaceae sections of the Flora of Tropical Africa (Prain 1917) and *Flora Capensis* (Hutchison and Rattray 1933).

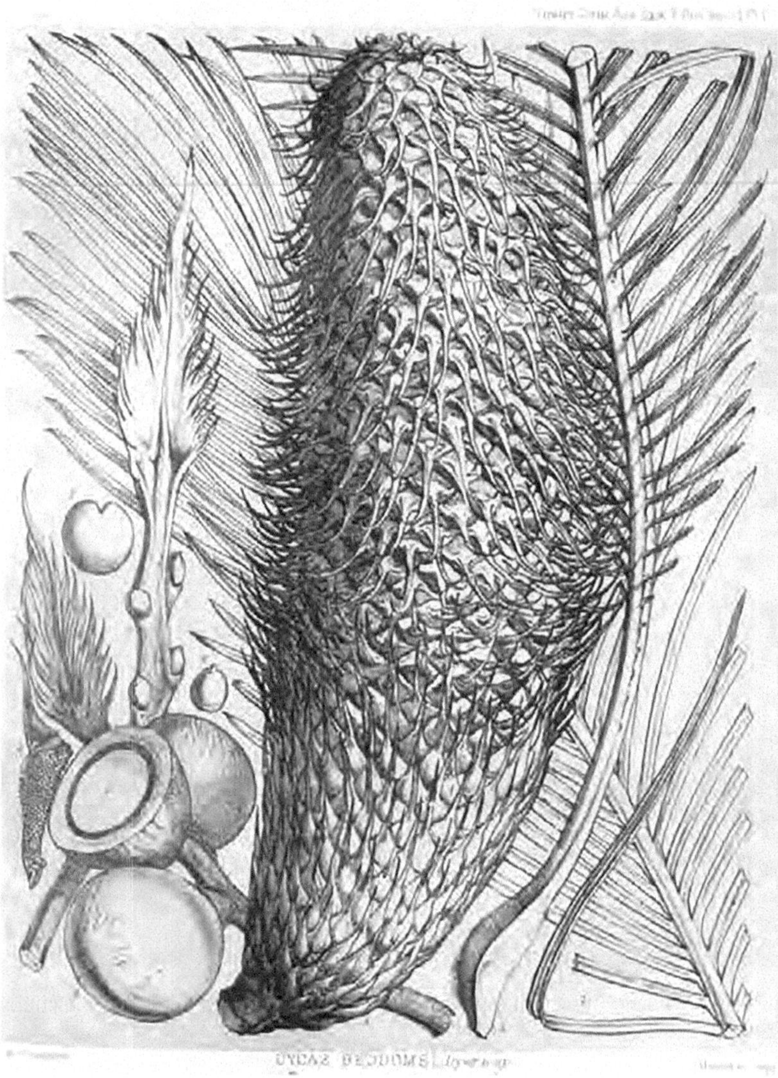

Fig. 5.51 Illustrations of *Cycas beddomei*, a species described by Thiselton-Dyer in the Transactions of the Linnean Society of London, Series 2(5): Table 17. 1885

Thiselton-Dyer was a Fellow of the Linnean Society, a Fellow of the Royal Society, a Fellow of the University of London, a Royal Commissioner to the 1900 Paris International Exhibition and the 1904 St Louis Exposition, a Botanical Advisor to the Secretary of State for the Colonies, a Member of the British Association for the Advancement of Science, and a Member of the Court of the University of Bristol. He was awarded the Clarke Medal of the Royal Society of New South Wales in 1892 and made Companion of the Order of Saint Michael and Saint George (KCMG) in 1899.

Fig. 5.52 The Western Australian cycad *Macrozamia dyeri* which Ferdinand von Mueller first called *Encephalartos dyeri* in recognition of the botanical work of Sir William Thiselton-Dyer. (Photo: William Archer)

Ferdinand von Mueller honored Thiselton-Dyer in naming the Western Australian cycad *Encephalartos dyeri*; that name was later re-assigned to *Macrozamia dyeri* (Fig. 5.52) by Charles Gardner. The Japanese botanist Takenosin Nakai created the genus *Dyerocycas* in 1943, but that name was sunk into synonymy with *Cycas*. Non-cycad taxa recognizing Thiselton-Dyer in eponymy are the small genus *Dyerophytum* (Plumbaginaceae) and the Asian tree genera *Dyera* (Apocyanaceae) and *Dyerella* (Dipterocarpaceae).

Thiselton-Dyer married Joseph Hooker's daughter, botanical illustrator Harriet Hooker, in 1877, and the couple had one son and one daughter. He died on December 23, 1929, and was buried in the churchyard of Saint Peter's, Bentham, survived by his wife, son, and daughter.

Cycad Names with Thiselton-Dyer's Authorship
Data from the *World List of Cycads*

Ceratozamia mexicana var. *robusta* (Miq.) Dyer (1884) = *C. robusta* Miq.
Ceratozamia mexicana var. *tenuis* Dyer (1884) = *C. tenuis* (Dyer) D.W.Stev. & Vovides.
Cycas beddomei Dyer (1885).
Cycas micholitzii Dyer (1905).
Cycas rumphii var. *bifida* Dyer (1902) = *C. bifida* (Dyer) K.D.Hill.
Dioon edule var. *latipinnium* Dyer (1884) = *D. mejiae* Standl. & L.O.Williams.
Dioon spinulosum Dyer ex Eichler (1883).
Zamia acuminata Oerst. ex Dyer (1884).
Zamia cycadifolia Dyer (1884) = *Z. loddigesii* Miq.
Zamia lawsoniana Dyer (1884).
Zamia picta Dyer (1884) = *Z. variegata* Warsz.

Thiselton-Dyer's Cycad Publications

Thiselton-Dyer, W.T. 1882. New garden plants: *Zamia monticola*, *Z. obliqua*. The Gardeners' Chronicle n.s. 17: 460–461.

Thiselton-Dyer, W.T. 1884. Cycadaceae. Pp. 190–195 in W.B. Hemsley, *Biologia Centrali-Americana*, Contributions to the Knowledge of the Flora and Fauna of Mexico and Central America, Vol. 3, Botany. R.H. Porter, London. [*Ceratozamia mexicana* var. *robusta* (Miq.) Dyer, *C. mexicana* var. *tenuis* Dyer, *Dioon edule* var. *latipinnium* Dyer, *Zamia acuminata* Oerst. ex Dyer, *Z. lawsoniana* Dyer, *Z. picta* Dyer].

Thiselton-Dyer, W.T. 1885. On a new species of *Cycas* from southern India. Transactions of the Linnean Society of London, Series 2(5): 85–86. [*Cycas beddomei* Dyer].

Thiselton-Dyer, W.T. 1888. Cycadaceae. Pp. 655–658 in J.D. Hooker (ed.), Flora of British India, Vol. 5. L. Reeve & Co., London.

Thiselton-Dyer, W.T. 1895–1933. *Flora Capensis*, Vols 4–7 and supplement 5(2).

Thiselton-Dyer, W.T. 1902. Cycadaceae. Journal of the Linnean Society, Botany 26: 560–561. [*Cycas rumphii* var. *bifida* Dyer].

Thiselton-Dyer, W.T. 1902–1913. Flora of Tropical Africa, Vols 4–6 and addenda.

Thiselton-Dyer, W.T. 1905. New or Noteworthy Plants. *Cycas micholitzii*. The Gardeners' Chronicle 3(38): 142–144. [*Cycas micholitzii* Dyer].

Gustav Wallis (1830–1878)

After brief spells as a gardener in Detmold and Munich, Wallis started working as a professional plant collector in 1856. His first venture in South America was a financial disaster, but after two years, he was commissioned by Jean Linden's Brussels nursery to hunt for orchids and other rare plants in Brazil and its neighboring countries. In 1866, Wallis came across a novel *Cattleya* at the confluence of the Rio Negro and the Amazon. He sent a large consignment of these plants to Linden who called the species *Cattleya eldorado*, the name later revised to *C. wallisii*. By the following year, Linden's nursery had more than 700 plants of the new species in bloom on display in Paris. However, the relationship between a demanding Linden and a desperate Wallis was a turbulent one, demonstrated in this extract from a lengthy discourse printed by Linden in the Gardener's Chronicle of 1875:

> After dragging on a miserable existence for several years, he [Wallis] sought my assistance. I found him poor, unknown and forsaken; and he left me with a name, a small fortune, a museum of curiosities of all sorts ... and honours. For my numerous kindnesses [he] has shown towards me the greatest ingratitude and bad feeling.

One can perhaps see why Wallis then switched allegiance to the English nurserymen, James Veitch and Sons, who sent him to the Philippines in 1870 principally to search for *Phalaenopsis* orchids. That venture was financially unsuccessful, and Wallis agreed to relocate to Colombia in South America where, between 1872 and 1874, he discovered other novel orchids, two giant-leaved *Anthurium* species and two species of *Zamia*, details of which follow.

Wallis's contract with Veitch ended in 1874, but he continued to collect plants independently. Unfortunately, his health began to suffer from the effects of yellow fever and malaria. Although he recovered temporarily in Panama, he suffered a relapse together with dysentery and passed away, penniless and friendless, in the Sister of Charity hospital in Cuenca, Ecuador, on June 20, 1878. Apart from his recognition in the cycad that we discuss below, Wallis's name is acknowledged in Regel's bromeliad genus *Wallisia* and in 14 *wallisii* epithets among the many species that he introduced into cultivation in Europe.

What of the cycads discovered by Wallis? What we gather from the literature archives is that in 1873, Wallis rediscovered the *Zamia* population that Berthold Seemann had found at Cabo Corrientes on the Colombian west coast in 1847 and then located a further two *Zamia* species in the mountainous parts of Antioquia, sending various samples of these cycads to James Veitch's nursery in London and to the Berlin Botanical Gardens.

Alexander Braun (1875a, b) used Wallis's material and field notes to describe *Zamia montana*, *Z. obliqua*, and *Z. wallisii* (Figs. 5.53, 5.54, 5.55, 5.56, and 5.57). It appears that the German botanist August Eichler filed leaflets of at least two of these species in the Berlin herbarium and provided tracings of these leaflets to Thiselton-Dyer; these drawings are apparently still present in the Kew herbarium. Only one possible *Zamia wallisii* specimen was salvaged after the destruction of the Berlin herbarium in World War II; that was relocated to the Swedish Museum of Natural History. Although it is greatly fragmented, it stands as the lectotype (Lindström 2009) for *Zamia wallisii* (Fig. 5.54).

While primarily interested in studying Colombian palms (Bernal et al. 1989), another of Veitch's contractors, Wilhelm Kalbreyer, found more of the mountain-dwelling *Zamia* species between 1877 and 1880 and sent samples to Europe. James Veitch's nursery provided Kew with *Zamia wallisii* specimens from which Joseph Hooker published a lithograph (Fig. 5.55) of the plant in Curtis's Botanical Magazine of 1890. William Bull's nursery in Chelsea, London, commissioned the plant dealers Edward Shuttleworth and John Carter to harvest a consignment of *Zamia*

Fig. 5.53 The German plant collector, Gustav Wallis, who introduced over 1000 plant species to Europe, many of which were named after him. (From Morren 1879)

montana plants from Antioquia in 1882; some of these went to Kew allowing William Thiselton-Dyer to write a description in the Gardeners' Chronicle in 1882. Unfortunately, plants of neither species adapted to cultivation in Europe, and no plants with provenance to the Colombian collections survived.

After these nineteenth-century collections and shipments, there was little further knowledge of the two mountainous cycad species and their wild populations for many years. The situation was exacerbated by the military conflict between the Colombian government and rebel forces that made visits to the plant localities a hazardous venture. It was more than a hundred years later that Colombian botanists Rodrigo Bernal and colleagues rediscovered a population of *Zamia wallisii*. By following the route that Kalbreyer had taken through the area, they found the plants in September 1983 at an altitude between 1200 and 1500 m in the Municipio de Frontino, Antioquia, and filed the specimen *Bernal, Galeano, and Turner 735* (Fig. 5.56) at the National Herbarium of Colombia.

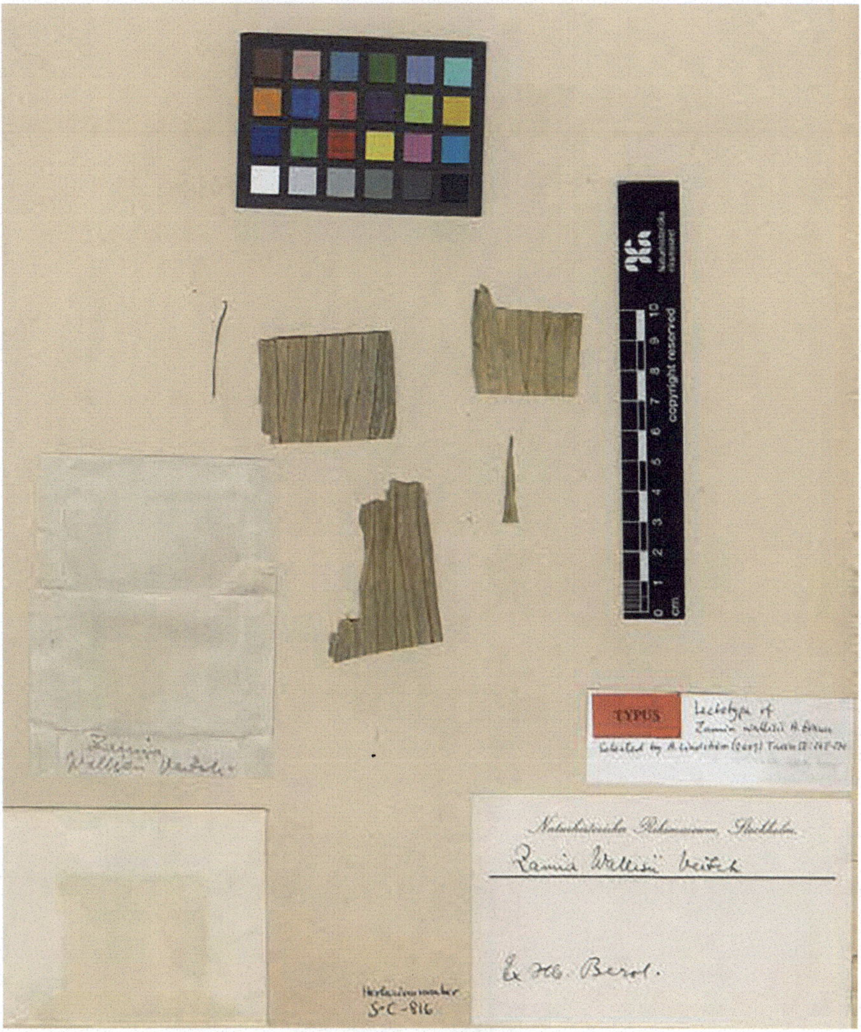

Fig. 5.54 The original *Zamia wallisii* specimen salvaged after the destruction of the Berlin herbarium in 1943 War II and relocated to the Swedish Museum of Natural History. Although it is greatly fragmented, it stands as the lectotype for *Zamia wallisii*

Four years later, Sergio Sabato, Dennis Stevenson, Knut Norstog, and his son Eric along with Rodrigo Bernal were able to visit the population and collect material for botanic gardens and herbaria (Norstog 1985). The voucher *Stevenson* et al. *582* at the New York Botanical Garden Herbarium is a suggested neotype for the species name.

The rediscovery of *Zamia montana* is similar in many respects to that described for *Z. wallisii*. In November 1986, again over a hundred years after the original discovery, a team of Colombian botanists found a population of *Zamia montana* in high-altitude cloud forests in Antioquia. The team comprised Dario Sánchez, Carlos

Fig. 5.55 A color plate of *Zamia wallisii* by John Fitch with lithography by Vincent Brooks, Day and Son, published in Curtis's Botanical Magazine of 1890. The illustration, Plate 7103, is from a plant that was given to Kew by James Veitch and Sons in 1888. This species has the largest leaflets of any cycad, measuring as much as 60 cm in length and 25 cm in width

Fig. 5.56 The voucher *Bernal, Galeano, and Turner 735* at the National Herbarium of Colombia (COL) testifies to the rediscovery of *Zamia wallisii* in September 1983, one hundred and ten years after Wallis first found the population in the Municipio de Frontino, Antioquia

Fig. 5.57 Knut Norstog in Antioquia, Colombia, in 1987, assisting with the collection of herbarium material and live plants of *Zamia wallisii* for the New York Botanical Garden, the Fairchild Tropical Botanical Garden, and the Naples Botanical Garden. (Photo: Dennis Stevenson)

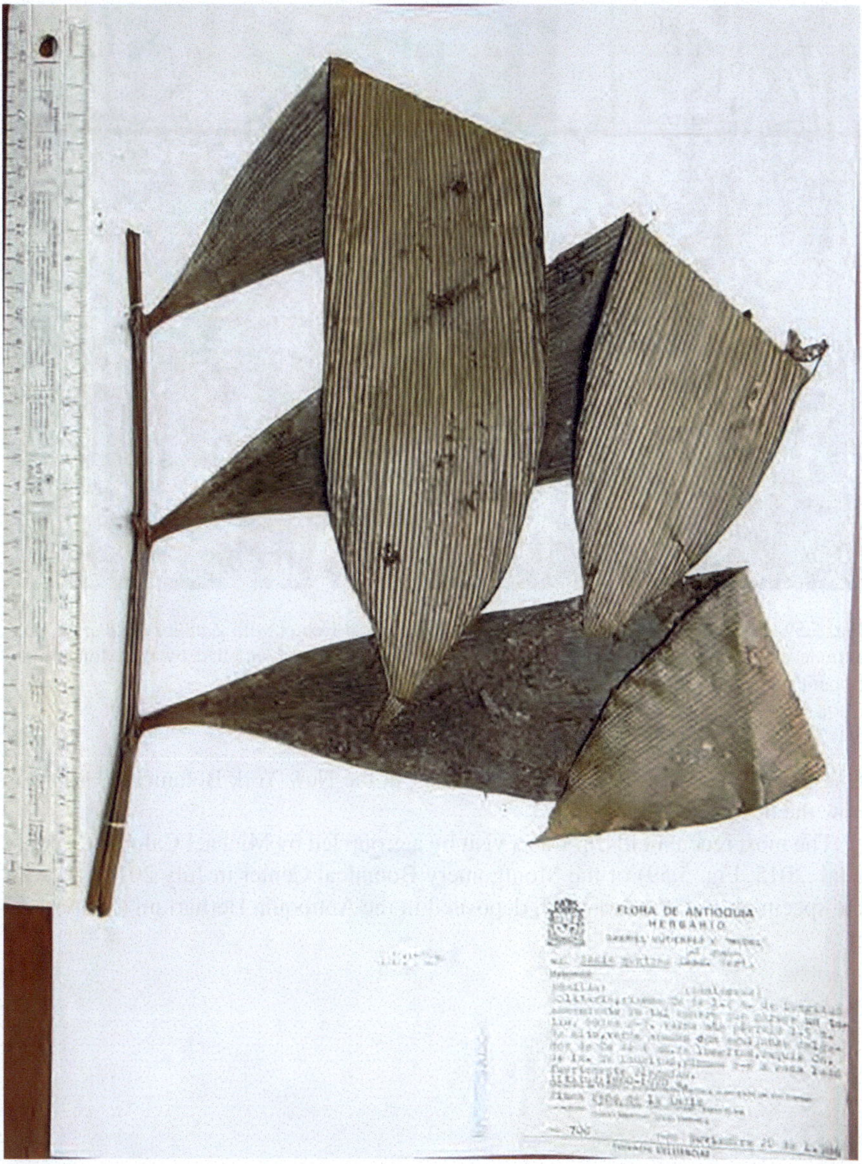

Fig. 5.58 The voucher *Sánchez* et al. *700* at the University of Antioquia, which represents the relocation of *Zamia montana* in November 1986, one hundred and thirteen years after its original discovery by Gustav Wallis

Orrego, Stelia Sylva, Gladys Mártinez, Diego Restrepo, and Gloria Galeano, and their collections are represented by specimens *Sánchez* et al. *700* (Fig. 5.58) and *900* at the University of Antioquia. The region attracted other researchers who made collections of *Z. montana* between 1986 and 1991. One of these was a visit in September 1987 by James Zarucchi, a botanist working for Missouri Botanic

Fig. 5.59 Michael Calonje of the Montgomery Botanical Center with *Zamia montana* in high-altitude cloud forest in Antioquia, Colombia, July 2014. (Photo supplied by the Montgomery Botanical Center)

Gardens; the specimen *Zarucchi* et al. *5724* at the New York Botanical Gardens is now the designated neotype.

The most recent field trip was a visit by a group led by Michael Calonje (Calonje et al. 2015; Fig. 5.59) of the Montgomery Botanical Center in July 2014 and with the specimen, *F.J. Roldán 4725,* deposited in the Antioquia Herbarium (HUA).

Chapter 6
The Early Twentieth Century

Charles Joseph Chamberlain (1863–1943)

Born on February 23, 1863, near Sullivan, Ohio, Charles Joseph Chamberlain (Fig. 6.1) was educated at Oberlin College, Ohio, the oldest coeducational liberal arts college in the U.S.A., and at the University of Chicago, an institution that was to become the prominent feature of his life. He was the first recipient of a Ph.D. degree in Botany from the University, where he was appointed Associate Professor in 1911 and full Professor in 1915. He was instrumental in establishing glasshouses (Fig. 6.2) in Garfield Park in Chicago to house the cycads he brought back from his many field trips around the world.

Chamberlain was innovative in applying zoological microscopic techniques to the study of plant material, especially in his work on cycad anatomy and morphology. In addition to organizing and directing the botanical laboratories at the University, he created the world's foremost greenhouse collection of living cycads, containing specimens from Cuba, Mexico, Australia, and South Africa in a plant collection that remained unsurpassed until a decade after his death.

During the first two decades of the 1900s, Chamberlain traveled extensively—by ship, by rail, and even on horseback. Funded by grants from the Botanical Society of America and the University of Chicago, he went first to Florida; then on to Cuba, seeing *Zamia* and *Microcycas* in the wild; and then twice to Veracruz, Mexico, where he saw *Dioon* and *Ceratozamia*. For his first trip to Veracruz, he had a letter of introduction from the President of the University of Chicago to the U.S. Ambassador to Mexico, who then wrote to the Governor of Veracruz, whose influence made it possible for the local train to make unscheduled stops so Chamberlain could make collections in the vicinity of Chavarillo. Connections in those days paved the way for fundamental discoveries in cycad biology.

Those early excursions were followed by Chamberlain's major and challenging round-the-world trip, commencing by boat from the U.S.A. to the Sandwich Islands,

R. Osborne, D. Wm. Stevenson, *The Cycad Botanists*, Memoirs of The New York Botanical Garden 123, https://doi.org/10.1007/978-3-031-79082-9_6

Fig. 6.1 Charles Chamberlain (left) and William Land (right), a friend and colleague from the University of Chicago. Chamberlain was a renowned sharpshooter who trained marksmen in WWI, hence the medals. (A photograph given to Dennis Stevenson by Chamberlain's grandson, Charles Allsopp)

onwards to Fiji, New Zealand, and then Australia. Arriving in Sydney, he lost no time in exploring the local cycads and other aspects of flora and fauna. He then traveled almost the full length of the Queensland coast, from Brisbane in the south to Cairns in the north. During that time, he saw all four genera of Australian cycads. After that, he went once again by boat, this time via Perth to South Africa, where he spent time in Natal, the Eastern Cape, and Cape Provinces, becoming familiar with *Encephalartos* and *Stangeria*, before leaving for London and then home to Chicago.

In total, Chamberlain saw all known cycad genera and some 30 different species, making field notes, taking photographs with 3.25 × 4-inch glass plates that he later hand-colored (Fig. 6.3a, b), collecting material for microscopic studies, and arranging for live plants to be shipped to Chicago for the greenhouse collection. Unfortunately, Chamberlain's living plant collections ultimately ended up in a private garden, and the associated collection records were lost. Another irony is that the *Zamia monticola* that he described "was grown from a seed collected on the steep mountain side about six or eight miles west of Jalapa (now Xalapa), opposite the extinct crater of Naolinco." It was never seen again until it was realized that it was not from Mexico, but rather endemic to Guatemala when finally collected again

Fig. 6.2 Charles Chamberlain with part of his cycad collection in the Garfield Park glasshouse, University of Chicago. (A photograph given to Dennis Stevenson by Chamberlain's grandson, Charles Allsopp)

by the German botanist Harald Förther in 2001. Careful records are important; Chamberlain had in fact received seeds and other material from Guatemala.

"The Living Cycads" (Fig. 6.4a) makes for fascinating reading (Osborne 1985). This volume (Chamberlain 1919) was so popular that Chamberlain used royalties from the book sales to subsidize his continued research work. In Part One, he recounts his worldwide travels in search of cycads. He struck up warm and ongoing

Fig. 6.3 (**a**). *Bowenia serrulata* at Byfield near Rockhampton on the Queensland central coast. A hand-colored slide taken by Chamberlain during his trip to Australia, as used in his 1912 paper elevating the taxon from subspecific to specific rank. (**b**). *Zamia pygmaea* in habitat. A similar hand-colored slide taken by Chamberlain during a trip to Cuba and named by him at the time as *Zamia kickxii*

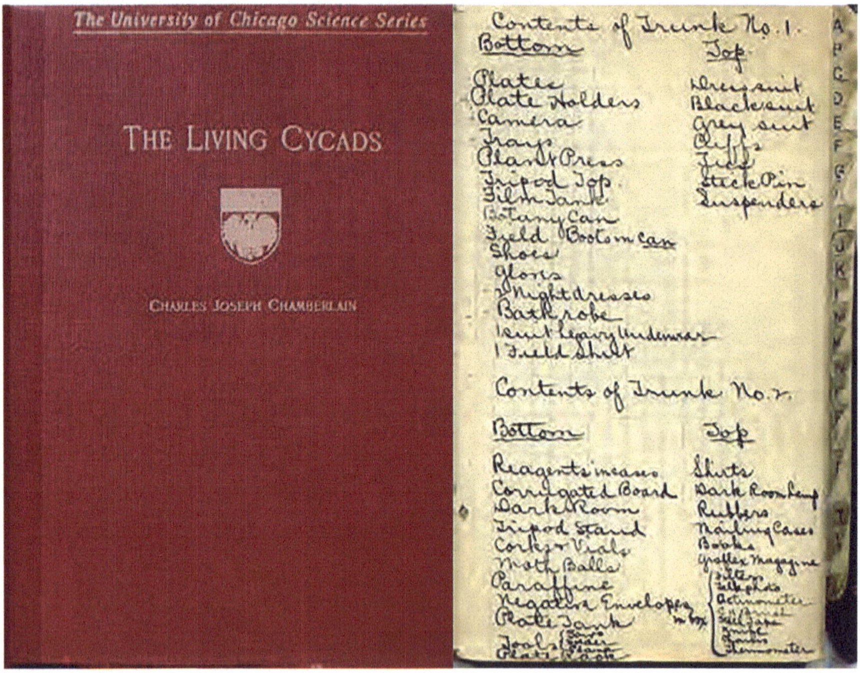

Fig. 6.4 (**a**). The cover of Chamberlain's popular book "The Living Cycads" (1919). (**b**). The handwritten inventory of the contents of Chamberlain's two trunks packed for his trip to Australia and South Africa. Note the formal attire packed in the upper layer of trunk No.1. From the notebook given to Dennis Stevenson by Chamberlain's grandson, Charles Allsopp

friendships with many people in all the countries he visited and clearly spent time in the company of high-ranking officials and other persons of influence; in anticipation of such occasions, he had packed formal attire in his baggage (Fig. 6.4b). Part Two deals with morphology and anatomy, while Part Three covers evolutionary and phylogenetic aspects. An interesting aspect of his fieldwork, and the lantern slides, is that he gave popular lectures to garden clubs and horticultural societies that gave him an honorarium. These funds helped to subsidize both his fieldwork and laboratory research.

During his second visit to Mexico, he described in The Living Cycads the forest flora at Tierra Blanca, on the Veracruz-Oaxaca border, while searching for *Dioon spinulosum* in…

> a forest of Spanish cedar, mahogany, ceiba, various kinds of rubber and occasionally ebony. Orchids, bromeliads, ferns and other plants weigh down the branches of the trees until they break off, so that one may collect this epiphytic vegetation without the labor of climbing.

In Australia, Chamberlain (1919) was much taken with *Macrozamia moorei* (Fig. 5.31)...

> which is abundant at Springsure, about two hundred miles west of Rockhampton, just on the Tropic of Capricorn. It has a massive trunk, seldom more than eight or ten feet in height, but often fifteen inches or even two feet in diameter. The grass was dry and brown but the cycads looked fresh and vigorous with dark-green leaves and a wonderful display of cones.

In South Africa, he described how he was taken by James Wylie's Zulu guide to see the single extant specimen of *Encephalartos woodii*…

> about ten miles from Mtunzini in the midst of the *Stangeria* and *Encephalartos brachyphyllus* [= *E. ngoyanus*].

In another, he tells of the excitement of finding *Encephalartos latifrons* in the Trappes Valley near Grahamstown.

Knut Norstog (1995) wrote interestingly of Chamberlain in an unpublished manuscript entitled "C. J. Chamberlain and the Ladies" that, in Chamberlain's various books, he customarily cited male writers in the conventional way by surname and initials (e.g., Lang, W. H.; Arnold, C. A.; Le Goc, M. J.; Worsdell, W. C.; etc.). However, more often than not, he referred to female authors by their more embellished titles (e.g., Browne, Lady Isabel; Langdon, Dr. LaDema; Dorety, Sister Helen Angela; Lamb, Sister Mary Alice, etc.). Norstog speculated that it is less likely that Chamberlain did this out of particular fondness for lady authors but more "that he felt that women weren't getting a fair shake in academia and in science and [he wanted] his contemporaries to notice that they also could do good science."

Chamberlain married Martha Life in 1888, and they had one daughter. After his wife passed away in 1931, he married another Martha, Martha Lathrop. He died in Chicago, Illinois on February 5, 1943, just a few days before his 80th birthday. The Philippine cycad, *Cycas chamberlainii* (now *C. riuminiana*), was named for him in 1925 (Brown and Kienholz, 1925) as was an Argentinian fossil genus *Neochamberlainia* (Artabe et al. 2010) and the Mexican species *Ceratozamia chamberlainii* (Fig. 6.5; Martínez-Domínguez et al. 2017).

Fig. 6.5 *Ceratozamia chamberlainii* in habitat in San Luis Potosi, Mexico. (Photo: Chip Jones)

Names of Cycads with Chamberlain's Authorship
Data from the *World List of Cycads*

Bowenia serrulata (W. Bull) Chamb. (1912).
Zamia monticola Chamb. (1926).
Zamia sylvatica Chamb. (1926) = *Z. loddigesii* Miq.

Chamberlain's Cycad Publications

Chamberlain, C. J. 1906. The ovule and female gametophyte of *Dioon*. Botanical Gazette 42(5): 321–358.
Chamberlain, C. J. 1906. Preliminary note on *Ceratozamia*. Botanical Gazette 43(2): 13.
Chamberlain, C. J. 1906. The ovule and female gametophyte of *Dioon*. Botanical Gazette 42(5): 321–358.
Chamberlain, C. J. 1908. Report on *Dioon* and *Ceratozamia*. Science 27: 333–334.
Chamberlain, C. J. 1909. Spermatogenesis in *Dioon edule*. Botanical Gazette 47: 215–236.
Chamberlain, C. J. 1909. *Dioon spinulosum*. Botanical Gazette 48(6): 401–413.
Chamberlain, C. J. 1910. Fertilization and embryogeny in *Dioon edule*. Botanical Gazette 50(6): 415–429.
Chamberlain, C. J. 1911. The adult cycad trunk. Botanical Gazette 52(2): 81–104.
Chamberlain, C. J. 1912. Morphology of *Ceratozamia*. Botanical Gazette 53: 1–19.

Chamberlain, C. J. 1912. Two species of *Bowenia*. Botanical Gazette 54: 419–423. [*Bowenia serrulata* (W.Bull) Chamb.].
Chamberlain, C. J. 1913. *Macrozamia moorei*, a connecting link between living and fossil cycads. Botanical Gazette 55(2): 141–154.
Chamberlain, C. J. 1915. A phylogenetic study of the cycads. Proceedings of the National Academy of Sciences 1: 86–90.
Chamberlain, C. J. 1916. *Stangeria paradoxa*. Botanical Gazette 61(5): 353–372.
Chamberlain, C. J. 1919. The Living Cycads. University of Chicago Press, Chicago.
Chamberlain, C. J. 1920. The living cycads and the phylogeny of seed plants. American Journal of Botany 7(4): 146–153.
Chamberlain, C. J. 1925. The origin of the cycads. Science n.s. 61: 73–77.
Chamberlain, C. J. 1926. Two new species of *Zamia*. Botanical Gazette 81(2): 218–227. [*Zamia monticola* Chamb., *Z. sylvatica* Chamb.].
Chamberlain, C. J. 1926. Hybrids in cycads. Botanical Gazette 81: 401– 418.
Chamberlain, C. J. 1932. Methods in Plant Histology. Fifth Revised Edition. University of Chicago Press, Chicago.
Chamberlain, C. J. 1935. Gymnosperms: Structure and Evolution. University of Chicago Press, Chicago. Facsimile edition 1961, Johnson Reprint Corporation, New York; 1966, Dover, New York.
Chamberlain, C. J. n.d. A taxonomic monograph of the cycads. Unpublished monograph in the library of Fairchild Tropical Botanical Garden, with a copy at the New York Botanical Garden.
Coulter, J. M. & C. J. Chamberlain. 1901. Morphology of the Spermatophytes. Part 1, Gymnosperms. University of Chicago Press, Chicago.
Coulter, J. M. & C. J. Chamberlain. 1903. The embryology of *Zamia*. Botanical Gazette 35: 184–194.

John Hutchinson (1884–1972)

John Hutchinson (Fig. 6.6) was born in the small village of Wark-on-Tyne, Northumberland, on April 7, 1884, son to the Head Gardener of a country estate. On his 20th birthday in 1904, Hutchinson joined the Royal Botanic Gardens, Kew, as a gardener, but within one year, his taxonomic and drawing skills were noticed, and he was transferred to the Herbarium, initially as personal assistant to Otto Stapf (whom he remembered with affection and gratitude). Hutchinson worked in the Indian and the African sections at Kew until 1936 when he became Keeper of the Museums, a position he held until his retirement in 1948. He continued voluntary duties at Kew until his death in 1972.

Hutchinson undertook two major collecting trips to South Africa, which he recounts in detail in his "A Botanist in Southern Africa" (1946). Between 1928 and 1929, he toured the Cape Province, Namaqualand, and the South African Provinces then called the Eastern Cape, Natal, and the Transvaal, variously in the company of fellow botanists Neville Pillans, Rudolf Marloth, Robert Dyer, and General Jan Smuts. On a second visit from June to September 1930, he joined the Smuts expedition to the Victoria Falls, Northern Rhodesia (now Zambia), Lake Tanganyika, and the Belgian Congo (now the Democratic Republic of Congo), later returning to

Fig. 6.6 Dr. John Hutchison in the Kew Herbarium. Image from the Biographical Memoirs of Fellows of the Royal Society 21: 345–365. 1975

Natal with Inez Verdoorn and then again visiting the Eastern Cape. Jan Christian Smuts, a South African prime minister, philosopher, and botanist, developed a close friendship with Hutchinson. The four-page foreword by Smuts in "A Botanist in Southern Africa" (Fig. 6.7) was inspired by their friendship.

Although he undoubtedly saw many cycads during these visits, Hutchinson published a taxonomic description of only one species, *Encephalartos kosiensis* (Hutchinson 1932), and that was based on leaf material sent to him after his return to Kew. The plant material had been collected near the Kosi Bay in northern Natal. Hutchinson realized at the time that this cycad "is no doubt allied to *Encephalartos ferox* from Mozambique"; only much later, when the distribution of *E. kosiensis* was found to extend into and along coastal Mozambique, all populations were considered to fall within the concept of *E. ferox*. The seed cones (Fig. 6.8) of this species are among the most colorful and geometrically striking of all cycads.

Importantly, in the major cycad monographs, Hutchinson, assisted by John Dalziel, produced the section Cycadaceae for Volume 1 of the "Flora of West Tropical Africa" (1927–1928) and then, in 1933, co-authored the section Cycadaceae for *Flora Capensis* with Dr. George Rattray in South Africa with much of this work based on William Thiselton-Dyer's notes. This update, more than a hundred years after Carl Thunberg's first treatment, reviewed the taxonomy of 13 species of *Encephalartos* and the monotypic *Stangeria eriopus* and was a predecessor to Dyer and Verdoorn's 1966 treatment of the Zamiaceae in the "Flora of Southern Africa."

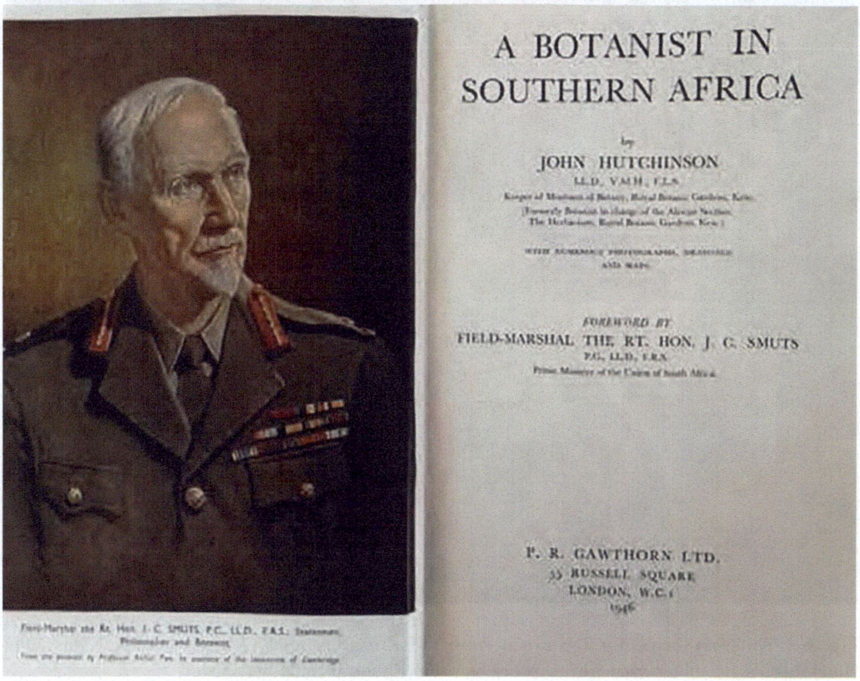

Fig. 6.7 The title pages of Hutchison's "A Botanist in Southern Africa" (1946) with an image of Field Marshall Jan Christian Smuts

Fig. 6.8 An early name for the cycad from the Kosi Bay area of northern Natal was John Hutchinson's *Encephalartos kosiensis* (=*E. ferox*). (Photo: Dennis Stevenson)

Foremost among his many writings in the area of plant systematics was Hutchinson's "The Families of Flowering Plants" proposing a radical revision of the earlier classification systems of Joseph Hooker, Adolf Engler, and Karl Prantl. This two-volume treatise was first released in 1926 and 1934 but was followed by a second edition in 1959 and a third printed posthumously. Hutchinson retired in 1948 but continued working on the phylogeny of flowering plants in his three volumes "The Genera of Flowering Plants" (1964, 1967, and also posthumously). His numerous and varied other writings are listed in a detailed bibliography by Carolyn Pope that accompanied Brenan's 1974 tribute.

John Hutchinson was awarded an Honorary Doctor of Laws from the University of St. Andrews, the Veitch Memorial Medal and Loder Cup of the Royal Horticultural Society, a Victoria Medal in horticulture, the Linnean Gold, and Darwin-Wallace Centenary Silver Medals and was appointed to the Order of the British Empire. He was elected Fellow of the Royal Society, the Association for Tropical Biology and Conservation, the American Amaryllis Society, and the Philadelphia Academy of Sciences and was a Member of the Botanical Society of Edinburgh and the Botanical Society of America. He passed away on September 2, 1972, at the age of 88 while still working in the Herbarium at Kew and was survived by his wife Lilian, two sons, and three daughters. His name is commemorated in *Hutchinsonia*, a genus in the Rubiaceae published by his friend and colleague, Walter Robyns, in 1928.

Name of the Cycad with Hutchinson's Authorship
Data from the *World List of Cycads*

Encephalartos kosiensis Hutch. (1932) = *E. ferox* Bertol.*fil.*

Hutchinson's Cycad Publications

Hutchinson, J. 1932. *Encephalartos kosiensis* Hutch. Bulletin of Miscellaneous Information 1932: 512. [*Encephalartos kosiensis* Hutch.]

Hutchinson, J. & G. Rattray. 1933. Cycadaceae. Pp. 24–44 in *Flora Capensis* 5(2). [Other families dealt with include Euphorbiaceae, Moraceae, Myricaceae & Proteaceae.]

Hutchinson, J. & J. M. Dalziel.1927–1928. Flora of West Tropical Africa. First edition: two vols. Second edition, with R. W. J. Keay & F. N. Hepper, three vols, 1954–1972.

David Prain (1857–1944)

David Prain (Fig. 6.9) was born on July 11, 1857, in Fettercairn, Scotland, to a saddler father, also David, and Mary Thomson. He attended the local parish school, then Aberdeen Grammar School, before entering the Universities of Aberdeen and Edinburgh to study medicine. In 1884, he accepted a posting as a botanist/physician to the Indian Medical Service. Three years later, he was appointed Curator to the

Fig. 6.9 Sir David Prain. (Photo from the archives of the Hunt Institute)

Calcutta Herbarium, and in 1898, he became Director of the Royal Botanic Garden of Calcutta and the Botanical Survey of India—as well as becoming Professor of Botany at the Medical College of Calcutta (now more correctly written in the Bengali name Kolkata). His research activities included studies on Indian hemp, wheat, mustards, pulses, and indigo. Among his publications was the volume "Bengal Plants," describing nearly 3000 species, published in 1903 at his own expense.

In 1903, Prain accompanied Sir Francis Younghusband on a plant-collecting expedition to the Sikkim-Tibet boundary. He also collected material in Burma (now Myanmar), Ceylon (now Sri Lanka), and the Andaman and Nicobar Islands.

In 1905, Prain returned to England to succeed William Thiselton-Dyer as Director, at Kew. Retaining this position for 17 years, he initiated a medicinal garden in the grounds of Cambridge Cottage, installed a large tank for aquatic plants, and acquired the Japanese gateway from the Japan-British exposition of 1910. His willingness to delegate decision-making was much appreciated after the somewhat autocratic leadership style of Thiselton-Dyer. During his tenure at Kew, Prain authored numerous publications while serving as editor of *Curtis's Botanical Magazine* and *Index Kewensis*. He contributed to parts of *Flora Capensis*, *Flora of Indochina*, *Flora of the Sundribuns*, and *Flora of Tropical Africa*.

Prain's cycad interests commenced with brief references to *Cycas pectinata*, *C. revoluta*, and *C. rumphii* in his "Bengal Plants" of 1903. Later, in *Curtis's Botanical Magazine*, he published annotated and attractive lithographs, prepared by Vincent Brooks, Day and Sons, of *Cycas micholitzii* (Plate 8242, 1909; Fig. 6.10), *Encephalartos barteri* (Plate 8232, 1909), and *E. hildebrandtii* (Plates 8592 and 8593, 1915). This followed a series of similar *Encephalartos* lithographs published during the editorship of Sir Joseph Hooker over the period 1856–1891.

Prain's only complete cycad description is that of *Encephalartos gratus* and was based on plant material, photographs, and detailed notes sent to Kew from Mount Mlangi (now Mulanje) in southeastern Nyasaland (now Malawi) by colonial officials John Mahon, John McClounie, and E. W. Davy over the period 1899–1916 and published in the *Kew Bulletin* of 1916. His section on Cycadaceae based largely on William Thiselton-Dyer's notes gives a usefully detailed review of *Cycas thouarsii* and the seven species of *Encephalartos* known from the area at that time (*E. barteri*, *E. ferox*, *E. gratus* (Fig. 6.11), *E. hildebrandtii*, *E. laurentianus*, *E. poggei*, and *E. septentrionalis*).

Among his many duties, David Prain is remembered for his services as Chair of the John Innes Horticultural Institution, President of the Linnean Society, and Vice President of the Royal Horticultural Society. He was awarded the Linnean Medal, the RHS Victoria Medal of Honour, and a Veitch Memorial Medal. He was made a Companion Order of St. Michael and St. George and received a knighthood in 1912. He died at Whyteleafe, Surrey, on 16 March 1944 (Burkill 1944; Desmond 2004). The genus *Prainea* (Moraceae) was naed for him in 1888.

Fig. 6.10 A lithograph of *Cycas micholitzii* prepared by Vincent Brookes, Day and Sons Ltd., and published with notes on the species by Prain as Plate 8242 in Curtis's Botanical Magazine of 1909

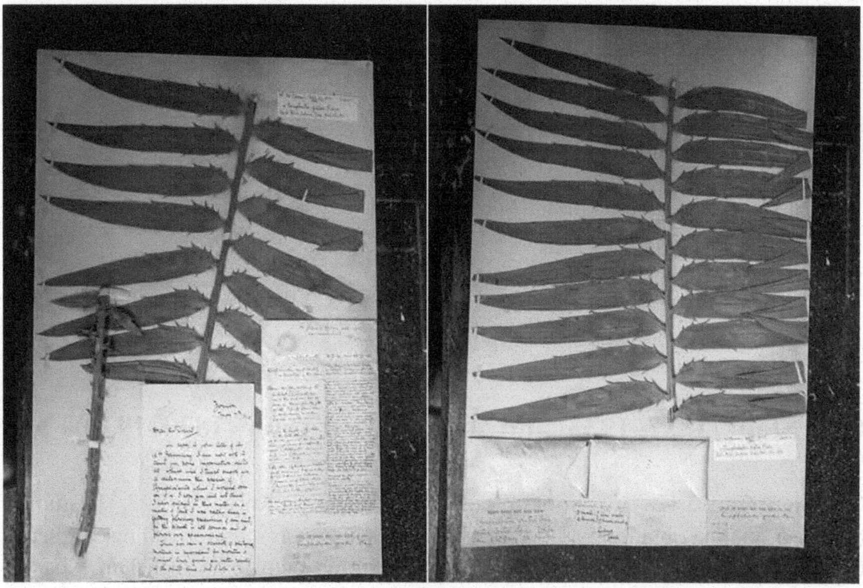

Fig. 6.11 Two of the three sheets which comprise David Prain's 1916 type for *Encephalartos gratus,* together with some of the correspondence between Prain in Kew and E. W. Davy, then Assistant Director of Agriculture in Zomba, Nyasaland (now Malawi)

Name of the Cycad with Prain's Authorship

Data from the *World List of Cycads*

Encephalartos gratus Prain (1916).

Prain's Cycad Publications

Prain, D. 1903. Cycadaceae. Bengal Plants 2: 745–746.

Prain, D. 1909. *Cycas micholitzii*. Curtis's Botanical Magazine 135: Plate 8242.

Prain, D. 1909. *Encephalartos barteri*. Curtis's Botanical Magazine 135: Plate 8232.

Prain, D. 1915. *Encephalartos hildebrandtii*. Curtis's Botanical Magazine 141: Plates 8592–8593.

Prain, D. 1916. Diagnoses Africanae: *Encephalartos gratus*. Kew Bulletin of Miscellaneous Information 7: 180–182. [*Encephalartos gratus* Prain].

Prain, D. 1917. Cycadaceae. Vol. 6(2). Pp. 344–354 in D. Oliver (ed.), Flora of Tropical Africa.

Prain's Cycad Publications

Prain, D. 1903. Cycadaceae. Bengal Plants 2: 745–746.
Prain, D. 1909. *Cycas micholitzii*. Curtis's Botanical Magazine 135: Plate 8242.
Prain, D. 1909. *Encephalartos barteri*. Curtis's Botanical Magazine 135: Plate 8232.
Prain, D. 1915. *Encephalartos hildebrandtii*. Curtis's Botanical Magazine 141: Plates 8592–8593.
Prain, D. 1916. Diagnoses Africanae: *Encephalartos gratus*. Kew Bulletin of Miscellaneous Information 7: 180–182. [*Encephalartos gratus* Prain].
Prain, D. 1917. Cycadaceae. Vol. 6(2). Pp. 344–354 in D. Oliver (ed.), Flora of Tropical Africa.

Otto Stapf (1857–1933) and Joseph Burtt Davy (1870–1940)

Otto Stapf (Fig. 6.12a) was born in Bad Ischl, Austria, on April 23, 1857. He studied botany in Vienna obtaining a Ph.D. and then staying on as lecturer and researcher at the University. In 1885, Jakob Polak, personal physician to Nasr al-Din, Shar of Persia, sponsored a nine-month expedition for Stapf to the southern and western parts of Persia (now Iran) where Stapf discovered and subsequently published reports on many new plant species. After a disagreement with his supervisor in Vienna, Stapf moved to England in 1890, took up the post of Keeper of the Royal Botanic Gardens Herbarium at Kew, and also became naturalized as a British citizen.

Stapf's career continued with many travels and numerous publications in systematic botany to his credit. He was recognized as a global expert on grasses, with a great many publications on their economic botany and their systematics, including his contribution of the section on the grasses in William Thiselton-Dyer's edition of *Flora Capensis* and his many systematic entries in the *Flora of Tropical Africa*. A legacy of Stapf's is his monograph on *Ephedra*, which has remained largely intact since its publication in 1889. He was editor of several issues of *Curtis's Botanical Magazine* and co-author with Georg Pritzel in two volumes of *Index Londinensis*.

Otto Stapf passed at the age of 76 on August 3, 1933, while on a summer trip to Austria. His name is commemorated in the grass genera *Neostapfia*, *Neostapfiella*, *Ottochloa*, *Stapfia*, and *Stapfiola* and in the genera *Stapfiella* (Passifloraceae) and *Stapfiophyton* (Melastomataceae), as well as in the German geobotanical journal *Stapfia*. He was a Fellow of the Royal Society, Botanical Secretary to the Linnean Society of London, a Member of the Austrian Academy of Sciences, and a Member of the German Botanical Society.

Joseph Burtt Davy (Fig. 6.12b) was born at Findern, Derbyshire, England, on August 20, 1870, and was educated at Ilkley School in West Yorkshire. He worked for two years as an administrative assistant in the Director's office at Kew until, seeking a more agreeable climate, he moved to the U.S.A. in 1892. There he took a teaching position at the University of California and later became Assistant Curator in the Department of Agriculture in Washington. In 1903, Burtt Davy moved to

Fig. 6.12 (**a**). Otto Stapf, Welcome Collection. (**b**). Joseph Burtt Davy. Image from Gardeners' Chronicle series 3, 75: 266. 1924

South Africa, taking up a government post as agrostologist and botanist in Pretoria. During this time, he gathered some 14,000 plant specimens in the Transvaal Province for what would ultimately become the South African National Herbarium. After ten years, Burtt Davy left the government service to go farming in the Transvaal Province but later returned to England to spend another five years working on the first two parts of his "*Manual of the Flowering Plants and Ferns of the Transvaal with Swaziland, South Africa*, 1926–1932" (Fig. 6.13) a major work illustrated with detailed line drawings by his wife Alice Davy.

In 1925, Burtt Davy obtained a Ph.D. from Cambridge and then joined the Imperial Forestry Institute at Oxford as a lecturer in tropical forest botany. He was awarded a D.Phil. degree by Oxford University in 1937. After a second retirement in 1939, he continued working on a third and fourth part of the Transvaal flora but died at Birmingham on August 20, 1940, before the work could be completed. Joseph Burtt Davy's name is commemorated in the genus *Burttdavya* (an East African tree in the Rubiaceae) and in numerous specific epithets including *Aloe davyana* (Asphodelaceae), *Acacia davyi* (Fabaceae), *Hibiscus burtt-davyi* (Malvaceae), and *Ficus burtt-davyi* (Moraceae).

It was at Kew, during the 1920s, when Stapf and Burtt Davy collaborated on their descriptions of *Encephalartos laevifolius*, *E. lanatus*, *E. paucidentatus*, *E. striatus*, and *E. transvenosus* for inclusion in their *Manual of the Flowering Plants* (1926). While most of the descriptions are straightforward, there are two taxa that present minor difficulties.

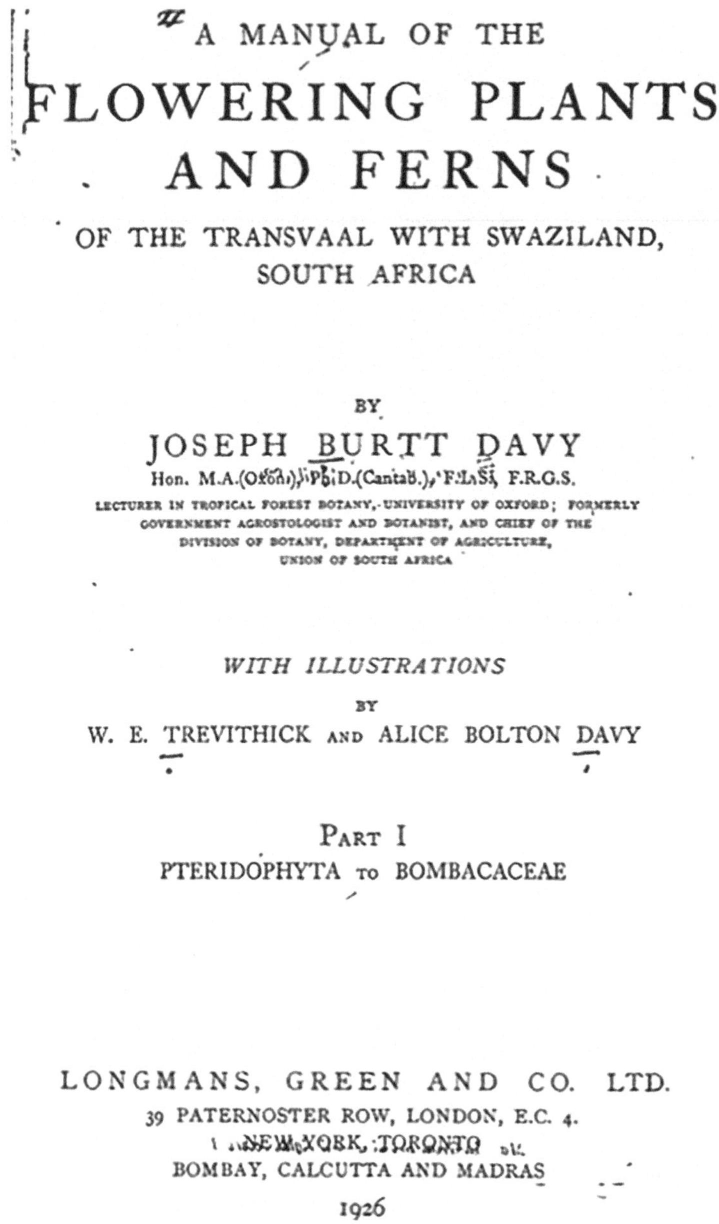

A MANUAL OF THE

FLOWERING PLANTS AND FERNS

OF THE TRANSVAAL WITH SWAZILAND, SOUTH AFRICA

BY

JOSEPH BURTT DAVY

Hon. M.A.(Oxon.), Ph.D.(Cantab.), F.L.S., F.R.G.S.

LECTURER IN TROPICAL FOREST BOTANY, UNIVERSITY OF OXFORD; FORMERLY GOVERNMENT AGROSTOLOGIST AND BOTANIST, AND CHIEF OF THE DIVISION OF BOTANY, DEPARTMENT OF AGRICULTURE, UNION OF SOUTH AFRICA

WITH ILLUSTRATIONS

BY

W. E. TREVITHICK AND ALICE BOLTON DAVY

PART I

PTERIDOPHYTA TO BOMBACACEAE

LONGMANS, GREEN AND CO. LTD.

39 PATERNOSTER ROW, LONDON, E.C. 4.

NEW YORK, TORONTO

BOMBAY, CALCUTTA AND MADRAS

1926

Fig. 6.13 The frontispiece to Part 1 of Joseph Burtt Davy's Manual of the Flowering Plants and Ferns of the Transvaal with Swaziland, South Africa (1926) in which he and Otto Stapf published descriptions of *Encephalartos laevifolius*, *E. lanatus*, *E. paucidentatus*, *E. striatus*, and *E. transvenosus*

The first relates to the locality of their type *Legat H5185* for *Encephalartos paucidentatus*, said to have been collected on the farm Breslau, near Pont Drift, on the Limpopo River. Grobbelaar (2002a, b) pointed out that this must be in error as that species occurs about 450 km to the southeast in a restricted area near Barberton on the Mpumalanga-Eswatini border. A second matter concerns the taxon that Stapf and Burtt Davy called *Encephalartos striatus*, based apparently on plants offered for sale in Pretoria by a local farmer. Several authors have attempted to address the nomenclature and the origin for that taxon; the present consensus is that this cycad is simply a rather vigorous form of *E. villosus* that occurs in the Umbeluzi (now Mbuluzi) valley between Swaziland (now Eswatini) and Mozambique. *Encephalartos striatus* is presently listed as a synonym for *E. villosus* on the *World List of Cycads*. Together, Stapf and Burtt Davy contributed to a better understanding of *Encephalartos* and helped set the stage for later studies of the genus.

Another interesting issue was the determination of the holotype of *E. lanatus* (Fig. 6.14). The original label, from the Colonial Herbarium of the Transvaal Department of Agriculture, indicates that it was collected in April 1911 on a farm at Middelburg by Mr. O. C. Weeber and was identified subsequently as *E. friderici-guilielmi*. A second note refers to this as being similar to a British Museum 1885 specimen from Lydenburg, while another partially hidden annotation mentions *E. cycadifolius*. The final note is the determination as the holotype of *E. lanatus* by Stapf and Burtt Davy in the 1920s.

Fig. 6.14 The holotype from Transvaal for *Encephalartos lanatus* at Kew

Names of Cycads with Stapf and Burtt Davy's Authorship
Data from the *World List of Cycads*

Encephalartos laevifolius Stapf & Burtt Davy (1926).
Encephalartos lanatus Stapf & Burtt Davy (1926).
Encephalartos paucidentatus Stapf & Burtt Davy (1926).
Encephalartos striatus Stapf & Burtt Davy (1926) = *E. villosus* Lem.
Encephalartos transvenosus Stapf & Burtt Davy (1926).

Stapf and Burtt Davy's Cycad Publications

Burtt Davy, J. 1912. A first check-list of the flowering plants and ferns of the Transvaal and Swaziland. Annals of the Transvaal Museum 3(3): 119–182.

Stapf, O. & J. Burtt Davy. 1926. Manual of the Flowering Plants and Ferns of the Transvaal with Swaziland, South Africa. Vol. 1: 40, 98–99. [*Encephalartos laevifolius* Stapf & Burtt Davy, *E. lanatus* Stapf & Burtt Davy, *E. paucidentatus* Stapf & Burtt Davy, *E. striatus* Stapf & Burtt Davy, *E. transvenosus* Stapf & Burtt Davy].

John Medley Wood (1827–1915)

John Medley Wood (Fig. 6.15) was born in Mansfield, England, on December 1, 1827. He spent seven years at sea before joining his father, solicitor James Wood, who had emigrated to Durban in 1852 and who had become deputy sheriff of the Natal Colony. John purchased a property at the mouth of the Umdhloti (now Mdloti) River, just north of Durban, where he experimented with various crop plants (Brown 1915). In 1855, he married his stepmother's younger sister, Elizabeth Haygarth.

In 1868, after the death of his father, Medley Wood moved about 30 km inland to Inanda (now eNanda) where he continued farming and ran a trading store. Here, he developed an interest in the native flora, particularly ferns, mosses, and fungi, and he started corresponding with local and European botanists and recording his observations. In 1877, he co-authored the book "A popular description of the Natal ferns, designed for the use of amateurs."

In 1882, Medley Wood accepted the position of Curator of the Durban Botanic Gardens, subsequently becoming Director of the Gardens and the Natal Herbarium (now the kwaZulu-Natal Herbarium). The Botanic Gardens at that time was focused on plants of economic value to the Natal Colony, and Medley Wood was instrumental in developing the so-called Uba sugar cane, a virus-resistant cultivar, which was to become of paramount importance to the Colony's agricultural industry.

Apart from his work in the Gardens and the Herbarium, Medley Wood wrote and supervised the production of six volumes of "Natal Plants," a "Guide to the Trees and Shrubs of the Natal Garden," a "Handbook to the Flora of Natal," and a "Revised

Fig. 6.15 John Medley Wood. Image from the Annals of the Bolus Herbarium 2: 33. 1916

List of the Flora of Natal." Under his guidance, the Botanic Gardens evolved into a leading international garden with the acquisition of many indigenous shrubs (Strey 1977). He also found time to describe 57 Natal plants, mostly in collaboration with Maurice Evans.

But what of cycads? Medley Wood was certainly familiar with all the Natal species, and the Durban Botanical Gardens already had an incipient collection of South African and exotic cycads by the turn of the century. But it is for the discovery and salvage of the single specimen of the only known specimen of *Encephalartos woodii* that Medley Wood will be forever remembered. Medley Wood (1908) recalled the chance finding of this cycad:

> In the year 1895 I was on a botanical collecting trip with wagon and oxen in Zululand, and having reached a spot where the country was very rough, I stayed for several days botanizing in the vicinity, and in doing so found a solitary clump of *Encephalartos*, consisting of four stems, the tallest of which was 18 ft. high with proportionate girth of stem, and with a few offsets at the base; the stems were all male, and not another plant of the species could be found in the vicinity.

In 1903, Medley Wood sent his assistant James Wylie (who later succeeded him as Curator of the Gardens) to fetch some of the offsets. Three of these, about a meter in height, were planted in the Gardens where, after two years of dormancy, they commenced vigorous growth (Fig. 6.16).

Medley Wood commented that, in relation to the other cycads in the Gardens

> …these three, as far as the foliage is concerned, are, in my opinion, not only the handsomest of all, but are striking different from all the others.

However, it was evident that longitudinal sections of the two taller stems had been hacked off, apparently for use by Zulu medicine men, and that continued such usage would jeopardize the plant's existence. For that reason, Medley Wood instructed Wylie to salvage the remaining material from the Zululand site in March 1907. These two trunks remain as magnificent features on the old conservatory site of the Durban Botanic Gardens (Figs. 6.17 and 6.18).

Despite the fact that only one clump (Fig. 6.19) was ever found in the wild and that clump no longer exists, the plants in the Durban Botanic Gardens have grown exceptionally well, and numerous offsets have been separated, rooted, and distributed throughout the world. Subsequent generations from those plants have likewise been propagated, and the species is now well known in cultivation, despite being "extinct in the wild." There is a slight possibility that a female plant can be produced, either spontaneously or via biological engineering (Osborne and Gorelick 2007) but that remains to be achieved.

Not all is that clear with respect to *Encephalartos woodii* specimens in public and private collections. Firstly, a series of morphologically similar plants, also all male in gender, have their origin in the Kranzkoof area near Durban. These have become colloquially known as "pseudowoodii" plants, and a morphometric assessment has shown them to be intermediate between *Encephalartos woodii* and the closely related *E. natalensis* (Osborne and Baijnath 1995). Many of these plants have been traded on the international markets as genuine species. Furthermore, a

Fig. 6.16 One of the three smaller plants brought from the Zululand site to the Durban Botanic Garden by James Wylie in 1903. Medley Wood reported on this in his 1906–1907 annual report using the name "*Encephalartos altensteini* var. *bispinosa*"

ENCEPHALARTOS WOODII.

Fig. 6.17 A photograph of one of the larger *Encephalartos woodii* plants in the Durban Botanic Garden, with John Medley Wood standing proudly alongside, as published in an account of the species in the 1914 issue of Kew's *Bulletin of Miscellaneous Information*. (Photo by William Davenport Haygarth)

Fig. 6.18 A trio of geometrically striking large pollen cones at the apex of one of Medley Wood's original *Encephalartos woodii* specimens at the Durban Botanic Gardens. (Photo: Roy Osborne)

Fig. 6.19 The only known photographic record of the original male plant of *Encephalartos woodii* at the site of its discovery in Zululand's Ngoye Forest. In the photograph is James Wylie (right) with an employee of the Natal Government Railways who assisted in getting the trunks to the Durban Botanic Gardens. This image, taken in 1907, was provided by the kwaZulu-Natal Herbarium

great many plants have been produced by pollinating female cones of closely related *Encephalartos* species with *E. woodii* pollen, and that process has been repeated many times introducing greater amounts of "woodii" genes into each successive generation. These plants too have been sold as the true species. The only authentic *E. woodii* plants are those where provenance can be traced back to Medley Wood's original material in the Durban Botanic Gardens.

Nomenclatural note: Medley Wood referred to the taxon in 1907 as *Encephalartos altensteini var. bispinosa* (the varietal epithet later being mistranscribed by Schuster (1932) to *bispinna*, but neither epithet correctly matching the gender for the genus). A brief account of the plant was given in The Gardeners' Chronicle of April 25,

1908, based on plants exhibited at the Ghent Exhibition by Messrs. F. Sander and Sons, who used the provisional name *Encephalartos woodii* (Pearson 1908). That name was ratified soon afterward in a second description by Medley Wood on June 27 in the same journal. Finally, a photograph of the plant at the Ghent Exhibition was published in The Gardeners' Chronicle on May 2, 1908, and is considered to be the neotype for the name *Encephalartos woodii* Sander *ex* J.M.Wood.

Medley Wood was elected an Associate of the Linnean Society in 1887 and awarded a D.Sc. degree by the University of Cape Town in 1913. He died on September 26, 1915, at the age of 87 having lived through the rule of five British sovereigns. His name is commemorated in the genera *Woodia* (Apocynaceae), *Woodiella* (a fungus of the order Helotiales), a large number of species epithets, and in the small but attractive Medwood Gardens alongside the Durban City Hall. Professor John Bews, himself a distinguished botanist, named Medley Wood the "Father of Natal Botany."

Name of the Cycad with Medley Wood's Authorship
Data from the *World List of Cycads*

Encephalartos altensteini *var. *bispinosa*** J.M. Wood (1907). *orthographically corrected to *altensteinii*. **Schuster (1932) incorrectly transcribed this as "*bispinna*" from which a gender correction to *bispinnus* is given in the *World List of Cycads*.

Encephalartos woodii Sander *ex* J.M. Wood (1908).

Medley Wood's Cycad Publications

Medley Wood, J. 1906–1907. Annual report on Natal Botanic Gardens and Colonial Herbarium: 8–9, with illustration. [*Encephalartos altensteini* var. *bispinosa* J.M.Wood].

Medley Wood, J. 1908 (27 June). Colonial Note: *Encephalartos woodii*. The Gardeners' Chronicle, series 3, 43: 414. [*Encephalartos woodii* Sander *ex* J.M.Wood].

Chapter 7
Contemporary

Luigi Califano (1901–1976)

Luigi Califano (Fig. 7.1) was born on October 12, 1901, in Salerno, Italy. He lost his father, a magistrate, at an early age, and his paternal uncle provided for his schooling and encouraged his passion for plants and the natural world.

Califano's undergraduate and postgraduate studies in medicine were carried out at the Institute of General Pathology at the University of Naples (*Università degli Studi di Napoli Federico II*), under the direction of Francesco Pentimalli, Pietro Rondoni, Gino Galeotti, and Vittorio Scaffidi, the latter of whom launched him on a long-term academic career. His research work focused on three major areas of biology and experimental medicine: the metabolism of tissues in normal and pathological conditions, biological oxidations, and physiology and bacterial biochemistry. His dedication and enthusiasm led to his appointments as Professor of Microbiology, Professor of General Pathology, and Dean of the Faculty of Medicine and Surgery at the University of Naples. He was a Member of the National Research Council and a Member of the Council of Public Education.

Quite apart from the demands of his medical teaching and research, Califano had a consuming passion for plants and assembled an important personal collection of rare and unusual specimens in greenhouses that he constructed on the terrace of his home in the Naples City Center. Early in his career, Califano established collaborative relations with the Naples Botanical Garden (*Orto Botanico di Napoli*). In 1967, he worked with Director Aldo Merola to facilitate the construction of a 5000 m^2 new glasshouse, which was appropriately named the Califano Greenhouse (*Serre Califano*; Fig. 7.2) in recognition of his contributions (Merola 1976–1977). The complex comprised several controlled environment chambers each holding different plant collections and each having independent heating and humidifying systems (Sibilio 2014–2015).

R. Osborne, D. Wm. Stevenson, *The Cycad Botanists*, Memoirs of The New York Botanical Garden 123, https://doi.org/10.1007/978-3-031-79082-9_7

Fig. 7.1 Professor Luigi Califano with a specimen of *Tillandsia*, a bromeliad genus in which he had a great interest and a fine collection

The new greenhouse complex prompted the garden's staff to undertake numerous field trips, principally to Latin America, but also including Zaire, South Africa, Australia, and elsewhere. The first of these, to Florida, Mexico, and Costa Rica in 1969, was organized and funded by Luigi Califano. This was followed by a series of trips between 1971 (Figs. 7.3 and 7.4) and 1994 involving the Neapolitan staff members Paolo De Luca, Sergio Sabato, Aldo Moretti, Aldo Merola, Gesualdo Siniscalco Gigliano, and Luciano Gaudio, accompanied on occasions by botanists and students from other Italian and international institutions. Califano joined the botanists on the 1969, 1971, and 1974 trips. An interesting and illustrated account of this field work is given by De Luca et al. (1995).

Fig. 7.2 Part of the Califano Glasshouse complex at the Naples Botanical Garden. (Photo: Giancarlo Sibilio)

Fig. 7.3 The Naples Botanical Garden expedition to Mexico in 1971. On the back of the truck are Sergio Sabato (second from the left) and Paolo De Luca (third from the left), while Luigi Califano is the passenger seated in the front. (Photo from the archives of the Naples Botanical Garden)

Fig. 7.4 Luigi Califano (left) with Paolo De Luca (center) and the German botanist Werner Rauh during a trip to Abra Anticona in Peru in 1971. (Photo from the archives of the Naples Botanical Garden)

Sadly, Luigi Califano passed away on January 14, 1976. His family announced that his entire plant collection, including 102 cycad specimens in eight genera, was to be bequeathed to the Naples Botanical Garden. More recently, the present Director, Paolo Caputo, has further enriched the cycad collection through collaborative exchanges and purchases from specialized plant nurseries. At present, the Naples Garden has nearly 600 mature cycad specimens comprising 110 species in all ten genera, something that is only surpassed today by the diligence of the Montgomery Botanical Center. Among the many prize specimens in the Califano Greenhouse (Fig. 7.5) is one of the world's finest ex situ collections of *Dioon* species. Other parts of the glasshouse complex are homes to plants such as *Welwitschia*, *Ephedra*, *Gnetum*, *Sanseviera*, insectivorous species, succulents, bromeliads, and ferns.

In another section of this book, we tell of the research work of Professor Sergio Sabato, a staff member of the university, also closely associated with the Naples Botanical Garden, and who accompanied Califano on two of the field trips to Mexico.

Luigi Califano was married to Pia Carusio, and the couple had three children, Giuseppe, Salvatore, and Francesco. All three became full Professors at the Universities of Naples, Florence, and Rome, respectively. The contribution of Luigi Califano to both medicine and to the Naples Botanical Garden cannot be underestimated. Apart from his name being recognized in the Califano Glasshouse, the German botanist Werner Rauh named the bromeliad *Tillandsia califanii* for him in

Fig. 7.5 Large potted *Cycas* specimens in the Califano Glasshouse complex at the Naples Botanical Garden. (Photo: Giancarlo Sibilio)

1971. Then, when Paolo De Luca and Sergio Sabato (1979) found a novel cycad near Teotitlán del Camino (now Teotitlán de Flores Magón) on the Oaxaca/Puebla border in October 1974, they were pleased to be able to name it *Dioon califanoi* (Fig. 7.6). The type specimen, *De Luca and Sabato s.n.*, is filed in the herbarium of the Department of Plant Biology of the University of Naples.

Fig. 7.6 *Dioon califanoi* in the mountains above Teotitlán de Flores Magón on the Oaxaca/Puebla border, Mexico. Notice the *Tillandsia* on the upper branch of the tree, one of the Mexico oaks, probably *Quercus oleoides*. (Photo: Wes Field)

Paolo De Luca (1944–2021)

Paolo De Luca (Fig. 7.7) was born in Naples on July 27, 1944, and grew up in Posillipo, an affluent residential quarter at the western end of the Bay of Naples. He attended the local classical high school and then enrolled at the University of Naples where he graduated in biological sciences. Immediately after graduating, he took on various academic roles at the university, carrying out research work on the taxonomy and ecology of algae from thermal and acid environments, the taxonomy and distribution of tropical Bromeliaceae, and the taxonomy of Cucurbitaceae and thermophilic ferns of the Tertiary era. Thereafter, he dedicated himself to the study of the Cycadales, examining practically almost all aspects of their biology. In order to study representative species in habitat, he carried out numerous botanical expeditions over the period 1969–1978 to various parts of North and Central America and Asia.

After becoming a full-time Professor of Botany at the University of Naples, in 1981, De Luca was appointed Director of the Naples Botanical Garden (*Orto Botanico di Napoli*), an institution belonging to the University and also housing the Department of Botany in which all the university's botanists carried out teaching and research activities. At that time, the garden was in decline both physically and in terms of research productivity. De Luca was able to secure significant government funding to improve the grounds, restore the orangery, add new glasshouses, renew the old medicinal garden, and establish a prominent research ethos.

Fig. 7.7 Paolo De Luca with a specimen of *Dioon califanoi* in the Greenhouse Califano of the Naples Botanical Garden. (Photo by Giancarlo Sibilio)

Staff-wise, he recruited Sergio Sabato from Florence and supported Aldo Moretti, Gesualdo Siniscalco Gigliano, and other staff members including the young Paolo Caputo, current Director of the Botanical Garden, bringing expertise and expanding the research team as well as hiring horticulturalists to tend the living cycad collections. It was also thanks to Paolo De Luca that the Botanical Garden became the seat for most of the research conducted by the botanists of the University of Naples and, also, of other Italian and several foreign universities.

To promote the educational role, De Luca ensured that all garden signage was correct and informative and facilitated the development of a museum with exhibits for the public and school children to learn about ethnobotany, evolution, and plant diversity. James Mickle and Dennis Stevenson helped with the design and development of the museum exhibits and content. Not only was De Luca a diplomat and an institutional thinker, but he also paid attention to detail. For but one example, he had a mold constructed from one of the original decorative pots on top of the posts along the wall facing Via Foria and then used that to replace all the damaged ones.

In our reports on Luigi Califano and Sergio Sabato, we have mentioned the numerous cycad-oriented field trips in which the Neapolitan workers were involved. Paolo De Luca was a member of the very first of those expeditions, in 1969, and he himself organized six more Mexican field trips from 1979 to 1984. It was these trips that led to a much greater knowledge of the biology of the genera *Ceratozamia*, *Dioon*, *Microcycas*, and *Zamia*, both in terms of publications and acquisition of living material for the Naples Botanical Garden. De Luca was co-author of the formal descriptions of seven species of *Dioon* and of that of *Zamia vazquezii*. A visit to Zaire in 1987 by Aldo Moretti, Jean-Pierre Sclavo, and François Malaisse led to a greater understanding of the central African cycads and the publication of *Encephalartos delucanus* and *E. sclavoi*. A comprehensive report on the twenty-five years of these expeditions was printed in the journal "*Encephalartos*" (De Luca et al. 1995).

Together with Jean-Pierre Sclavo and Dennis Stevenson, De Luca played a key role in the organization of the First International Conference on Cycad Biology (Fig. 7.8), held at Beaulieu-sur-Mer, France, in 1987. It was the first opportunity that many members of the international cycad community had the opportunity to meet De Luca and his team of co-workers from Naples. Very appropriately, he gave the opening address on the "Historical perspective on cycads from antiquity to the present" (De Luca 1990). That was followed later by a presentation from his research colleagues in which the Conference host, Jean-Pierre Sclavo, was honored in the formal description of the species *Encephalartos sclavoi*.

Paolo De Luca passed away on December 4, 2021, after suffering from Creutzfeldt-Jakob disease for nearly a year. He is survived by his wife Maria Teresa and daughters, Adriana and Federica. Two cycads are named for him: *Encephalartos delucanus* (Fig. 7.9) from Tanzania as published by François Malaisse, Jean-Pierre Sclavo, and Catherine Crosiers in 1992 and *Ceratozamia delucana* (Fig. 7.10) from Veracruz (and later found also in Puebla), Mexico, as described by Mario Vázquez Torres, Aldo Moretti, and César Carvajal-Hernández in 2013.

Fig. 7.8 Delegates to Cycad 1987, the First International Conference on Cycad Biology, Beaulieu-sur-Mer, France, in April 1987, included (left to right): Jean-Pierre Sclavo, Mario Vázquez Torres, Paolo De Luca, Sergio Sabato, Gesualdo Siniscalco Gigliano, Aldo Moretti, and Dennis Stevenson. (Photo: Roy Osborne)

Fig. 7.9 *Encephalartos delucanus* in habitat, Katavi Province, Tanzania. (Photo: D. Moyer, from the World List of Cycads)

Fig. 7.10 *Ceratozamia delucana* in cloud forest habitat in Veracruz, Mexico, with Mario Vázquez Torres, principal author of the species. (Photo: Chip Jones)

Names of Cycads with De Luca's Authorship
Data from the *World List of Cycads*

Dioon califanoi De Luca and Sabato (1979).
Dioon caputoi De Luca, Sabato & Vázq. Torres (1980).
Dioon holmgrenii De Luca, Sabato & Vázq. Torres (1981).
Dioon merolae De Luca, Sabato & Vázq. Torres (1981).
Dioon rzedowskii De Luca, A. Moretti, Sabato & Vázq. Torres (1980).
Dioon tomasellii De Luca, Sabato & Vázq. Torres (1984).
Dioon tomasellii var. *sonorense* De Luca, Sabato & Vázq. Torres (1984) ≡.
D. sonorense (De Luca, Sabato & Vázq.Torres) Chemnick, T.J.Greg. & Salas-Mor.
Encephalartos sclavoi A. Moretti, D.W. Stev. & De Luca (1990).
Zamia vazquezii D.W. Stev., Sabato & De Luca (1998).

De Luca's Cycad Publications

De Luca, P. 1990. An historical perspective on cycads from antiquity to the present. Pp. 1–7 in D. W. Stevenson (ed.), The Biology, Structure and Systematics of the Cycadales. Proceedings of the First International Conference on Cycad Biology, Beaulieu-sur-Mer, France, 1987. Memoirs of the New York Botanical Garden 57.

De Luca, P., V. La Valva & S. Sabato. 1980. Spermatogenesis and tissue formation in cycad pollen grains. Caryologia 33(2): 261–265.

De Luca, P., P. Medeghini Bonatti, A. Moretti, S. Sabato & G. Siniscalco Gigliano. 1981. *Ultrastruttura dei canali mucipari delle Cycadales e dati chemiotassonomici sulla composizion e del le mucillagini.* Giornale Botanico Italiano 115: 397–398.

De Luca, P., A. Moretti & A. Balduzzi. 1995. Twenty-five years of field studies at the Botanical Garden of Naples, Italy. Encephalartos 42: 11–18.

De Luca, P., A. Moretti & S. Sabato. 1979. Regeneration in megagametophytes of cycads. Giornale Botanico Italiano 113: 129–143.

DeLuca, P., A. Moretti & S. Sabato. 1980. *Distribuzione et assonomia del genere Dioon (Zamiaceae).* Giornale Botanico Italiano 114(3–4): 129–130.

De Luca, P., A. Moretti, S. Sabato & G. Siniscalco Gigliano.1980. The ubiquity of cycasin in cycads. Phytochemistry 19(10): 2230–2231.

De Luca, P., A. Moretti, S. Sabato & M. Vázquez Torres. 1982. A comparative study of cycad mucilages. Phytochemistry 21(7): 1609–1611.

De Luca, P., A. Moretti, S. Sabato & M. Vázquez Torres. 1980. *Dioon rzedowskii* (Zamiaceae), a new species from Mexico. Brittonia 32(2): 225–229. [*Dioon rzedowskii* De Luca, A.Moretti, Sabato & Vázq.Torres].

De Luca, P., A. Moretti, G. Siniscalco Gigliano, P. Caputo, S. Cozzolino, L. Gaudio, D. W. Stevenson, E. T. Wurtzel & R. Osborne. 1995. Molecular systematic of cycads. Pp. 131–137 in P. Vorster (ed.), Proceedings of the Third International Conference on Cycad Biology, Pretoria, South Africa, 1993, Cycad Society of South Africa, Stellenbosch.

De Luca, P. & S. Sabato. 1978. Nomenclature and taxonomy of *Dioon mejiae* Standl. & Williams (Zamiaceae). Delpinoa 20: 45–50.

De Luca, P. & S. Sabato. 1979. In vitro spermatogenesis of *Encephalartos* Lehm. Caryologia 32(2): 241–245.

De Luca, P. & S. Sabato. 1979. *Dioon califanoi* (Zamiaceae), a new species from Mexico. Brittonia 31(1): 170–173. [*Dioon califanoi* De Luca & Sabato].

De Luca, P. & S. Sabato. 1980. *Osservazioni sur adicic or alloidia pogame.* Giornale Botanico Italiano 114: 112.

De Luca, P. & S. Sabato. 1980. Regeneration of coralloid roots on cycad megagametophytes. Plant Science Letters 18(1): 27–31.

De Luca, P., S. Sabato, A. Balduzzi & R. Nazzaro. 1980. Coralloid root regeneration on *Macrozamia* megagametophytes. Giornale Botanico Italiano 114(6): 271–275.

De Luca, P., S. Sabato & D. W. Stevenson. 1984. Comments opposed to the proposal to conserve the spelling *Dioon* against *Dion* (Zamiaceae). Taxon 33(4): 728–729.

De Luca, P., S. Sabato & M. Vázquez Torres. 1970. *Dioon purpusii* Rose (Zamiaceae), a mis-known species. Delpinoa n.s. 20: 31–35.

De Luca, P., S. Sabato & M. Vázquez Torres. 1980. *Dioon caputoi* (Zamiaceae), a new species from Mexico. Brittonia 32(1): 43–46. [*Dioon caputoi* De Luca, Sabato & Vázq. Torres].

De Luca, P., S. Sabato & M. Vázquez Torres. 1981. *Dioon merolae* (Zamiaceae), a new species from Mexico. Brittonia 33(2): 179–185. [*Dioon merolae* De Luca, Sabato & Vázq. Torres].

De Luca, P., S. Sabato & M. Vázquez Torres. 1981. *Dioon holmgreniii* (Zamiaceae), a new species from Mexico. Brittonia 33(4): 552–555. [*Dioon holmgrenii* De Luca, Sabato & Vázq.Torres].

De Luca, P., S. Sabato & M. Vázquez Torres. 1982. Distribution and variation of *Dioon edule* (Zamiaceae). Brittonia 34(3): 355–362.
De Luca, P., S. Sabato & M. Vázquez Torres. 1984. *Dioon tomasellii* (Zamiaceae), a new species with two varieties from western Mexico. Brittonia 36(3): 223–227. [*Dioon tomasellii* De Luca, Sabato & Vázq.Torres, *Dioon tomasellii* var. *sonorense* De Luca, Sabato & Vázq.Torres].
Sabato, S. & P. De Luca. 1985. Evolutionary trends in *Dion* (Zamiaceae). American Journal of Botany 72(9): 1353–1363.
Stevenson, D. W., A. Moretti & P. De Luca 1990. A new species of *Encephalartos* (Zamiaceae) from Tanzania. Pp. 156–161 in D. W. Stevenson (ed.), The Biology, Structure and Systematics of the Cycadales. Proceedings of the First International Conference on Cycad Biology, Beaulieu-sur-Mer, France, 1987. Memoirs of the New York Botanical Garden 57. [*Encephalartos sclavoi* A.Moretti, D.W.Stev. & De Luca].
Stevenson, D. W., S. Sabato, A. Moretti & P. De Luca. 1998. What is *Zamia fischeri* Miquel? Delpinoa 37–38: 9–17. [*Zamia vazquezii* D.W.Stev., Sabato & De Luca].

Robert Allen Dyer (1900–1987) and Inez Clare Verdoorn (1896–1989)

Robert Allen Dyer (Fig. 7.11) was born on September 21, 1900, in Pietermaritzburg, Natal (now KwaZulu-Natal), and was educated at Merchiston and Michaelhouse schools in that province. He studied at the University of Natal (now the University of KwaZulu-Natal), graduating with a M.Sc. He was employed as a research scientist with the Natal Sugar Association for two years before being appointed Curator to the Albany Museum, Grahamstown in 1925. He spent the years 1931–1934 as a South African botanical liaison officer at the Royal Botanic Gardens, Kew, before achieving a D.Sc. from the University of Natal in 1937. He then took the post of Head of the National Herbarium in Pretoria, later Chief of the Division of Botany of the Department of Agriculture and Forestry, retiring in 1963 but staying on as an associate until just before his death in 1987. He was instrumental in founding the National Botanical Garden in Pretoria (Vorster 1987; Codd 1989).

Apart from his cycad work (Dyer 1965a, b), he was a specialist in the Euphorbiaceae and Asclepiadaceae, with an interest in various other succulents as well as the Amaryllidaceae, co-authoring a mammoth two-volume monograph entitled "The Succulent Euphorbieae" with Alain White and Boyd Sloane in 1941. He was the author or co-author of some 450 publications, co-editor with Leslie Codd and Brian Rycroft of the "Flora of South Africa," and editor of both the "Flowering Plants of South Africa" (Dyer 1975–1976) and "Bothalia." After retirement, he devoted his time to producing "The Genera of South African Flowering Plants."

Robert Dyer's work was justifiably recognized in numerous positions and awards: he was a Fellow of the Royal Society of South Africa and the American Cactus and Succulent Society and President of Section C of the South African Association for the Advancement of Science, the South African Biological Society, and the Pretoria Horticultural Society. He was awarded the Senior Captain Scott

Fig. 7.11 Robert Allen Dyer (left) and Inez Clare Verdoorn (right), prominent South African botanists responsible for the description of 17 species of *Encephalartos*. Images kindly supplied by the South African National Biodiversity Institute

Medal of the South African Biological Society, the Gold Medal of the South African Association of Botanists, and the Herbert Medal of the American Amaryllis Society and was granted an honorary D.Sc. by the University of Witwatersrand in 1976. The genus *Radyera* (Malvaceae) and the species *Aridaria dyeri* and *Hereroa dyeri* (Aizoaceae) are named for him as is the cycad *Encephalartos dyerianus* (Fig. 7.12). This name was based upon specimens collected on a granite hill north of Mica in Limpopo Province, South Africa, by J. Smit in 1966. It was originally identified as *E. eugene-maraisii* and later used by John Lavranos and Douglas Goode (1988) as the holotype (Fig. 7.13) in recognition of Dyer's work. He passed away in Johannesburg on October 26, 1987, at the age of 87.

Inez Clare Verdoorn (Fig. 7.11) was born on June 15, 1896, and was the niece of the much-respected South African naturalist, poet, and writer Eugene Marais (1871–1936). She attended the Loreto Convent school in Pretoria and joined the National Herbarium in Pretoria as an assistant in 1917. She spent 1925–1926 as a South African liaison officer at Kew. She then returned to Pretoria to take charge of the National Herbarium, officially retiring in 1951 but staying on as an associate for many years thereafter.

Inez Verdoorn is the author or co-author of over 200 botanical publications. She has served as President of the South African Biological Society and President

Fig. 7.12 An impressive and aged multi-trunked specimen of *Encephalartos dyerianus* in its restricted habitat on a single granite hill in Limpopo Province, South Africa. (Photo: Wynand van Eeden)

Fig. 7.13 The two sheets that constitute the holotype *PRE 32864* of *Encephalartos dyerianus*. Images kindly supplied by the South African National Biodiversity Institute

(Section B) of the South African Association for the Advancement of Science and was awarded an honorary Ph.D. by the University of Natal in 1967. Her name is commemorated in the genus *Inezia* (Asteraceae) and the species *Aloe verdoorniae* (Asphodelaceae), *Senecio verdoorniae* (Asteraceae), and *Teclea verdoorniana* (Rutaceae). Volume 28 of The Flowering Plants of Africa was dedicated to her. She passed away at the age of 92 on April 2, 1989, surviving two years later than her colleague Robert Dyer.

Robert Dyer and Inez Verdoorn worked together at the National Botanical Research Institute for over nearly five decades. Their cycad publications remain the most important contributions to our knowledge of southern African cycads, with 15 major botanical contributions and the authorships of 17 species of *Encephalartos*. Both Dyer and Verdoorn spent a great deal of their time studying southern African cycads—and many other plants—in the field and are personally responsible for the collection and preparation of many of the type specimens in the Pretoria Herbarium.

Names of Cycads with Dyer and Verdoorn's Authorships
Data from the *World List of Cycads*

Encephalartos arenarius R.A. Dyer (1956).
Encephalartos chimanimaniensis R.A. Dyer & I. Verd. (1969).
Encephalartos concinnus R.A. Dyer & I. Verd. (1969).
Encephalartos cupidus R.A. Dyer (1971).
Encephalartos eugene-maraisii I. Verd. (1945).
Encephalartos eximius I. Verd. (1954) = *E. cycadifolius* (Jacq.)Lehm.
Encephalartos heenanii R.A. Dyer (1972).
Encephalartos humilis I. Verd. (1951).
Encephalartos inopinus R.A. Dyer (1964).
Encephalartos lebomboensis I. Verd. (1949).
Encephalartos munchii R.A. Dyer & I. Verd. (1969).
Encephalartos natalensis R.A. Dyer & I. Verd. (1951).
Encephalartos ngoyanus I. Verd. (1948).
Encephalartos princeps R.A. Dyer (1965).
Encephalartos pterogonus R.A. Dyer & I. Verd. (1969).
Encephalartos trispinosus (Hook.) R.A. Dyer (1965).
Encephalartos umbeluziensis R.A. Dyer (1951).

Dyer and Verdoorn's Cycad Publications

Dyer, R. A. 1951. *Encephalartos umbeluziensis*. Flowering Plants of Africa 28, Plate 1100. [*Encephalartos umbeluziensis* R.A.Dyer].
Dyer, R. A. 1956. A new cycad from the Cape Province. Journal of South African Botany 22(1): 1–4, Plates 1–3. [*Encephalartos arenarius* R.A.Dyer].
Dyer, R. A. 1964. Notes and new records of African plants, Cycadaceae. Bothalia 8(2): 169–170. [*Encephalartos inopinus* R.A.Dyer].

Dyer, R. A. 1965. The Cycads of Southern Africa. Bothalia 8(4): 405–515.
Dyer, R. A. 1965. New species and notes on type specimens of South African *Encephalartos*. Journal of South African Botany 31(2): 111–121. [*Encephalartos princeps* R.A.Dyer, *E. trispinosus* (Hook.) R.A.Dyer].
Dyer, R. A. 1966. Stangeriaceae. Pp. 1–3 in L. E. Codd, B. De Winter & H. B. Rycroft (eds), Flora of Southern Africa Vol.1.
Dyer, R. A. 1971. A further new species of cycad from the Transvaal. Bothalia 10(2): 379–383. [*Encephalartos cupidus* R.A.Dyer].
Dyer, R. A. 1972. A new species of *Encephalartos* from Swaziland. Bothalia 10(4): 539–546. [*Encephalartos heenanii* R.A.Dyer].
Dyer, R. A. & I. Verdoorn. 1951. Cycadaceae. *Encephalartos natalensis*. Bothalia 6(1): 205–211. [*Encephalartos natalensis* R.A.Dyer & I.Verd.].
Dyer, R. A. & I. C. Verdoorn.1966. Zamiaceae. Pp. 3–34 in L. E. Codd, B. De Winter & H. B. Rycroft (eds), Flora of Southern Africa Vol.1.
Dyer, R. A. & I. Verdoorn. 1969. *Encephalartos manikensis* and its near allies. Kirkia 7: 147–158. [*Encephalartos chimanimaniensis* R.A.Dyer & I.Verd., *E. munchii* R.A.Dyer & I.Verd., *E. pterogonus* R.A.Dyer & I.Verd.].
Verdoorn, I. C. 1945. A new species of *Encephalartos* from the Waterburg. Journal of South African Botany 11(1): 1–3. [*Encephalartos eugene-maraisii* I.Verd.].
Verdoorn, I. C. 1949. *Encephalartos ngoyanus*. Flowering Plants of Africa 27: Plates 1053–1054. [*Encephalartos ngoyanus* I.Verd.].
Verdoorn, I. C. 1949. *E. lebomboensis*. Flowering Plants of Africa 27: Plates 1078–1079. [*Encephalartos lebomboensis* I.Verd.].
Verdoorn, I. C. 1951. Cycadaceae. *Encephalartos humilis*. Bothalia 6(1): 220–221, Plate 3. [*Encephalartos humilis* I.Verd.].
Verdoorn, I. C. 1954. Cycadaceae. *Encephalartos eximius*. Bothalia 6(2): 426–428, Plates 2–4. [*Encephalartos eximius* I.Verd.].

Cynthia Giddy (1933–1998)

Cynthia Giddy nee Theron (Fig. 7.14) was born in Cradock in the Eastern Cape Province, South Africa, on September 20, 1933, to parents Christian, a journalist, and Johanna Theron, a teacher. She was the oldest of three daughters and an adopted sister. She attended Rocklands Junior School in Cradock and Jan van Riebeeck High in Cape Town and then completed an Honours degree in Social Anthropology and Psychology at Rhodes University, Grahamstown, graduating at the end of 1955. Rhodes University was also the *alma mater* for several other "cycad people" such as John Donaldson, Nat Grobbelaar, Roy Osborne, Carin Swart, and Gimme Walter.

Cynthia married Edward Noel (Ted) Giddy in 1956 after they had completed their studies and when they were both in their early twenties. They spent time in Port Elizabeth and went briefly overseas to London and then back to Barkly East and Maclear in the Eastern Cape, before moving to the Natal midlands. Cynthia taught Afrikaans, while Ted joined the Entomology Department of the then University of Natal in Pietermaritzburg.

Cynthia rose to prominence in her Umlaas Road Nursery (Fig. 7.15), at Camperdown, about 20 km from Pietermaritzburg in KwaZulu-Natal, which she and her husband ran until 1996. [The nursery was later taken over as "The Cycad

Fig. 7.14 A happy snapshot of Cynthia Giddy, provided by her daughters

Centre" by Danie Nel and Avis Meresman.] The nursery boasted a fine collection of cycad specimens used for offset and seed propagation. Cynthia was generous with her time and knowledge and spent many hours educating the numerous and frequent nursery visitors, both local and from overseas. Cynthia often demonstrated techniques of artificial pollination that she had learned from Bruce Bursey, a cycad grower in the Eastern Cape. We believe that Cynthia was the first person to produce a crop of *Encephalartos natalensis* × *E. woodii* hybrids.

Fig. 7.15 Ted and Cynthia Giddy at their Umlaas Road Nursery, KwaZulu-Natal, South Africa, in 1991. (Photo: Edwin Brown)

Parallel with her horticultural interests, Cynthia was an advocate for cycad conservation, becoming actively, and sometimes controversially, involved with the South African Nature Conservation authorities at national and provincial levels. She collaborated with the Natal Parks Board in implementing cycad conservation legislation and assisted in their staff training sessions. She would regularly appear as an expert witness in cycad trafficking court hearings, and she promoted her cycad conservation ethos in newspaper and magazine articles and on television. Her catchcry of "conservation through cultivation" (Giddy 1990) was accepted to some extent but created concern from those who favored habitat conservation rather than the treatment of single species. Her encouragement of community cycad nurseries (Giddy 1995) was admirable in theory, but expectations that third-world communities could develop horticultural and marketing skills proved somewhat over-ambitious—both in South Africa and internationally. She served on the CITES Threatened Plants Committee (Fig. 7.16), representing the Cycad Specialist Group at meetings in Malawi (1991), Thailand (1992), and Mexico (1994), and she was Chair of the Cycad Specialist Group for the 1994–1996 triennium.

Cynthia's passion for cycads led to her publishing the successful book "Cycads of South Africa" (1974, 1984; Fig. 7.17), a text well illustrated with photographs and enhanced with line drawings by Barbara Jeppe. One of the first cycad-specific books designed for the interested layman, this was a popularization of the scientific publications on the South African cycad flora by Dr. Robert Allan Dyer (1965a, b) and by Drs. Dyer and Inez Verdoorn (1966), researchers from the South African

Fig. 7.16 Cynthia Giddy inspecting *Zamia furfuracea* in Veracruz, Mexico, after attending a meeting of the CITES Threatened Plants Committee. (Photo: Uwe Schippmann)

Botanical Research Institute. Cynthia opens and closes the preface to her book with these emotive sentences:

> It has been my privilege to stand in the swirl of cloud and mist among the age-old stems of the Modjadji Palms [*Encephalartos transvenosus*] on the hill below the Kraal of the Rain Queen and feel something of the magic of rain in Africa. The cycads do not belong to us. They were here before us. They belong to the generations not yet born.

Cynthia was a prominent participant at the first three International Conferences on Cycad Biology, at each of which she spoke eloquently on matters of cycad conservation. At the first of these meetings, in Nice, France, in 1987, and sponsored by Jean-Pierre Sclavo, she surprised the attendees by delivering her opening address in fluent French. She shared, with Nat Grobbelaar and Roy Osborne, the planning and administrative responsibilities for the third conference in Pretoria in 1993, one of the most successful of all these events.

Besides her activities with the IUCN Cycad Specialist Group and the CITES Threatened Plants Committee, she was a very prolific spokesperson promoting the cause of cycad conservation in the popular press via numerous articles (Fig. 7.18) and editorials over many years.

Apart from her cycad interests, Cynthia grew and hybridized aloes; the *Aloe* "Rooikappie" was one of her successes, although the related plant sold as *Gasteria croucheri* "Cynthia Giddy" (≡ *Aloe croucheri*) was probably originally wild collected. Much later, *Aloe* "Cynthia Giddy" (Fig. 7.19) was named in recognition of Cynthia's horticultural endeavors when it was produced at the Rancho Soledad

Fig. 7.17 Front cover of the second edition (1984) of Cynthia Giddy's popular book "Cycads of South Africa"

THE GREAT TREE ROBBERY

Improbable as it may sound, plant-poachers are at the centre of South Africa's latest smuggling rings

By Patricia McCracken

Fig. 7.18 One of many newspaper and magazine articles in which Cynthia Giddy promoted the cause of cycad conservation. (From the South African magazine, "Personality," October 23, 1989 (McCracken 1989))

Fig. 7.19 *Aloe* "Cynthia Giddy," an attractive, spotted, clumping aloe named for Cynthia by the Rancho Soledad Nursery, North San Diego County in California. Image from the Aloes database of the National Gardening Association, https://garden.org

Nursery in California; it is thought that this is a hybrid of *A. lateritia* with *A. cameronii*. Cynthia was also one of the first to propagate the sought-after yellow *Clivia miniata* var. *citrina*, her products being marketed as "Natal Yellow" and "Cynthia's Best." A further horticultural interest was her collection of *Epidendrum* orchids.

Among her more eclectic interests, Cynthia had a passion for making dolls' houses which she decorated with elaborate furnishings and exquisitely dressed miniature dolls. Ted and Cynthia were also keen animal lovers and bulldog and pug breeders. A further hobby shared by Ted and Cynthia was their passion for genealogy; they were proud of a long ancestry showing a linear descendancy from 1820 settler stock. Added to this are their shared interests in philately, gourmet cooking, photography, community work, and a zest for travel.

Cynthia was an erudite, well-educated, and outspoken woman. She had a lack of tolerance for bureaucracy. Her fiery nature and shortness of temper led to her falling out with several of the more conservative cycad biologists, but she was nevertheless admired for her passion and tenacity. Roy Osborne recalls that several times during their telephone conversations, Cynthia would slam the phone down in anger when their respective points of view differed, but then they would get together for a friendly lunch at the farm the following weekend. A recollection in the tributes published in the journal *Encephalartos* reads that, after Cynthia had severely chastised a cycad collector, she was asked "You know you actually have a good heart, why do you do these things?" to which Cynthia replied, "Professional and famous people are not always popular, so therefore I am keeping up with my reputation."

Very tragically, and only two years after they had retired from the demands of the nursery business, Ted and Cynthia were fatally injured in a collision with a truck near Bathurst in the Eastern Cape in 1998, passing away respectively on May 15 and June 16. Their tombstone (Fig. 7.20) reflects much in the few but appropriate words: "Eccentrics – Passionate Conservationists – Teachers –They Dared to be Different." Ted and Cynthia were survived by their four daughters: Caitlin, Janet, Louise, and Laure (Osborne 1998).

Giddy's Cycad Publications

Giddy, C. 1974. Cycads of South Africa. Purnell & Sons, Cape Town. Second revised edition 1984, C. Struik Publishers, Cape Town.

Giddy, C. 1990. Conservation through cultivation. Pp. 89–93 in D. W. Stevenson (ed.), The Biology, Structure and Systematics of the Cycadales. Proceedings of the First International Conference on Cycad Biology, Beaulieu-sur-Mer, France, 1987. Memoirs of the New York Botanical Garden 57.

Giddy, C. 1990. Focus on ... *Encephalartos ngoyanus* I.Verd. Encephalartos 23: 3–9.

Giddy, C. 1991. The Cycad Specialist Group. Species 17: 72–73.

Giddy, C. 1992. Focus on ... *Encephalartos cerinus* Lavranos & D.L.Goode. Encephalartos 32: 4–7.

Giddy, C. 1993. Cycad conservation legislation — does it work in South Africa? Pp. 329–334 in D. W. Stevenson & K. J. Norstog (eds), The Biology, Structure and Systematics of the Cycadales. Proceedings of the Second International Conference on Cycad Biology, Townsville, Australia, 1990. Palm & Cycad Societies of Australia Ltd., Milton, Queensland.

Giddy, C. 1995. Cycad conservation — a third world perspective. Pp. 19–25 in P. Vorster (ed.), Proceedings of the Third International Conference on Cycad Biology, Pretoria, South Africa, 1993. Cycad Society of South Africa, Stellenbosch.

Fig. 7.20 The tombstone commemorating the lives of Ted and Cynthia Giddy, St. John's Anglican Churchyard, Bathurst, Eastern Cape, South Africa. (Photo provided by the daughters of Ted and Cynthia)

Nathanaël (Nat) Grobbelaar (1928–2015)

Nathanaël (Nat) Grobbelaar (Fig. 7.21) was born in Port Elizabeth, South Africa, on July 27, 1928. He attended Pearson High School and was awarded a scholarship as the best matriculant in his year. His academic studentship is nothing less than outstanding. He commenced studies at Rhodes University (the *alma mate*r also of John Donaldson, Cynthia Giddy, and Roy Osborne) where he obtained a B.Sc. cum *laude* in botany and chemistry, followed by postgraduate studies at Pretoria University with a CSIR bursary where he was awarded a M.Sc. in plant biochemistry, also cum *laude*, in 1952. He was then granted a Rockefeller Fellowship to study at Cornell University, New York, completing in 18 months a doctoral degree supervised by Professor F. C. Steward. That was followed by a 6-month postdoctoral research

Fig. 7.21 Nathanaël (Nat) Grobbelaar. Photo published as part of the centenary celebration of the University of Pretoria when they honored the "100 most brilliant minds" of their staff and students

fellowship at Cal Tech, California, under the mentorship of Professor Fritz Went, and finally a research visit to King's College in London, England.

On returning to South Africa in 1955, he accepted a post with the Department of Plant Physiology and Biochemistry (later the Department of Botany) at the University of Pretoria, a 35-year tenure during which he held the Bertha Stoneman Chair from 1959 to 1985.

His initial research interest was in the biosynthesis of amino acids, including pipecolic acid (piperidine-2-carboxylic acid, an important physiological regulator), which led to projects on nitrogen assimilation by cyanobacteria and other organisms and more general studies on metabolic regulation in plants. This work lead to numerous publications and international acclaim. Nat's interest in nitrogen fixation and plant toxin physiology led to the University of Pretoria's Schweickerdt Herbarium assembling a unique specialist collection of legumes, many of which were collected by Nat on numerous field excursions.

Subsequently, Nat developed a passion for understanding all aspects of the biology of cycads. Under his administration, the University of Pretoria Garden became home to a diverse and impressive collection of native southern African cycads, a vital research resource for future generations of students. We asked Nat's wife if she could tell us what catalyzed Nat's enthusiasm for cycads; here's what she told us (Hanneke Grobbelaar, pers. comm., May 2012).

> A colleague from Bristol University came to visit us and while walking through the garden, he remarked that he was surprised that Nat has not done any research on cycad nitrogen fixation, seeing that we have quite an abundance of these plants in South Africa. Nat immediately became interested, not so much in collecting, but more in the plants' physiology. I, on the other hand, started to collect a few plants in our home garden and then a cousin of Nat's brought him a medium sized *Encephalartos lehmannii* from his late father's farm in the Groot Winterhoek mountains northwest of Uitenhage in the Eastern Cape. Nat recognised this plant as the one his father was always trying to eradicate because the angora goats didn't eat it! Afterwards he started collecting seedlings … and it went from there.

During his academic career, Nat was the author or co-author of more than 185 articles in scientific and popular journals, many of them cycad-oriented. He coined a number of morphological terms where he saw a necessity for clarity, for example, cycad "corruptules," "corruptels," "omnules," and "omnels" (to distinguish between fertilized and unfertilized ovules and the internal seed kernel or "cleaned seed" of horticulture). Although these distinctions are scientifically important, the terms have not been widely accepted by cycad biologists. Nat also attended several International Conferences on Cycad Biology, presenting lectures on such themes as the vegetative propagation of cycads using excised leaves and coning behavior in *Encephalartos transvenosus*. He also shared with Cynthia Giddy and Roy Osborne the organization and planning of the third conference, held in Pretoria in 1993 and one of the most successful of all these events. But Nat is best known to cycad enthusiasts for his much-appreciated volume "Cycads — with special reference to the southern African species," first published in Afrikaans, its immediate success leading to its translation into English (Grobbelaar 2002a, b; Fig. 7.22).

Nat was a Founding Member of the South African Association of Botanists and the first Editor of the *South African Journal of Botany* and most deservedly awarded an honorary life membership and the Society's Gold Medal. He served as President of the Biological Society of South Africa and was awarded their Junior and Senior Captain Scott medals. He was a Member of the South African Academy of Arts and Sciences and held their Havenga prize for biology. He served on the Executive Council of the International Union for Biological Sciences (part of UNESCO), for nine years. He was a Member of the American Association for the Advancement of Science, the *Association pour l'étude taxonomique de la flora d'Afrique tropicale*, the American Society of Plant Physiologists, and the Botanical Society of America.

Fig. 7.22 Front cover of the 2002 English edition of Grobbelaar's "Cycads — with special reference to the southern African species"

He was President of the Cycad Society of South Africa from 1990 to 1993 and was granted honorary life membership thereafter. Nat is also fondly remembered by many of us who were privileged to visit his magnificent private collection of cycads and enjoy the generous hospitality of his family at their beautiful home *Tarentaal* (the Afrikaans word for guinea fowl) just north of Pretoria.

Nat passed away on January 4, 2015, and was survived by his wife Hanneke, three sons, one daughter, and eight grandchildren. We conclude by reprinting the final paragraph from the obituary by his friends and colleagues, Marion Meyer and Kobus Eloff (2015):

"His logical, rational and enthusiastic approach to research was an exceptional gift. His honest, open and straightforward opinion on any matter was often a talking point amongst his colleagues. He frequently played pranks on his colleagues and could recount these episodes with much humour. Nat's favourite quote—by Louis Pasteur—was probably:

In the field of observation, chance favours only the prepared mind."

Nathanaël Grobbelaar was an exceptional leading mind, role model, and inspiration for many natural scientists, young or old, and we cherish his memory.

Grobbelaar's Cycad Publications

Chang, D. C. N., N. Grobbelaar & J. Coetzee. 1988. SEM observations of cyanobacteria-infected cycad coralloid roots. South African Journal of Botany 54(5): 491–495.

Grobbelaar, N. 1989. Bees and cycads. Encephalartos 19: 43–44.

Grobbelaar, N. 1989. Disintegration of *Encephalartos* strobili. South African Journal of Botany 55(6): 581–585.

Grobbelaar, N. 1990. Spininess of cycad pinnae. Encephalartos 23: 28.

Grobbelaar, N. 1993. Senescence patterns in the female cones of *Encephalartos*. Pp. 159–161 in D. W. Stevenson & K. J. Norstog (eds), The Biology, Structure and Systematics of the Cycadales. Proceedings of the Second International Conference on Cycad Biology, Townsville, Australia, 1990. Palm & Cycad Societies of Australia Ltd., Milton, Queensland.

Grobbelaar, N. 1995. Cycads: their corrupules, omnules and omnels. Encephalartos 41: 36–37.

Grobbelaar, N. 1995. Vegetative propagation of *Encephalartos* species using excised leaves. Pp. 85–89 in P. Vorster (ed.), Proceedings of the Third International Conference on Cycad Biology, Pretoria, South Africa, 1992. Cycad Society of South Africa, Stellenbosch.

Grobbelaar, N. 1998. Unusual female cone development in *Encephalartos longifolius*. Encephalartos 53: 5–6.

Grobbelaar, N. 1998. *Encephalartos lanatus* sex ratio. Encephalartos 53: 8.

Grobbelaar, N. 1999. Growth of female cones and their omnules after the cones have become receptive to pollination. Pp. 276–279 in C. Chen (ed.), Biology and Conservation of Cycads. Proceedings of the Fourth International Conference on Cycad Biology, Panzhihua, China, 1998. International Academic Publishers, Beijing.

Grobbelaar, N. 1999. Coning frequency, gender ratio, and pollination of *Encephalartos transvenosus* (Zamiaceae) at the Modjadji Nature Reserve, South Africa, and the germination of this cycad's seed. Pp. 309–318 in C. Chen (ed.), Biology and Conservation of

Cycads. Proceedings of the Fourth International Conference on Cycad Biology, Panzhihua, China, 1998. International Academic Publishers, Beijing.

Grobbelaar, N. 2001. *Encephalartos middelburgensis*: the Avontuur variant. Encephalartos 67: 24.

Grobbelaar, N. 2002. *Broodbome, met speciale verwysing na die Suider-Afrikaanse soorte* (Afrikaans version). Cycads, with special reference to the southern African species (English version edited by R. Osborne). Published by the author, Pretoria.

Grobbelaar, N., W. E. Hatting & J. Marshall. 1986. The occurrence of coralloid roots on the South African species of the Cycadales and their ability to fix nitrogen symbiotically. South African Journal of Botany 52: 467–471.

Grobbelaar, N. & J. Marshall. 1993. Specificity of the cycad-cyanobiont symbiosis. Pp. 162–164 in D. W. Stevenson & K. J. Norstog (eds), The Biology, Structure and Systematics of the Cycadales. Proceedings of the Second International Conference on Cycad Biology, Townsville, Australia, 1990. Palm & Cycad Societies of Australia Ltd., Milton, Queensland.

Grobbelaar, N., J. J. M. Meyer & J. Burchmore. 1989. Coning and sex ratios of *Encephalartos transvenosus* at the Modjadji Nature Reserve. South African Journal of Botany 55: 79–82.

Grobbelaar, N., W. E. Scott, W. E. Hatting & J. Marshall. 1987. The identification of the coralloid root endophytes of the southern African cycads and the ability of the isolates to fix dinitrogen. South African Journal of Botany 53: 111–118.

Grobbelaar, N., J. M. Strauss & E. G. Groenewald. 1971. Non-leguminous seed plants in southern Africa which fix nitrogen symbiotically. Plant & Soil (special volume): 325–335.

Grobbelaar, N. & I. van der Westhuijzen. 1995. Stomatal densities on the leaves of some African cycads. Encephalartos 41: 17–19.

Marshall, J., T. C. Huang & N. Grobbelaar. 1987. Comparative morphological and physiological studies on cyanobionts of *Encephalartos transvenosus*. South African Journal of Botany 55: 574–580.

Marshall, J., N. Grobbelaar, J. Coetzee & R. Osborne. 1989. Pollen morphology of the Cycadales — with special reference to *Encephalartos* species. Pollen et Spores 31(3–4): 229–249.

Osborne, R. & N. Grobbelaar. 2004. What controls coning in *Cycas revoluta*? Pp. 142–149 in A. J. Lindstrom (ed.), The Biology, Structure and Systematics of the Cycadales. Proceedings of the Sixth International Conference on Cycad Biology, Chonburi, Thailand, 2002. Nong Nooch Tropical Botanical Garden, Chonburi, Thailand.

Osborne, R., N. Grobbelaar & P. L. D. Vincent. 1993. A numerical phenetic study of the genus *Encephalartos* Lehm. Pp. 279–280 in D. W. Stevenson & K. J. Norstog (eds), The Biology, Structure and Systematics of the Cycadales. Proceedings of the Second International Conference on Cycad Biology, Townsville, Australia, 1990. Palm & Cycad Societies of Australia Ltd., Milton, Queensland.

Osborne, R., N. Grobbelaar & P. Vorster. 1988. South African cycad research — progress and prospects. South African Journal of Science 84: 891–896.

Robbertse, H. P. J., N. Grobbelaar & E. du Toit. 2011. Origin of the coleorhiza in cycad seedlings and its structural homology with that of the Poaceae. Botanical Review 77: 1–10.

Vogel, J. C., H. van der Merwe & N. Grobbelaar. 1995. The use of radiocarbon for determining the growth rate of arborescent cycads. Pp. 115–119 in P. Vorster (ed.), Proceedings of the Third International Conference on Cycad Biology, Pretoria, South Africa, 1993. Cycad Society of South Africa, Stellenbosch.

Kenneth David Hill (1948–2010)

Kenneth (Ken) David Hill (Figs. 7.23 and 7.24) was born in Armidale, New South Wales, Australia, on August 6, 1948, and grew up on a local farm. His mother, Nola, described him as "a collector and classifier, a voracious reader and interested in plants from a young age." After his schooling at Armidale High, he attended the University of New England where he obtained an Honours degree in Science followed by a Master's degree in Geology—while at the same time taking additional courses in botany.

Fig. 7.23 Staff photograph of Ken Hill, March 2003, Royal Botanic Gardens Sydney

Fig. 7.24 Ken Hill working on photographic slides while Australian Botanical Liaison Officer at Kew in 1997–1998. (Photographer unknown)

After a brief spell as an exploration geologist in northern Australia and Sudan, in 1983, Ken took the position of Technical Officer at the National Herbarium of New South Wales, part of the Royal Botanic Gardens, Sydney, where his first project was to work with Lawrie (L. A. S.) Johnson and Don Blaxell on a revision of the classification of eucalypts. Their team was ultimately responsible for several major taxonomic changes and numerous relevant publications. One such example was the far-reaching and somewhat controversial decision by Hill and Johnson (1995) to erect the genus *Corymbia* to encompass 113 species of bloodwoods, ghost gums, and spotted gums as a genus distinct from *Eucalyptus*. Ken also started a lifelong interest and study of cycads at that time. In 1986, he took the post of botanist, continued his work on eucalypts for the Flora of Australia, and also accepted the challenge of writing the treatment of Gymnosperms for the Flora. He was present at the Second International Conference of Cycad Biology in Townsville, Australia, and on the subsequent field trips. That was the beginning of his career-long relationship with cycad societies, cycad conferences, and cycad fraternity in general.

From 1997 to 1998, Ken was seconded to the position of Botanical Liaison Officer at the Royal Botanic Gardens at Kew in England, sharing the series of these overseas appointments with other senior cycad botanists. John Maconochie, for example, held the post some two decades earlier.

Back in Australia, he took the role of Senior Research Scientist at the Royal Botanic Gardens, Sydney, a post he retained until 2004 when, so very sadly, his health started a progressive and irreversible deterioration, and he was obliged to take early retirement. Ken's career is wonderfully summarized by RBG staff colleague Leonie Stanberg (2010), who wrote the following paragraphs:

> During his relatively short botanical career, Ken demonstrated a broad range of skills, knowledge and interests which he drew upon to achieve many exceptional things. He published extensively on the systematics, ecology and biogeography of eucalypts (most notably the bloodwoods) and cycads. He was also involved in the identification, naming, description and subsequent research on the Wollemi Pine. He became an internationally well-respected botanist whose calibre, value and application of research work has spoken for itself, as evidenced by its acceptance and use by the botanical community at large. Ken played an international ambassadorial role for the Royal Botanic Gardens Sydney with his cycad work and also with his initiating and leading role in the uptake and development of bioinformatics at RBG Sydney. Combining his well-developed computing skills with an impressive ability to be original and innovative, he took the lead in delivering a number of electronic products that have greatly benefited the staff of RBG Sydney as well as the botanical and general community. Most notable among these are 'The Cycad Pages,' 'NSWdata' (the first specimen database for NSW), 'NSW FloraOnline,' EucaLink' and 'WattleWeb' (later developed into 'PlantNET').

Apart from his outstanding contributions to our knowledge of cycads, Ken Hill was perhaps the world's leading authority on eucalypts. As the last in a continuous line of resident eucalypt specialists on the staff at NSW, Ken continued the long tradition of expertise in eucalypts that was maintained for over a century. He travelled widely in Australia studying eucalypts (Fig. 7.25) and cycads (Fig. 7.26) and in later years in the botanically very poorly known Asia/Pacific region in pursuit of *Cycas.* Ken's extensive fieldwork combined with his skill and expertise in the field, enabled the discovery of many new species, solved numerous problems of identification and nomenclature, and significantly expanded our knowledge of these two groups. He was a long-standing member of the IUCN/SSC Cycad Specialist Group and played a key role in many of the IUCN Red List assessments of Asian and Australian cycads."

Ken was extremely well known and well liked in the global cycad community and attended nearly all the International Conferences on Cycad Biology, producing an astonishing number of important papers on cycad taxonomy and being the principal author of the texts "Cycads of Australia" (Hill and Osborne 2001; Fig. 7.27) and "Cycads of Vietnam" (Osborne and Gorelick 2007). His prolific work on cycads includes scientific descriptions of 60 new *Cycas* taxa and four new species of *Macrozamia*, an achievement unsurpassed in the history of cycad science (Stanberg 2012). The Proceedings of the Eighth International Conference on Cycad Biology,

Fig. 7.25 Here he inspects *Eucalyptus macrocarpa* in Mount Annan Botanic Gardens, New South Wales, in 1996. (Photo: Leonie Stanberg)

Fig. 7.26 Ken Hill inspecting a *Cycas xipholepis–C.tuckeri* intergrade near Coen in Australia's Cape York Peninsula, July 1994. (Photo: Leonie Stanberg)

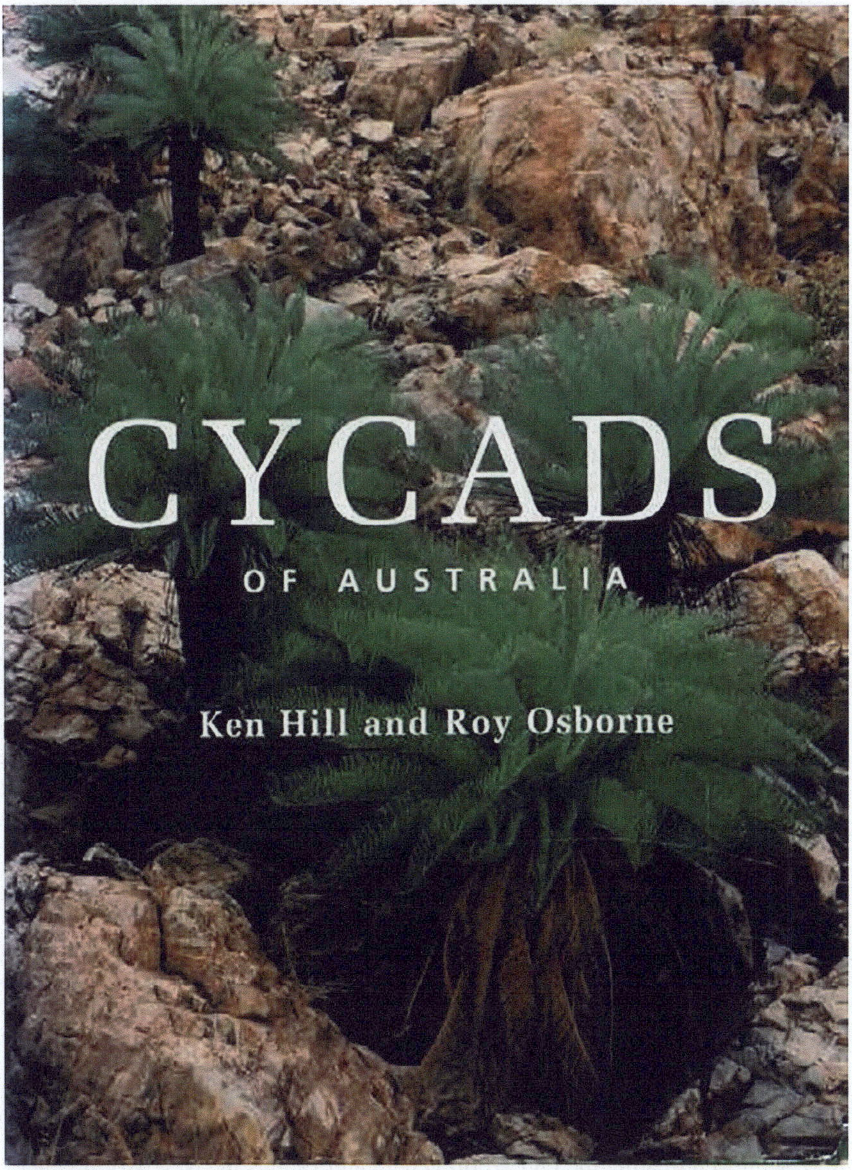

Fig. 7.27 Front cover of Hill and Osborne's "Cycads of Australia" (2001), Kangaroo Press, Sydney, featuring *Cycas pruinosa* in habitat near Lake Argyle in Western Australia

held in Panama City in 2008, were dedicated to Ken's memory (Stevenson et al. 2012).

Besides eucalypts and cycads, Ken was involved in the Wollemi Pine discovery (Jones et al. 1995) and was one of the three authors of the new species *Wollemia nobilis* (Fig. 7.28), an extremely rare member of the Araucariaceae restricted to a

Fig. 7.28 Ken Hill at the Royal Botanic Gardens, Sydney in 1994 comparing fresh material of the then newly discovered Wollemi pine with a fossil species in the Araucariaceae. (Photo: Peter Rae, Sydney Morning Herald)

remote series of narrow, steep-sided, sandstone gorges in Wollemi National Park in New South Wales, Australia.

With respect to Ken's physical attributes, Roy Osborne recalls:

> It was after the International Cycad Conference in South Africa in July 1993 that Ken was amongst the 40-odd attendees who took the opportunity of a post-conference trip that terminated in Cape Town. In that city, the delegates were treated on their last day to the popular cable car ride to the top of Table Mountain. Ken dismissed that as altogether too easy and exclaimed that he would run up the mountain instead. He did so, and, astonishingly, reached the summit well before the rest of the party made their more sedate arrival in the cable car.

Ken's interests were by no means confined to matters botanical. He was knowledgeable on a wide variety of other subjects: ornithology, genealogy, philately, motorcycles, technology in general, world travel, gardening, music, and movies. Ken's final objective in his cycad endeavors was to produce a modern monograph of the cycads to replace that of Schuster's (1932) *Pflanzenreich* treatment. As part of this process, he established "The Cycad Pages," an online facility which has since been removed from the Internet. Together with Dennis Stevenson, they divided the effort, with Ken treating the Old World genera and Dennis the New World genera. Similarly, they divided topics, Ken doing phytogeography, taxonomy, history, etc. and Dennis anatomy morphology, chemistry, ethnobotany, etc. Very unfortunately, Ken did not live to see this work completed when his illness overtook him and he passed away in Sydney on August 4, 2010. He was survived by his wife, Lesley Greenwood, his daughters Zoe and Phoebe, his mother Nola, and his brother Don. The memory of our special friend will stay with us for many more years.

Names of Cycads with Hill's Authorship
Data from the *World List of Cycads.*

Cycas aculeata K.D. Hill & T.H. Nguyên (2004).
Cycas aenigma K.D. Hill & A. Lindstr. (2008).
Cycas apoa K.D. Hill (1994).
Cycas arenicola K.D. Hill (1993).
Cycas arnhemica K.D. Hill (1994).
Cycas arnhemica subsp. *muninga* Chirgwin & K.D. Hill (1996) = *C. arnhemica* K.D. Hill.
Cycas arnhemica subsp. *natja* K.D. Hill (1996) = *C. arnhemica* K.D. Hill.
Cycas badensis K.D. Hill (1996).
Cycas bifida (Dyer) K.D. Hill et al. (2004).
Cycas bougainvilleana K.D. Hill (1994).
Cycas brachycantha K.D. Hill, T.H. Nguyên & P.K. Lôc (2004).
Cycas brunnea K.D. Hill (1992).
Cycas campestris K.D. Hill (1994).
Cycas canalis K.D. Hill (1994).
Cycas canalis subsp. *carinata* K.D. Hill (1994) = *C. canalis* K.D. Hill.
Cycas candida K.D. Hill et al. (2004).
Cycas chamaoensis K.D. Hill (1999).
Cycas clivicola K.D. Hill (1999).
Cycas clivicola subsp. *lutea* K.D. Hill (1999) = *C. clivicola* K.D. Hill.
Cycas collina K.D. Hill, T.H. Nguyên & P.K. Lôc (2004).
Cycas condaoensis K.D. Hill & S.L. Yang (2004).
Cycas curranii (J. Schust.) K.D. Hill (1995).
Cycas dolichophylla K.D. Hill, T.H. Nguyên & P.K. Lôc (2004).
Cycas elephantipes A. Lindstr. & K.D. Hill (2003).
Cycas falcata K.D. Hill (1999).
Cycas fugax K.D. Hill, T.H. Nguyên & P.K. Lôc (2004).
Cycas indica A. Lindstr. & K.D. Hill (2007).
Cycas lacrimans A. Lindstr. & K.D. Hill (2008).
Cycas lindstromii S.L. Yang, K.D. Hill & T.H. Nguyên (1997).
Cycas litoralis K.D. Hill (1999) = *C. edentata* de Laub.
Cycas maconochiei Chirgwin & K.D. Hill (1996).
Cycas maconochiei subsp. *lanata* K.D. Hill (1996).
Cycas maconochiei subsp. *viridis* K.D. Hill (1996).
Cycas media subsp. *banksii* K.D. Hill (1996).
Cycas media subsp. *ensata* K.D. Hill (1996).
Cycas megacarpa K.D. Hill (1992).
Cycas micronesica K.D. Hill (1994).
Cycas montana A.Lindstr. & K.D. Hill (2009).
Cycas nitida K.D. Hill & A. Lindstr.(2008).

Cycas nongnoochiae K.D. Hill (1999).
Cycas ophiolitica K.D. Hill (1992).
Cycas orientis K.D. Hill (1992).
Cycas pachypoda K.D. Hill et al. (2004).
Cycas petraea A. Lindstr. & K.D. Hill (2003).
Cycas platyphylla K.D. Hill (1992).
Cycas pranburiensis S.L. Yang, W. Tang, K.D. Hill & Vatch. (1999).
Cycas saxatilis K.D. Hill & A. Lindstr. (2008).
Cycas semota K.D. Hill (1996).
Cycas silvestris K.D. Hill (1992).
Cycas simplicipinna (Smitinand) K.D. Hill (1995).
Cycas sundaica Miq. *ex* A. Lindstr. & K.D. Hill (2009).
Cycas taitungensis C.F. Shen, K.D. Hill, C.H. Tsou & C.J. Chen (1994).
Cycas tansachana K.D. Hill & S.L. Yang (1999).
Cycas tropophylla K.D. Hill & P.K. Lôc (2004).
Cycas tuckeri K.D. Hill (1996).
Cycas vespertilio A. Lindstr. & K.D. Hill (2008).
Cycas xipholepis K.D. Hill (1996).
Cycas yorkiana K.D. Hill (1996).
Cycas zeylanica (J. Schust.) A. Lindstr. & K.D. Hill (2002).
Macrozamia elegans K.D. Hill & D.L. Jones (1998).
Macrozamia johnsonii D.L. Jones & K.D. Hill (1992).
Macrozamia montana K.D. Hill (1998).
Macrozamia reducta K.D. Hill & D.L. Jones (1998).

Hill's Cycad Publications

Chen, C. J., K. D. Hill & D. W. Stevenson. 2004. Comments on *Cycas* L., *Dyerocycas* Nakai and *Epicycas* de Laub. (Cycadaceae). Pp.57–68 in T. Walters & R. Osborne (eds), Cycad Classification — Concepts and Recommendations. CABI Publishing, Wallingford, U.K.

Hill, K. D. 1992. A preliminary account of *Cycas* (Cycadaceae) in Queensland. Telopea 5(1): 177–206. [*Cycas brunnea* K.D.Hill, *C. couttsiana* K.D.Hill, *C. megacarpa* K.D.Hill, *C. ophiolitica* K.D.Hill, *C. platyphylla* K.D.Hill, *C. silvestris* K.D.Hill].

Hill, K. D. 1993. A history of the taxonomy of Australian cycads. Pp. 250–259 in D. W. Stevenson & K. J. Norstog (eds), The Biology, Structure and Systematics of the Cycadales. Proceedings of the Second International Conference on Cycad Biology, Townsville, Australia, 1990. Palm & Cycad Societies of Australia Ltd., Milton, Queensland.

Hill, K. D. 1993. *Cycas arenicola* (Cycadaceae), a new species from the Northern Territory of Australia. Telopea 5(2): 419–422. [*Cycas arenicola* K.D.Hill].

Hill, K. D. 1993. The golden cycad of Fiji, and the correct name for the native Fijian cycad. Encephalartos 36: 11–13.

Hill, K. D. 1994. The *Cycas media* group (Cycadaceae) in New Guinea. Australian Systematic Botany 7(6): 527–541. [*Cycas campestris* K.D.Hill].

Hill, K. D. 1994. The *Cycas rumphii* complex (Cycadaceae) in New Guinea and the Western Pacific. Australian Systematic Botany 7(6): 543–567. [*Cycas apoa* K.D.Hill, *C. bougainvilleana* K.D.Hill, *C. micronesica* K.D.Hill].
Hill, K. D. 1994. Three new species of *Cycas* from the Northern Territory of Australia. Telopea 5(4): 693–701. [*Cycas arnhemica* K.D.Hill, *C. canalis* K.D.Hill, *C. canalis* subsp. *carinata* K.D.Hill, *C. orientis* K.D.Hill].
Hill, K. D. 1995. The genus *Cycas* (Cycadaceae) in the Indian region, with notes on the application and typification of the name *Cycas circinalis*. Taxon 44(1): 23–31.
Hill, K. D. 1995. The rediscovery of *Cycas macrocarpa*. Palms & Cycads 46: 18–23.
Hill, K. D. 1995. Infrageneric relationships, phylogeny and biogeography of the genus *Cycas* (Cycadaceae). Pp.139–162 in P. Vorster (ed.), Proceedings of the Third International Conference on Cycad Biology, Pretoria, South Africa, 1993. Cycad Society of South Africa, Stellenbosch. [*Cycas curranii* (J.Schust.) K.D.Hill, *C. simplicipinna* (Smitinand) K.D.Hill].
Hill, K. D. 1995. Taxonomic changes in the Australian cycads. Pp.193–223 in P. Vorster (ed.), Proceedings of the Third International Conference on Cycad Biology, Pretoria, South Africa, 1993. Cycad Society of South Africa, Stellenbosch.
Hill, K. D. 1996. A taxonomic revision of the genus *Cycas* (Cycadaceae) in Australia. Telopea 7(1): 1–64. [*C. arnhemica* subsp. *muninga* Chirgwin & K.D.Hill, *C. arnhemica* subsp. *natja* K.D.Hill, *C. badensis* K.D.Hill, *C. maconochiei* Chirgwin & K.D.Hill, *C. maconochiei* subsp. *lanata* K.D.Hill, *C. maconochiei* subsp. *viridis* K.D.Hill, *C. media* subsp. *banskii* K.D.Hill, *C. media* subsp. *ensata* K.D.Hill, *C. semota* K.D.Hill, *C. tuckeri* K.D.Hill, *C. xipholepis* K.D.Hill, *C. yorkiana* K.D.Hill].
Hill, K. D. 1996. Cycads in the Pacific. Pp. 267–274 in A. Keast & S. E. Miller (eds), The Origin and Evolution of Pacific Island Biotas, New Guinea to Eastern Polynesia: Patterns and Processes. SPB Academic Publishing, Amsterdam.
Hill, K. D. 1997. Cycads and conservation. Pp. 223–226 in K. D. Touchell & K. W. Dixon (eds), Conservation into the twenty-first Century. Proceedings of the Fourth International Botanic Gardens Conservation Congress. Kings Park and Botanic Garden, West Perth.
Hill, K. D. 1998. Flora of Australia 48. Gymnosperms — the paraphyletic stem of seed plants: 505–526, Pinophyta: 545–596, Cycadophyta: 597–661. CSIRO Australia. [*Macrozamia elegans* K.D.Hill & D.L.Jones, *M. montana* K.D.Hill, *M. reducta* K.D.Hill & D.J.Jones].
Hill, K. D. 1999. *Cycas falcata* (Cycadaceae), a new species from Sulawesi, Indonesia. Kew Bulletin 54(1): 209–213. [*Cycas falcata* K.D.Hill].
Hill, K. D. 1999. New discoveries in the Australian cycads. Pp. 75–97 in C. J. Chen (ed.), Biology and Conservation of Cycads. Proceedings of the Fourth International Conference on Cycad Biology, Panzhihua, China, 1996. International Academic Publishers, Beijing.
Hill, K. D. 1999. *Cycas* — an evolutionary perspective. Pp. 98–115 in C. J. Chen (ed.), Biology and Conservation of Cycads. Proceedings of the Fourth International Conference on Cycad Biology, Panzhihua, China, 1996. International Academic Publishers, Beijing.
Hill, K. D. 1999. *Les Cycadales*. Manureva 7: 34–37.
Hill, K. D. 2000. What is *Cycas circinalis*? The Cycad Newsletter 22(4): 7–9.
Hill, K. D. 2001. A note on the typification of *Cycas miquelii* Warb. Encephalartos 65: 39–40.
Hill, K. D. 2001. Cycads — the effects of taxonomy and biogeography on conservation. Pp. 131–138 in L. G. Shaw, L. S. L. Chua & K. C. Khoo (eds), Taxonomy — the cornerstone of biodiversity. Proceedings of the Fourth International Flora Malesiana Symposium. Forestry Research Institute Malaysia, Kuala Lumpur.

Hill, K. D. 2003. Regional overview: Australia. Chapter 4. Pp. 20–24 in J. Donaldson (ed.), Status Survey and Conservation Action Plan, Cycads. IUCN/SCC Cycad Specialist Group. Gland, Switzerland.
Hill, K. D. 2004. *Cycas candida* (Cycadaceae), a new species from Queensland together with an extension of the range and amended description of *Cycas yorkiana*. Telopea 10(2): 607–611. [*Cycas candida* K.D.Hill].
Hill, K. D. 2004. Character evolution, species recognition and classification concepts in Cycadaceae. Pp. 23–44 in T. Walters & R. Osborne (eds), Cycad Classification — Concepts and Recommendations. CABI Publishing, Wallingford, U.K.
Hill, K. D. 2004. An analysis of morphological characters used in classification of the Cycadaceae. Pp. 6–29 in A.J. Lindstrom (ed.), The Biology, Structure and Systematics of the Cycadales. Proceedings of the Sixth International Conference on Cycad Biology, Chonburi, Thailand, 2002. Nong Nooch Tropical Botanical Garden, Chonburi, Thailand.
Hill, K. D. 2005. Diversity and evolution of gymnosperms. Chapter 3. Pp. 25–44 in R. J. Henry (ed.), Plant Diversity and Evolution — Genotypic and Phenotypic Variation in Higher Plants. CABI Publishing, Wallingford, U.K.
Hill, K. D. 2008. The genus *Cycas* (Cycadaceae) in China. Telopea 12(1): 71–118.
Hill, K. D., M. W. Chase, D. W. Stevenson, H. G. Hills & B. Schutzman. 2003. The families and genera of cycads: a molecular phylogenetic analysis of cycadophyte based on nuclear and plastid DNA sequences. International Journal of Plant Science 164(6): 933–948.
Hill, K. D. & C. J. Chen. 1994. On *Cycas taiwaniana* Carruthers (Cycadaceae) and the cycads of south-eastern China. Acta Phytotaxonomica Sinica 32(6): 538–548.
Hill, K. D., C. J. Chen & P. K. Lôc. 2003. Regional overview: Asia. Chapter 5. Pp. 25–30 in J. Donaldson (ed.), Status Survey and Conservation Action Plan, Cycads. IUCN/SSC Cycad Specialist Group. Gland, Switzerland.
Hill, K. D., H. T. Nguyên & P. K. Lôc. 2004. The genus *Cycas* (Cycadaceae) in Vietnam. Botanical Review 70(2): 134–193. [*Cycas aculeata* K.D.Hill & T.H.Nguyên, *C. bifida* (Dyer) K.D.Hill, *C. brachycantha* K.D.Hill, T.H.Nguyên & P.K. Lôc, *C. collina* K.D.Hill, T.H.Nguyên & P.K. Lôc, *C. condaoensis* K.D.Hill & S.L.Yang, *C. dolichophylla* K.D.Hill, T.H.Nguyên & P.K. Lôc, *C. fugax* K.D.Hill, T.H.Nguyên & P.K. Lôc, *C. pachypoda* K.D.Hill, *C. tropophylla* K.D.Hill & P.K. Lôc].
Hill, K. D. & R. Osborne. 2001. Cycads of Australia. Kangaroo Press, Sydney, Australia.
Hill, K. D. & L. C. Stanberg. 1999. Epicuticular waxes and their taxonomic significance in the cycads. Pp. 159–174 in C. J. Chen (ed.), Biology and Conservation of Cycads. Proceedings of the Fourth International Conference on Cycad Biology, Panzhihua, China, 1996. International Academic Publishers, Beijing.
Hill, K. D. & L. C. Stanberg. 2009. Notes on *Cycas truncata* de Laub. and related matters. Telopea 12(3): 447–450.
Hill, K. D., D. W. Stevenson & R. Osborne. 2004. The World List of Cycads. Appendix 1. Pp. 219–235 in T. Walters & R. Osborne (eds), Cycad Classification — Concepts and Recommendations. CABI Publishing, Wallingford, U.K.
Hill, K. D., D. W. Stevenson & R. Osborne. 2004. The World List of Cycads. Botanical Review 70(2): 274–298. Also Pp. 195–212 in A.J. Lindstrom (ed.), The Biology, Structure and Systematics of the Cycadales. Proceedings of the Sixth International Conference on Cycad Biology, Chonburi, Thailand 2002. Nong Nooch Tropical Botanic Garden, Chonburi, Thailand.
Hill, K. D., D. W. Stevenson & R. Osborne. 2007. The World List of Cycads. Pp. 154–212 in A.P. Vovides, D.W. Stevenson & R. Osborne (eds), Proceedings of the Seventh International Conference on Cycad Biology, Xalapa, Mexico, 2005.

Hill, K. D. & P. Vatcharakorn. 1998. The cycads of Thailand. Pp.178–190 in D. R. Hodel (ed.), The Palms and Cycads of Thailand. Nong Nooch Tropical Botanical Garden, Chonburi, Thailand.

Hill, K. D. & S. L. Yang. 1999. The genus *Cycas* (Cycadaceae) in Thailand. Brittonia 51(1): 48–73. [*Cycas chamaoensis* K.D.Hill, *C. clivicola* K.D.Hill, *C. clivicola* subsp. *lutea* K.D.Hill, *C.litoralis* K.D.Hill, *C. nongnoochiae* K.D.Hill, *C. tansachana* K.D.Hill & S.L.Yang].

Hill, K. D. & S. L. Yang. 1999. The cycads of Thailand. Pp. 33–40 in C. J. Chen (ed.), Biology and Conservation of Cycads. Proceedings of the Fourth International Conference on Cycad Biology, Panzhihua, China, 1996. International Academic Publishers, Beijing.

Jones, D. L. & K. D. Hill. 1992. *Macrozamia johnsonii*, a new species of *Macrozamia* section *Macrozamia* (Zamiaceae) from northern New South Wales. Telopea 5(1): 31–34. [*Macrozamia johnsonii* D.J.Jones & K.D.Hill].

Kokubugata, K., K. D. Hill & K. Kondo. 2000. Chromosome numbers and karyotype of two *Cycas* species (Cycadaceae,Cycadales). Annals of the Tsukuba Botanical Garden 19: 19–23.

Kokubugata, K., K. D. Hill, G. W. Wilson, K. Kondo & L. M. Randall. 2001. A comparison of chromosome number and karyotype in somatic chromosomes of Stangeriaceae (Cycadales). Edinburgh Journal of Botany 58(3): 475–481.

Kokubugata, K., K. D. Hill & K. Kondo. 2002. Chromosome characters of two *Cycas* species (Cycadaceae, Cycadales). Annals of the Tsukuba Botanical Garden 21: 139–141.

Kokubugata, K., K. D. Hill & K. Kondo. 2002. Comparison of 5S ribosomal DNA sites on somatic chromosomes of eight species in *Cycas* (Cycadaceae). Chromosome Science 6(2): 41–46.

Kokubugata, K., K. D. Hill & K. Kondo. 2002. Ribosomal rDNA distribution in somatic chromosomes of *Stangeria eriopus* (Stangeriaceae, Cycadales). Brittonia 54(1): 1–5.

Lindstrom, A. J. & K. D. Hill. 2002. Notes of the species of *Cycas* (Cycadaceae) from Sri Lanka and the islands of the Andaman Sea. Novon 12(2): 237–240. [*Cycas zeylanica* (J.Schust.) A.Lindstr. & K.D.Hill].

Lindstrom, A. J. & K. D. Hill. 2003. New species and new records of *Cycas* (Cycadaceae) from Thailand. Brittonia 54: 298–304. [*Cycas elephantipes* A.Lindstr. & K.D.Hill, *C. petraea* A.Lindstr. & K.D.Hill].

Lindstrom, A. J. & K. D. Hill. 2007. The genus *Cycas* (Cycadaceae) in India. Telopea 11(4): 463–488. [*Cycas indica* A.Lindstr. & K.D.Hill].

Lindstrom, A. J., K. D. Hill & L. C. Stanberg. 2008. The genus *Cycas* (Cycadaceae) in the Philippines. Telopea 12(1): 119–145. [*Cycas aenigma* K.D.Hill & A.Lindstr.,*C. lacrimans* A.Lindstr. & K.D.Hill, *C. nitida* K.D.Hill & A.Lindstr., *C. saxatilis* K.D.Hill & A.Lindstr., *C. vespertilio* A.Lindstr. & K.D.Hill].

Lindstrom, A. J., K. D. Hill & L. C. Stanberg. 2009. The genus *Cycas* (Cycadaceae) in Indonesia. Telopea 12(3): 385–418. [*Cycas montana* A.Lindstr. & K.D.Hill, *C. sundaica* Miq. *ex* A. Lindstr. & K.D.Hill].

Osborne, R., M. A. Calonje, K. D. Hill, L. Stanberg & D. W. Stevenson. 2012. The World List of Cycads. Pp. 480–508 in D. W. Stevenson, R. Osborne & A. S. Taylor Blake (eds), Proceedings of the Eighth International Conference on Cycad Biology, Panama City, Panama, 2008. Memoirs of the New York Botanical Garden 106.

Osborne, R., K. D. Hill, H. T. Nguyen & L. Phan Ke. 2007. Cycads of Vietnam. Published by R. Osborne & W. van Eeden, Brisbane and Cape Town.

Osborne, R., D. W. Stevenson & K. D. Hill. 1995. A World List of Cycads. Pp. 55–64 in P. Vorster (ed.), Proceedings of the Third International Conference on Cycad Biology, Pretoria, South Africa, 1993. Cycad Society of South Africa, Stellenbosch.

Osborne, R., D. W. Stevenson & K. D. Hill. 1999. The world list of cycads. Pp. 224–239 in C. J. Chen (ed.), Biology and Conservation of Cycads. Proceedings of the Fourth International Conference on Cycad Biology, Panzhihua, China, 1996. International Academic Publishers, Beijing.

Shen, C. F., K. D. Hill, C. H. Tsou & C. J. Chen. 1994. *Cycas taitungensis* C.F.Shen, K.D.Hill, C.H.Tsou & C.J.Chen, *sp. nov.*, a new name for the widely known cycad species endemic in Taiwan. Botanical Bulletin Academia Sinica 35: 133–140. [*Cycas taitungensis* C.F.Shen, K.D.Hill, C.H.Tsou & C.J.Chen].

Steyn, E. M. A., G. F. Smith & K. D. Hill. 1999. Stangeriaceae. Species Plantarum: Flora of the World 2: 1–9. Australian Biological Resources Study, Canberra.

Yang, S. L., K. D. Hill & H. T. Nguyên. 1997. *Cycas lindstromii* (Cycadaceae), a new species from southern Vietnam. Novon 7: 213–215. [*Cycas lindstromii* S.L.Yang, K.D.Hill & T.H.Nguyên].

Yang, S. L., W. Tang, K. D. Hill & P. Vatcharacorn. 1999. *Cycas pranburiensis* (Cycadaceae), a new species from southern Thailand Brittonia 51(1): 44–47. [*Cycas pranburiensis* S.L.Yang, W.Tang, K.D.Hill & Vatch.].

Lawrence Alexander Sidney Johnson (1925–1997)

Lawrence (Lawrie) Alexander Sidney Johnson (Fig. 7.29) was born in Sydney, Australia, on June 26, 1925, the third child of Sid and Emily Johnson. He attended Beecroft public school and Parramatta High where he was considered a bright child who showed an interest in exploring the local bushland. Although his initial academic focus was on chemistry, it was during his studies at the University of Sydney that he became fascinated with botany, perhaps due to encouragement from Professor Eric Ashby (later Lord Ashby) and plant physiologist Bob (later Sir Rutherford) Robertson. An undergraduate analysis of the taxonomy of *Casuarina* developed into a major research project which brought him into contact with the staff of the New South Wales Herbarium. He graduated from the University of Sydney with First Class Honours in 1948.

After completion of his studies, Johnson joined the staff of the National Herbarium in Sydney and then the scientific arm of the Royal Botanic Gardens and part of the New South Wales Department of Agriculture. At the time, taxonomy was less than popular, and Sydney's herbarium was something of a scientific backwater. Johnson rose to the challenge and soon commenced improving the quality and diversity of the herbarium collection. At the same time, he developed a network with other botanists leading to joint research projects in taxonomy and other fields of botany.

His work extended from conservation and ecology, through theoretical systematics to the hands-on taxonomy of the families as diverse as the Myrtaceae (particularly the eucalypts *Eucalyptus*, *Corymbia*, and *Angophora)*, Casuarinaceae, Cyperaceae, Oleaceae, Proteaceae, Restionaceae, and Zamiaceae. His interest in eucalypts and cycads was to become an inspiration to Ken Hill who joined the herbarium staff in 1983.

Fig. 7.29 L. A. S. (Lawrie) Johnson. (From plantnet.rbgsyd.nsw.gov.au)

The number of herbarium vouchers he collected in Australia is an astonishing 18,449 (Data from Bionomia). Barbara Briggs (2000) commented that "alone or in collaboration with colleagues, he [Johnson] distinguished and described four new families of vascular plants, 33 new genera, 286 new species and he reclassified another 395 species." In her bibliography, she listed 145 scientific works in which Johnson was either lead author or co-author.

In 1962, Johnson was appointed as Australian Botanical Liaison Officer, a twelve-month posting to the Royal Botanic Gardens, Kew, a position also served later by John Maconochie in 1976 and Ken Hill in 1997. Ten years later, he became Director of the Sydney Royal Botanic Gardens. His new administrative responsibilities involved setting up the Royal Botanic Gardens and Domain Trust, initiating an active "Friends of the Royal Botanic Gardens" organization; undertaking various educational programs; sourcing funding for the new Robert Brown herbarium, administrative building, and new glasshouses; and establishing specialist satellite botanical gardens at Mount Annan and Mount Tomah.

Johnson was known for his irascible personality; he was highly critical of poor-quality management and what he perceived as political interference in the good work of the herbarium and the gardens. He wrote that "the Royal Botanical Gardens has been abominably treated by politicians and botanically totally ignorant managers .... a cruel perversion of sensible and decent management," harsh comments at one time jeopardizing his own employment at the Gardens. He was similarly forthright in his opinions on scientific methods; he held particularly strong views on phenetic and cladistic taxonomy (as detailed in David Hull's 1996 article in *Telopea*) as well as the morphology and terminology applied to inflorescences (Briggs and Johnson 1979), which fostered a symposium and debate at the 14th International Botanical Congress in 1987 in Berlin, which one of us (DWS) attended as an Assistant Professor at the time and upon whom it made a lasting impression as did the original paper.

Johnson's major contribution to cycad taxonomy was his 1959 treatment in the Proceedings of the Linnean Society of New South Wales. That work included descriptions of four new species: *Macrozamia communis*, *M. diplomera*, *M. lucida*, and *M. stenomera*, as well as two subspecies that were later upgraded to specific status, *M. pauli-guilielmi* subsp. *flexuosa* and *M. pauli-guilielmi* subsp. *plurinervia*. He also named the family Stangeriaceae in which he included only the monotypic South African genus *Stangeria* but which was later extended to include the two Australia *Bowenia* species. The Stangeriaceae subsequently became obsolete when the family was merged into the Zamiaceae on the basis of evidence from molecular studies.

Johnson's work was acknowledged in several significant ways. His research work earned him a D.Sc. degree from the University of Sydney in 1971. He was awarded the Clarke Medal of the Royal Society of New South Wales in 1979 and the Mueller Medal of the Australasian Association for the Advancement of Science in 1984. In 1986, he was elected a Fellow of the Australian Academy of Science, while in the following year, he was appointed a Member of the Order of Australia for services to botany. Johnson held honorary or corresponding memberships of the Linnean Society of London, the Botanical Society of America, and the American Society of Plant Taxonomists. His broad approach to botany is summed up by Barabara Briggs (1996) and Doug Benson (1996) in an issue of Telopea.

David Jones and Ken Hill commemorated Johnson's name in 1992 when they named the New South Wales cycad *Macrozamia johnsonii* (Fig. 7.30). His eponymy also includes taxa as diverse as in *Baumea johnsonii* (Cyperaceae), *Davidsonia johnsonii* (Smooth Davidson's Plum, Cunoniaceae), *Eucalyptus johnsoniana* (Johnson's Mallee, Myrtaceae), *Grevillea johnsonii* (Proteaceae), *Notelaea johnsonii* (Oleaceae), *Sclerolaena johnsonii* (Chenopodiaceae), *Typhonium johnsonianum*

Fig. 7.30 Impressive specimens of *Macrozamia johnsonii*, the cycad named in 1992 for Lawrie Johnson. (Photographed in Dalmorton State Forest, southwest of Grafton, New South Wales, by Yvonne Dalziel)

(Araceae), and the well-known "grass tree" *Xanthorrhoea johnsonii* (Asphodelaceae).

In 1996, Johnson had a surgical procedure to remove a melanoma. Sadly, this was unsuccessful, and by early the next year, he was diagnosed with multiple brain tumors. Stoically accepting the inevitable, he prepared a statement that was later read at his funeral by his friend Geoffrey Swain; it reviewed much of his life and offered thanks and encouragement to those who attended the final service. He passed away at Sydney's Royal North Shore Hospital on August 1, 1997, and was survived by his wife Merle and their five children, Christopher, Sylvia, Nicholas, Quentin, and Alexander (Sandy).

Dr. Barbara Briggs, a close colleague of Johnson's, commenced her 2000 biography of Johnson with the sentence:

> Johnson was a taxonomic botanist notable for the outstanding breadth of his interests and expertise, the rigour of his scientific approach, and the intensity with which he defended scientific conclusions and opinions.

Peter Raven, Director of the Missouri Botanical Garden, summed up Johnson's international reputation in a letter to Dr. Briggs:

> Our knowledge of Australian plants has been greatly improved as a result of the industrious, intelligent and forceful career of Lawrie Johnson. He has given us new insight into several of the most important groups of plants in Australia – ones that are leading components in the vegetation, and most interesting biogeographically. He was never afraid to take on difficult problems in systematics, and he made important contributions to our understanding of every group that he studied. Few have or could have accomplished so much.

Cycad Names with Johnson's Authorship
Data from the *World List of Cycads*

Macrozamia communis L.A.S. Johnson (1959).
Macrozamia diplomera (F.Muell.) L.A.S. Johnson (1959).
Macrozamia lucida L.A.S. Johnson (1959).
Macrozamia pauli-guilielmi subsp. *flexuosa* (C.Moore) L.A.S. Johnson (1959) ≡ *M. flexuosa* C.Moore.
Macrozamia pauli-guilielmi subsp. *plurinervia* L.A.S. Johnson (1959) ≡ *M. plurinervia* (L.A.S.Johnson) D.L.Jones.
Macrozamia stenomera L.A.S. Johnson (1959).

Johnson's Cycad Publications
Johnson, L. A. S. 1959. The families of cycads and the Zamiaceae of Australia. Proceedings of the Linnaean Society of New South Wales (Series 2) 84: 64–117. [*Macrozamia communis* L.A.S.Johnson, *M. diplomera* (F.Muell.) L.A.S.Johnson, *M. lucida* L.A.S.Johnson, *M. pauli-guilielmi* subsp. *flexuosa* (C.Moore) L.A.S.Johnson, *M. pauli-guilielmi* subsp. *plurinervia* L.A.S.Johnson, *M. stenomera* L.A.S.Johnson].
Johnson, L. A. S. 1961. Zamiaceae. Flora of New South Wales. Contributions from the New South Wales National Herbarium 1: 21–41.
Johnson, L. A. S. 1963. Cytological and taxonomic notes on Zamiaceae. Contributions from the New South Wales National Herbarium 3(4): 235–240.
Wilson, K. L. & L. A. S. Johnson. 1990. Cycadaceae. Pp. 362–377 in K. Kubitzki (ed.), The Families and Genera of Vascular Plants. Vol. 1. Pteridophytes and Gymnosperms. Springer-Verlag, Berlin.

Jacques Désiré Léandri (1903–1982)

Jacques Désiré Léandri (Fig. 7.31) was born in Sartène, Corsica, on July 12, 1903. Little is known of his early education, but he was apparently regarded as a diligent student, graduating with a science degree at age 17.

He joined the staff of the Paris Museum of Natural History (*Muséum national d'histoire naturelle)* in 1921 and continued postgraduate studies under the mentorship of Professors Gaston Bonnier, Paul Henri Lecomte, and Julien Costantin. In 1930, he completed a doctoral thesis on anatomical changes in the Thymeliaceae in response to their environment (Avmomin 1974).

Léandri then spent many years traveling, studying, and collecting plants, chiefly in Madagascar, but also in Morocco. He undertook a detailed study of the flora of Madagascar and the nearby Comoros, focusing especially on the Euphorbiaceae, a family in which he circumscribed 18 taxa. He also published extensively on the Urticaeae, Thymeliaceae, Chenopodiaceae, and many individual genera and species, being the author and co-author of some 150 scientific papers on Madagascan flora alone.

Fig. 7.31 Jacques Léandri at age 68. (A photo by P. Guex from the archives of the Natural History Museum in Paris)

Apart from his interest in Madagascan flora, he was also tasked with writing the monographs of the aquatic family Ceratophyllaceae, the Gnetaceae, and the Cycadaceae for the *Flore Génerále del'Indo-Chine*. In this work, we find Leandri's descriptions of two new cycads, summarized in the following paragraphs.

The cycad, *Cycas chevalieri* (Fig. 7.32), was first collected by François Fleury on behalf of the French botanist Auguste Chevalier in the Nghia Dan area of French

Fig. 7.32 *Cycas chevalieri* with an emerging pollen cone, Quang Tri Province, Vietnam. This cycad was first collected by François Fleury on behalf of Auguste Chevalier in 1841 but described by Leandri only in 1931. (Photo: Hiêp Nguyen)

Indochina (now Vietnam) in 1841. There was some doubt about this cycad's existence until it was formally described by Léandri nearly a century later in 1931. Léandri named it in recognition of Chevalier who had spent time in Vietnam and later became Professor of Botany at the Natural History Museum in Paris. The voucher *F. Fleury sub. A. Chevalier 32,612* in the Paris herbarium stands as the lectotype (Hill et al. 2004). The species has never been rediscovered at the site of its original finding, but several other populations have since been located in the provinces of Ha Tinh, Nghe An, Quang Binh, and Quang Tri in central Vietnam (Osborne et al. 2007).

In the early 1900s, the French botanist Eugène Poilane collected specimens of cycad from the southern coastal hills of Indochina (now southeast Vietnam). Léandri, seeing the similarity between this material and *Cycas pectinata*, named the taxon *C. pectinata* var. *elongata* in 1931 with the specimen *Poilane 9328* in the Paris herbarium being the lectotype. The taxon was elevated to the specific status as "*elonga*" by Ding Yue Wang in 1996 with the name corrected in *errata* to *Cycas elongata* (Fig. 7.33).

Léandri passed away in Paris on May 10, 1982, at the age of 78. His name is acknowledged in the Madagascan genera *Dendroleandria* (Malvaceae) and *Leandriella* (Acanthaceae) and the species *Rhynchosia leandrii* (Fabaceae) and *Blotia leandriana* (Euphorbiaceae).

Names of Cycads with Léandri's Authorship
Data from the *World List of Cycads*

Cycas chevalieri Leandri (1931).
Cycas pectinata var. *elongata* Leandri (1931) ≡ *C. elongata* (Leandri) D. Yue Wang.

Léandri's Cycad Publication

Leandri, J. 1931. Cycadaceae. Pp. 1085–1092 in P. H. Lecomte & F. Gagnepain (eds), *Flore générále del'Indo-Chine* 5(10). [*Cycas chevalieri* Leandri, *C. pectinata* var. *elongata* Leandri].

John Richard Maconochie (1941–1984)

John Richard Maconochie (Fig. 7.34) was born in Adelaide, South Australia, on June 13, 1941. John never knew his father who was apparently captured by the Japanese in the fall of Singapore in 1942 and never returned home. A tradition in the family is that the eldest son has always been named John Richard, on the basis of his unusual surname (clearly of Scottish origin). We speculate that the family may

Fig. 7.33 Dennis Stevenson in Phu Yen Province, Vietnam, inspecting a plant of *C. elongata*. (Photo: Hiêp Nguyen)

be distantly related to Alexander Maconochie (1787–1860), a penal reformer who worked in Tasmania, or to the family that created the infamous "Maconochie stew" of World War I.

He attended the University of Adelaide and, after obtaining an Honours degree in Botany, proceeded with postgraduate studies to submit a M.Sc. thesis in 1968

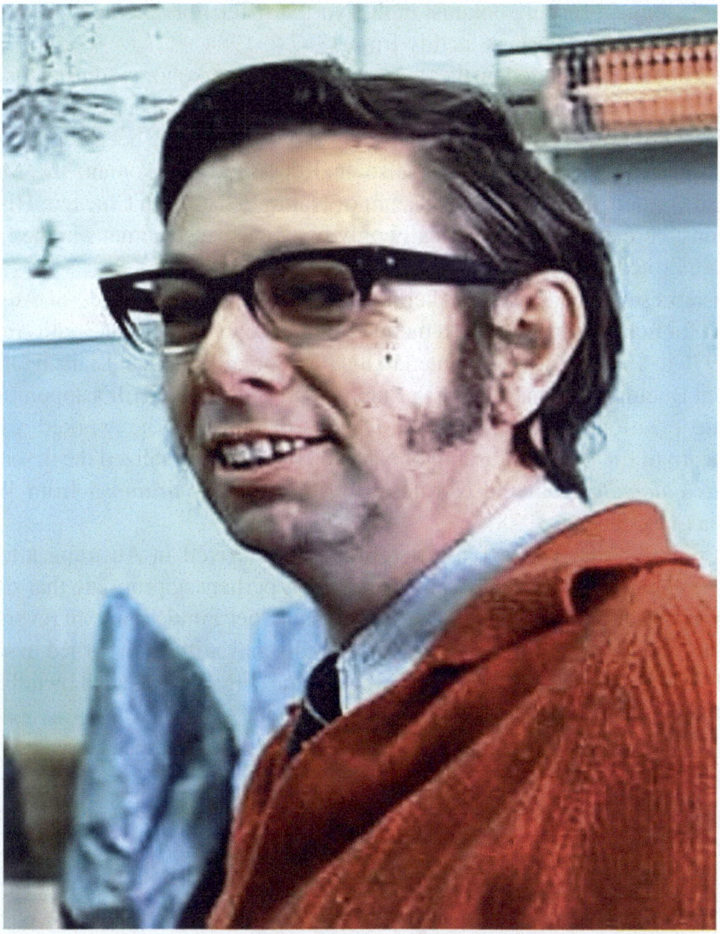

Fig. 7.34 John Maconochie in 1978. (From the files of the Australian National Botanical Gardens)

entitled "Studies on the leaf production of selected plant species in the temperate, southern arid and central arid regions of South Australia." This was the start of a lifelong research interest in plant ecology and physiology in arid landscapes.

In his mid-twenties, John was appointed as a botanist in charge of the Alice Springs Herbarium, taking over the position from George Chippendale in April 1967. Assisted by Des Nelson (2012), John immediately started an ambitious project to identify pasture grasses and monitor for possible invasive and detrimental species. That extended to mounting vouchers of plants poisonous to stock and eventually to a more comprehensive collection of local arid-zone plants. John was the author or co-author of numerous important publications in the field of arid zone ecology and physiology, floristic surveys, plants toxic to cattle and sheep, and ethnobotanical data relating to the Aboriginal people of the Northern Territory. John may be the longest-serving botanist in the Northern Territory, being employed continuously at the Alice Springs facility from 1967 to 1984.

John's interests extended further to encompass plant taxonomy; during his career, he collected or identified an astonishing total of 9527 specimens (data from Binomia). Of these, 460 were specimens of *Cycas*, which have been deposited in all the state herbaria of Australia and also in Kew, the Smithsonian, the Missouri Botanical Garden, and the Field Museum of Natural History in Chicago. His cycad collections were exceeded numerically only by extensive specimens in the families Fabaceae, Proteaceae, Poaceae, and Myrtaceae.

John represented the territory herbaria on the Council of Heads of Australian Herbaria for many years and was a member of the Royal Society of South Australia. From 1976 to 1977, he was appointed to the position of Botanical Liaison Officer at the Royal Botanic Gardens at Kew in England, preceding Ken Hill's appointment to the same position for two decades. During that period, he worked with Dr. W. T. Stearn of the British Museum of Natural History and finalized the descriptions of *Cycas calcicola* from the Northern Territory and *C. pruinosa* from Western Australia (Maconochie 1978).

John's wife, Aija, was a Latvian refugee who arrived in Australia after WW II. Her surname, Ozols, is Latvian for oak, so it is perhaps appropriate that she married a botanist. John nicknamed her "Slozo," using her family name in reverse. She was an accomplished drafts woman, and her skills may still be discerned in some of the buildings in Alice Springs including their home (Fig. 7.35) shaded by tall lemon-scented gums (*Corymbia citriodora*) planted by Aija.

Fig. 7.35 The Maconochie home in Alice Springs. (Photo taken in 2010 by Alex Nelson)

John also collaborated with the United Nations Food and Agriculture Organization, which took him on several excursions into parts of Africa and the Middle East. For example, in the early 1980s, he made some 800 herbarium collections, which effectively became the start of the National Herbarium of Oman (ON) in Muscat. But, very tragically, John was killed in a road accident while working in Somalia in January 1984, at the age of 42. He was survived by Aija and two sons, John (Max) and Paul.

John was the author or co-author of some 20 taxa including two species *Cycas calcicola* (Fig. 7.36) and *C. pruinosa*, and he also constructed a key to the species in the genus in Australia (1980). His name is acknowledged in Alice Springs in terms of Maconochie Road in the prestigious area of Mount Johns. His botanical eponymy extends to *Acacia maconochieana* (Mullan wattle, Fabaceae), *Euphorbia maconochieana* (Euphorbiaceae), *Hakea maconochieana* (Proteaceae), *Triumfetta maconochieana* (Malvaceae), and, of course, *Cycas maconochiei* (Fig. 7.37).

Fig. 7.36 Two specimens from the Perth Herbarium that comprise the isotype of *Cycas calcicola*, collected in April 1971 by Maconochie about 15 km north of Katherine in the Northern Territory

Fig. 7.37 Prolific stands of *Cycas maconochiei* in open savannah forest on Cox Peninsula, to the southwest of Darwin. (Photo: Bret and Yvonne Dalziel)

Names of Cycads with Maconochie's Authorship
Data from the *World List of Cycads.*

Cycas calcicola Maconochie (1978).
Cycas pruinosa Maconochie (1978).

Maconochie's Cycad Publications

Maconochie, J. R. 1978. Two new species of *Cycas* from Northern Australia. Journal of the Adelaide Botanical Gardens 1(3): 175–178. [*Cycas calcicola* Maconochie, *C. pruinosa* Maconochie].

Maconochie, J. R. 1980. Key to *Cycas*. In H. T. Clifford & J. Constantine (eds), Ferns, Fern Allies and Conifers of Australia. University of Queensland Press, St. Lucia, Queensland.

Ronald Melville (1903–1985)

Ronald Melville (Fig. 7.38) was born in Bristol, England, on March 12, 1903, the second of seven children of Edgar and Florence Melville. After an early childhood and schooling in Cardiff, he joined his uncle's pharmacy in that city, training and graduating as a pharmacist at the age of 22. In 1926, he was appointed pharmacist

Fig. 7.38 Dressed in tweeds and cardigan, Roland Melville was described as "a benevolent Dickensian figure" as he worked in his tiny office in Kew's herbarium (Arbose 1974)

at the Whitechapel Hospital in London while concurrently studying botany at Birbeck College (later the University of London). After obtaining his Ph.D. degree in plant physiology from a project on the growth of tomato plants, conducted in 1934 at the Cheshunt Experimental and Research Station, he joined the museum staff at the Royal Botanic Gardens at Kew.

By 1936, he had become an acknowledged expert in British elm trees. Apart from giving lectures to students at Kew, he worked on diverse teaching and research projects at the National Pinetum at Bedgebury, the Chelsea Polytechnic, and Birbeck College. During World War II, he assisted the British Ministry of Health by developing and promoting the use of rosehip syrup as a source of vitamin C—thus reducing the incidence of scurvy among children lacking fresh fruit in their diet. One of the first of his many botanical publications was his collaborative effort with John Hutchinson in the book "The Story of Plants and their Uses to Man" (Hutchinson and Melville 1948).

After 16 years at the Kew Museum, Melville transferred to the Herbarium to take charge of the Australasian Section, a post he held until his retirement. His "hands-on" knowledge was enhanced when he spent 1952–1953 at the Melbourne Botanical Garden on a grant from the Maud Gibson Trust. While in Australia, he collected more than 20,000 herbarium specimens, mostly in Victoria. Other destinations he visited to collect herbarium vouchers and to capture photographic records included New Zealand, Fiji, Hawai'i, Yugoslavia, Argentina, Israel, South Africa, and the Grand Canary Islands. His travels and biogeographical knowledge undoubtedly helped him in his postulates of migration routes of early angiosperms from the Southern to the Northern Hemisphere. He also spent some considerable effort in developing a somewhat controversial theory on the evolution of angiosperm floral components, eschewing the terms carpel and stamen and proposing the terms "androphylls," "gynophylls," and "androgynophylls" for the male, female, and bisexual structures, respectively (Melville 1960). Later, he extended these studies on leaf vein architecture to include the origin of the flowering plant leaf (Melville 1969).

Melville was the person to use the term "bulla" to describe the head of the cone scales, taking an analogy in classical literature for the boss of a shield (Fig. 7.39). Of particular interest to our cycad-oriented readers, in 1972–1973, Melville visited East Africa and met Peter Bailey of the Nairobi Museum in Kenya, an association which led to Melville (1957) describing the cycads *Encephalartos bubalinus* (Fig. 7.40), *E. hildebrandtii* var. *dentatus*, and *E. tegulaneus* (Fig. 7.41) in 1957 and writing up the Zamiaceae section in the *Flora of Tropical East Africa* in 1958.

Melville retired as Senior Principal Scientific Officer in early 1968 with over 100 scientific publications to his credit. But his longstanding interest in plant conservation, facilitated by a Royal Society grant, allowed him to maintain his office at Kew and commence drawing up the *IUCN Red Data Book* published in parts through the 1970s (Fig. 7.42) and listing all known threatened plants. Initially, there was little knowledge of the number of plants facing extinction based on observations from his own travels and from data constantly flowing into Kew. Melville initially estimated that up to 20,000 flowering plant species faced extinction. In the 1970s, he revised that figure upward. "Our information shows the problem to be much more serious," he said. "We believe that between 20,000 and 30,000 species — or one in every ten flowering plants — is endangered" (Arbose 1974).

Melville's various achievements included serving on the Royal English Forestry Society, the Systematic Association, the Council of the Botanical Exchange Club,

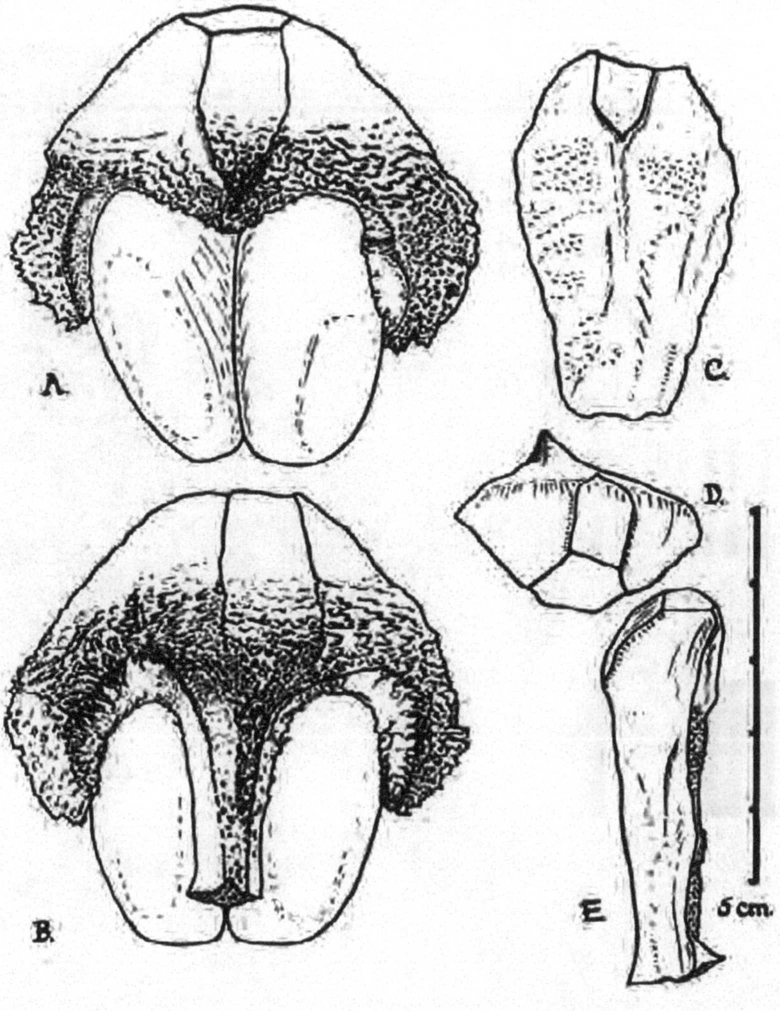

Fig. 7.39 View of megasporophylls (**a, b**) and microsporophylls (C–E) of *Encephalartos tegulaneus* that Melville used in his 1957 description of the species. Melville was the person to use the term "bulla" to describe the head of the cone scales, taking an analogy in classical literature for the boss of a shield

and the Richmond Scientific Society, holding leading roles with the Linnean Society, and becoming a Companion of the Imperial Service Order in 1938. The Australian tree *Acacia melvillei* (Yarran, Fabaceae) was named for him by Les Pedley, a wattle expert of the Queensland Herbarium, in 1978. Melville died at the age of 82 on August 6, 1985; he was survived by his wife Elsie (née Sharpington), two children, and four grandchildren.

Fig. 7.40 *Encephalartos bubalinus* in Narok, Kenya, a species described by Ronald Melville in 1957. (Photo: Andrew Cameron)

Fig. 7.41 *Encephalartos tegulaneus* in Samburu, Kenya, a species described by Ronald Melville in 1957. (Photo: Andrew Cameron)

In an obituary for Melville, his fellow research colleague at Kew, Martin Sands (1986), wrote

> [Ronald Melville] will be best remembered amongst botanists for his theories of angiosperm evolution. There is no doubt that, borne of a very active and enquiring mind, his controversial theories and ideas on this and other topics have always stimulated discussion, and will continue to be subjects for keen debate. He was exceptional in his depth of knowledge in a wide field of botanical disciples and all who know him during his many years at Kew, will remember him as a distinguished and valued colleague and with great affection as a friend.

Names of Cycads with Melville's Authorship
Data from the *World List of Cycads.*

Encephalartos bubalinus Melville (1957).
Encephalartos hildebrandtii var. *dentatus* Melville (1957) = *E. hildebrandtii* A.Braun & C.D.Bouché.
Encephalartos tegulaneus Melville (1957).

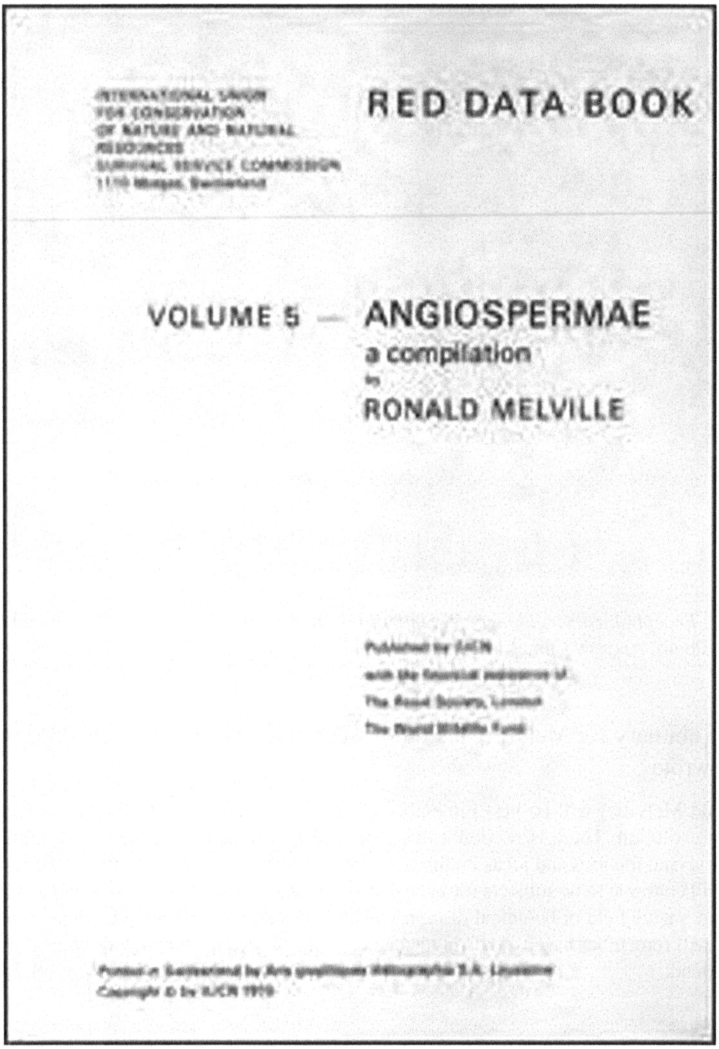

RED DATA BOOK

VOLUME 5 — ANGIOSPERMAE
a compilation
by
RONALD MELVILLE

Fig. 7.42 Ronald Melville spent much of his retirement years working on the IUCN's Red Data Book of Threatened Plant Species, Volume 5, Angiospermae, published in 1970

Melville's Cycad Publications

Melville, R. 1957. *Encephalartos* in Central Africa. Kew Bulletin 12(2): 237–257. [*Encephalartos bubalinus* Melville, *E. hildebrandtii* var. *dentatus* Melville, *E. tegulaneus* Melville].

Melville, R. 1958. Zamiaceae. Pp. 1–10 in Flora of Tropical East Africa.

Knut Jon Norstog (1921–2003)

In the history of cycadology, there are three particularly outstanding names: Friedrich Anton Miquel (1811–1871) in the mid-nineteenth century, Charles Joseph Chamberlain (1863–1943) in the first half of the twentieth century, and Knut Jon Norstog (1921–2003; Figs. 7.43 and 7.44) in the latter part of the twentieth century. What made these three so prominent was their eclectic interest and ability to be multidisciplinary in their research. Each of them also had very successful ventures into plant groups other than cycads. Each of them was an outstanding teacher and mentor. Of the three, Knut is the only one we have been privileged to know personally and even more privileged to call a friend and mentor.

Knut Norstog was born on June 11, 1921, on a farm near Walton, North Dakota. His father, Jon (1877–1942), was a poet and author foremost and a farmer secondarily. His mother, Inga, was superintendent of schools in McKenzie County. Knut's only sibling, Rolf, became an architect. Knut's early interest in

Fig. 7.43 Knut Norstog inspecting a specimen of *Zamia pseudoparasitica* in the greenhouse at Fairchild Tropical Botanical Garden. (Photo: Dennis Stevenson)

Fig. 7.44 Knut Norstog and his partner Priscilla Fawcett at the 16th International Botanical Congress in St. Louis, Missouri in 1999. (Photo: Andrew Vovides)

natural history came from his exposure to farming and exploring the local fauna and flora.

After graduating from Luther College in Decorah, Iowa, Knut entered the U.S. Army Air Corp and served in the European theater. Being part of the crew that filmed bombing runs, Knut was the photographer in the ventral ball turret, a fact that explains his partial deafness and his walking away from or hanging up the phone in the middle of a sentence, often his own. After being discharged, Knut went on to the University of Michigan where he completed his doctorate with mentor Carl LaRue. Interestingly, LaRue was a pioneer in tissue culture including experimental efforts with cycads and grasses; this initiated Knut's interest in both these plant groups.

Knut began his professorships at Wittenberg University in Ohio, then founded the Botany Department at the newly established University of South Florida, and eventually moved to Northern Illinois University. He was active in local scientific societies, in particular, the Ohio Academy of Science and the Illinois Academy of Science. He attended annual meetings of the former, and it was there, in 1968, that he presented a paper on reproduction in *Zamia*, a lecture that initiated Dennis Stevenson's interest in cycad biology and catalyzed an early 50-year mentorship and friendship. Knut had a long association with Fairchild Tropical Garden spanning some 35 years before his retirement and a move to Waterloo, Illinois, in 1989. He continued his botanical interest by becoming a consultant to the Missouri Botanical Garden.

Throughout his career, Knut showed a keen interest in fields as diverse as anatomy, ultrastructure, histochemistry, cytology, cycad reproduction, insect pollination and cycad symbiosis, tissue culture, and field biology. His interests were not confined to cycads but also extended to other plant groups including grasses. He kept abreast of new methods and approaches but did so judiciously and was not one to be accused of "hopping on the bandwagon." He was an enthusiastic supporter of and participant in the triennial International Conferences on Cycad Biology, acting as Co-editor for the Proceedings of the Australian meeting in 1990. The recognition of his eclecticism in botany led to his being the Editor-in-Chief of the American Journal of Botany for several years.

He was Chair of the IUCN's Cycad Specialist group from 1988–1990. He supervised and examined the work of numerous postgraduate students including the doctoral thesis of Roy Osborne and the postdoctoral work of Andrew Vovides. He was Editor-in-Chief of the American Journal of Botany for several years. Knut was made an Honorary Life Member of the Cycad Society of South Africa in 1988 (Osborne 1988).

Norstog's first book publication was "Plant Biology" and the associated "Laboratory Manual for Elementary Botany," both co-authored with Robert Long and published in 1976. He then co-authored another introductory general biology book, "Biology," with Andrew Meyerriecks in 1983. However, for those readers with a cycad interest, it was Knut Norstog and Trevor Nicholls' authoritative and comprehensive text "The Biology of the Cycads," published in 1997 by Cornell University Press (Fig. 7.45), that was immediately recognized as, and still remains, the fundamental reference work on the subject. Topics covered in that work included cycad morphology, anatomy, reproductive processes (Fig. 7.46), population dynamics, toxicity, cytology, cytogenetics, cycad fossils, and species descriptions. Knut explained to us that the text evolved in part from Charles Chamberlain's extensive notes, which he had inherited after Chamberlain's death. In the references section to "The Biology of the Cycads," there are 16 publications between 1965 and 1995 in

Fig. 7.45 Knut Norstog and Trevor Nicholls' authoritative and comprehensive text "The Biology of the Cycads" published in 1997 by Cornell University Press

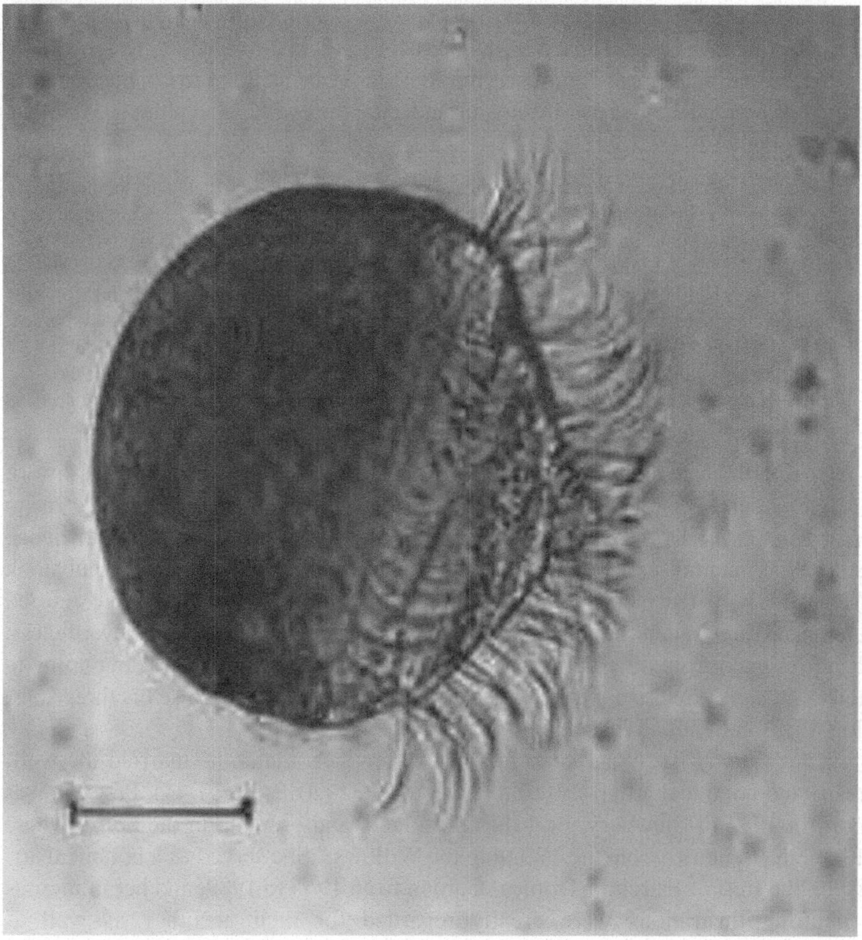

Fig. 7.46 A photomicrograph by Knut Norstog of a mature free-swimming spermatozoid (the male gamete) of *Zamia pumila*. Cycads and the Maidenhair Tree, *Ginkgo biloba*, are the only living seed plants that share this character. The scale bar represents 100 micrometers. (Photo: Knut Norstog)

which Norstog is the sole author and a further nine where he is the principal author with colleagues such as Priscilla Fawcett, Andrew Vovides, Dennis Stevenson, and others. Some of Knut's most salient and pioneering contributions to cycad biology were his publications on Robertsonian fusions and fissions in cycad chromosomes (Norstog, 1980) that set the stage for subsequent work by Aldo Moretti and Sergio Sabato (1984). Another was his establishment of insect pollination in cycads (Norstog et al., 1986; Norstog, 1987; Fawcett et al. 1995; Stevenson et al. 1998) as well as the ultrastructure of the male gametes (Norstog, 1967, 1974, 1990). Knut Norstog was above all eclectic, innovative, erudite, and synthetic in his thinking, research, and publications. He was in many ways our modern example, inspiration, and equivalent to Charles Joseph Chamberlain.

On several occasions, Knut used the Rosetta Stone analogy in connection with the contribution that cycads offer to our understanding of plant evolution. We quote from Norstog (2003):

> Cycads are the most ancient, still living seed plants with fossils that date to the late Carboniferous period some 300–325 million years ago. I have sometimes used the analogy of the Rosetta Stone for the fund of information stored within the living cycads and its importance to the interpretation of plant biology. You will recall that the Rosetta Stone was a slab found in 1799 near Rosetta, a town in Egypt near the mouth of the Nile, and which bore parallel inscriptions in Greek and Egyptian hieroglyphics. This discovery made it possible to decipher Ancient Egyptian writing and revealed many aspects of this great culture that were previously unknown. In a similar way, the very ancient structures and developmental pathways of cycads enable us to make connections between the early origins of seed plants and their present-day counterparts.

One particular incident summarizes Knut as a person. Panchanan Maheshwari, an internationally eminent plant embryologist from the University of Delhi, came to visit Knut at the University of South Florida because of his theoretical knowledge and experimental experience in embryology and tissue culture. Knut took Maheshwari to the faculty club on campus for dinner. When they were refused entrance, Knut called the university president and the club relented. Nevertheless, Knut resigned from his position the next day. In part, because he was not going to tolerate that behavior and, in part, because he was not going to raise his three sons, Jon, Erik, and Paul, in that environment.

Knut's first wife, Nona, died in 1980. A few years later, he married the well-respected botanical artist, Priscilla Fawcett (1932–2012; Fig. 7.44). Priscilla was born in 1932 in Blewbury, Berkshire, England, and grew up in the house where Kenneth Grahame wrote the "Wind in the Willows." She served as a botanical artist on the staff of Fairchild Tropical Garden from 1965 to 1989, and her numerous scientific illustrations were greatly regarded for their accuracy as well as their beauty.

Knut Norstog passed away after a short illness on June 8, 2003, and was survived by Priscilla and his sons Jon, Erik, and Paul. Knut always had a kind word for all, did not think ill of anyone, and was always positive in his outlook (Osborne 2003). All of us who knew him share the feeling that our lives were that much better and enriched for having known him. Truly, he was the very essence of a "teacher, scholar, friend and gentleman." In 1982, Dennis Stevenson acknowledged Knut's friendship when he named the Mexican cycad *Ceratozamia norstogii* (Figs. 7.46, 7.47, and 7.48), a species from Chiapas and Oaxaca. In addition, the Proceedings of the Fifth International Conference on Cycad Biology, held in Florida in 1999, were dedicated to Knut's memory (Stevenson 2004b).

Fig. 7.47 The rugged and wooded country near Fenia in Chiapas, Mexico, where *Ceratozamia norstogii* is located. (Photo: Roy Osborne)

Norstog's Cycad Publications

Ackery, P. R., R. J. Nash, E. A. Bell & K. J. Norstog. 1993. Sequestration of MAM-glycosides by insects. Pp. 3–7 in D. W. Stevenson & K.J. Norstog (eds), The Biology, Structure and Systematics of the Cycadales. Proceedings of the Second International Conference on Cycad Biology, Townsville, Australia, 1990. Palm & Cycad Societies of Australia Ltd., Milton, Queensland.

Chavez, V. M., R. E. Litz, P. A. Moon & K. J. Norstog. 1992. Somatic embryogenesis from leaf callus of mature plants of the gymnosperm *Ceratozamia mexicana* var. *robusta* (Miq.) Dyer (Cycadales). In Vitro Cell Development Biology 28: 59–63.

Chavez, V. M., R. E. Litz, P. A. Moon & K. J. Norstog. 1995. Somatic embryogenesis and organogenesis from leaf and embryo explants of *Ceratozamia* and *Zamia*. P. 91 (Abstract) in P. Vorster (ed.), Proceedings of the Third International Conference of Cycad Biology. Cycad Society of South Africa, Stellenbosch.

Chavez, V. M., R. E. Litz & K. J. Norstog. 1992. In vitro morphogenesis of *Ceratozamia hildae* and *C. mexicana* from megagametophytes and zygotic embryos. Plant Cell, Tissue and Organ Culture 30: 93–98.

Fawcett, P. K. S., K. J. Norstog & A. P. Vovides. 1995. Pollination of two species of *Zamia* in South Florida and its implication for the evolution of cycad strobili. Pp. 111–121 in D. D. Pant et al. (eds), Proceedings of the International Conference on Global Environment and Diversification of Plants through Geological Time. South Asian Publishers, Allahabad.

Fig. 7.48 *Ceratozamia norstogii* in habitat near Fenia, Chiapas, Mexico. (Photo: Lilí Martínez-Domínguez)

Monnier, M. & K. J. Norstog. 1984. Developmental aspects of immature *Zamia* embryos in culture. *Zeitschrift für Pflanzenphysiologie* 113: 105–116.

Monnier, M. & K. J. Norstog. 1986. Effects of *in ovulo* period of the differentiation and regulation of immature *Zamia* embryos cultured in vitro. Journal of Experimental Botany 37: 1633–1642.

Niklas, K. J. & K. J. Norstog. 1984. Aerodynamics and pollen grain depositional patterns on cycad megastrobili: implications on the reproduction of three cycad genera (*Cycas*, *Dioon* and *Zamia*). Botanical Gazette 145: 205–222.

Norstog, K. J. 1965. Induction of apogamy in megagametophytes of *Zamia integrifolia*. American Journal of Botany 52(10): 993–999.

Norstog, K. J. 1967. Fine structure of the spermatozoid of *Zamia* with special reference to flagellar apparatus. American Journal of Botany 54(7): 831–840.

Norstog, K. J. 1967. Some characteristics of growth and development in cycad tissue cultures. Pp. 4: 1–8 in Proceedings of the Fifth Conference on Cycad Toxicity, Fairchild Tropical Garden and University of Miami. Sponsored by the National Institute of Health.

Norstog, K. J. 1968. Fine structure of the spermatozoid of *Zamia*: observations on the microtubule system and related structures. Phytomorphology 18: 350–356.

Norstog, K. J. 1972. Role of archegonial neck cells of *Zamia* and other cycads. Phytomorphology 22(2): 125–130.

Norstog, K. J. 1974. Fine structure of the spermatozoid of *Zamia*: the *vierer Grüppe*. American Journal of Botany 61(5): 449–456.

Norstog, K. J. 1975. The motility of cycad spermatozoids in relation to structure and function. Biological Journal of the Linnean Society 7 (Supplement 1): 135–142.

Norstog, K. J. 1976. In search of some lost cycads. Fairchild Tropical Garden Bulletin 31(4): 5–11.

Norstog, K. J. 1977. The spermatozoid of *Zamia chigua* Seem. Botanical Gazette 138(4): 409–412.

Norstog, K. J. 1980. Chromosome numbers in *Zamia*. Caryologia 33: 419–428.

Norstog, K. J. 1980. Reptiles of the plant world — cycads. Fairchild Tropical Garden Bulletin 35(1): 25–27.

Norstog, K. J. 1981. Karyotypes of *Zamia chigua* (Cycadales). Caryologia 34(3): 225–260.

Norstog, K. J. 1982. Experimental embryology of gymnosperms. Pp. 204–216 In B. M. Johri (ed.), Experimental Embryology of Vascular Plants. Springer-Verlag, Berlin.

Norstog, K. J. 1985. Exploring for *Zamia wallisii*. Fairchild Tropical Garden Bulletin 40(1): 6–18.

Norstog, K. J. 1986. *Zamia chigua*, a case of mistaken identity? Fairchild Tropical Garden Bulletin 41(1): 6–13.

Norstog, K. J. 1986. The blepharoplast of *Zamia pumila* L. Botanical Gazette 147(1): 40–46.

Norstog, K. J. 1987. Cycads and the origin of insect pollination. American Scientist 75: 270–279.

Norstog, K. J. 1987. Cycad hybridization studies. Fairchild Tropical Garden Bulletin 42(3): 20–23.

Norstog, K. J. 1990. The spermatozoid of *Microcycas calocoma*: ultrastructure. Botanical Gazette 151(3): 275–284.

Norstog, K. J. 1990. Studies of cycad reproduction at Fairchild Tropical Garden. Pp. 63–81 in D. W. Stevenson (ed.), The Biology, Structure and Systematics of the Cycadales. Proceeding of the First International Conference on Cycad Biology, Beaulieu-sur-Mer, France, 1987. Memoirs of the New York Botanical Garden 57.

Norstog, K. J. 1992. Cycad phenology. Encephalartos 32: 11–13.
Norstog, K. J. 1993. Spermatogenesis in *Microcycas*: evolutionary significance of male gametes in seed plants. Pp. 270–278 in D. W. Stevenson & K. J. Norstog (eds), The Biology, Structure and Systematics of the Cycadales. Proceedings of the Second International Conference on Cycad Biology, Townsville, Australia, 1990. Palm & Cycad Societies of Australia Ltd., Milton, Queensland.
Norstog, K. J. 1995. Spermatogenesis in *Microcycas*: implications in the evolution of diplospermy in spermatophytes. Pp. 203–211 in D. D. Pant et al. (eds), Proceedings of the International Conference on Global Environment and Diversification of Plants through Geological Time. South Asian Publishers, Allahabad.
Norstog, K. J. 1995. C. J. Chamberlain and the ladies. Unpublished manuscript and correspondence with Selma Voth.
Norstog, K. J. 2003. Foreword, Pp. 3–8. in J. Donaldson (ed.), Cycads: Status, Survey, and Conservation Action Plan. IUCN/SSC Cycad Specialist Group. IUCN, Gland, Switzerland.
Norstog, K. J. & P. K. S. Fawcett. 1989. Insect-cycad symbiosis and its relation to the pollination of *Zamia furfuracea* (Zamiaceae) by *Rhopalotria molli*s (Curculionidae). American Journal of Botany 76(9): 1380–1394.
Norstog, K. J. & P. K. S. Fawcett. 1989. Insect pollination of cycads. Encephalartos 19: 38–41.
Norstog, K. J. & P. K. S. Fawcett. 1992. Mexican cycads. Encephalartos 30: 14–19.
Norstog, K. J. & P. K. S. Fawcett. 1993. Hurricane Andrew and Fairchild Tropical Garden. Encephalartos 33: 26–28.
Norstog, K. J., P. K. S. Fawcett, T. J. Nicolls, A. P. Vovides & E. Espinosa. 1995. Insect pollination of cycads: evolution and ecological considerations. Pp. 265–285 in P. Vorster (ed.), Proceedings of the Third International Conference on Cycad Biology, Pretoria, South Africa, 1993. Cycad Society of South Africa, Stellenbosch.
Norstog, K. J., P. K. S. Fawcett & A. P. Vovides. 1992. Beetle pollination of two species of *Zamia*: evolutionary and ecological considerations. The Palaeobotanist 41: 149–158.
Norstog, K. J., P. K. S. Fawcett, A. P. Vovides & M. V. Dunn. 1993. Toxicity and insect-cycad symbiosis in *Zamia furfuracea*. Pp. 42–50 in D. W. Stevenson & K. J. Norstog (eds), The Biology, Structure and Systematics of the Cycadales. Proceedings of the Second International Conference on Cycad Biology, Townsville, Australia, 1990. Palm & Cycad Societies of Australia Ltd., Milton, Queensland.
Norstog, K. J. & R. W. Long. 1976. Plant Biology. W. B. Saunders Co., Philadelphia.
Norstog, K. J. & R. W. Long. 1976. Plant Biology: A Laboratory Manual for Elementary Botany. W.B. Saunders Co., Philadelphia.
Norstog, K. J. & A. J. Meyerriecks. 1983. Biology. Charles E. Merrill, Columbus, Ohio.
Norstog, K. J. & T. J. Nicholls. 1997. The Biology of the Cycads. Cornell University Press, Ithaca. New York.
Norstog, K. J. & T. J. Nicholls. 1999. Some comments on the taxonomy and systematics of cycads. Pp. 92–97 in C. J. Chen (ed.), Biology and Conservation of Cycads. Proceedings of the Fourth International Conference on Cycad Biology, Panzhihua, China, 1996. International Academic Publishers, Beijing.
Norstog, K. J., E. M. Gifford & D. W. Stevenson, 2004. Comparative development of the spermatozoids of cycads and *Ginkgo biloba*. Botanical Review 70(1): 5–15.
Norstog, K. J. & R. Overstreet. 1965. Some observations of the gametophytes of *Zamia integrifolia*. Phytomorphology 15(1): 46–49.
Norstog, K. J. & E. Rhamstine. 1967. Isolation and culture of haploid and diploid cycad tissue. Phytomorphology 17: 374–381.

Norstog, K. J. & D. W. Stevenson. 1980. Wind? Or insects? The pollination of cycads. Fairchild Tropical Garden Bulletin 35(1): 28–30.
Norstog, K. J., D. W. Stevenson & K. J. Niklas. 1986. The role of beetles in the pollination of *Zamia furfuracea* L.*fil*. (Zamiaceae). Biotropica 18(4): 300–306.
Stevenson, D. W., K. J. Norstog & D. V. Molsen. 1966. Midribs of cycad pinnae. Brittonia 48(1): 67–74.
Stevenson, D. W., K. J. Norstog & P. K. S. Fawcett. 1998. Pollination biology of cycads. Pp. 277–294 in S. Owens & P. Rudall, Reproductive Biology: in systematics, conservation and economic botany. Royal Botanic Gardens, Kew.
Vovides, A. P., K. J. Norstog, P. K. S. Fawcett, M. W. Duncan, R. J. Nash & D. V. Molsen. 1993. Histological changes during maturation in male and female cones of *Zamia furfuracea* L.*fil*. (Zamiaceae, Cycadales) and their significance in relation to pollination biology. Botanical Journal of the Linnean Society London 111: 241–252.

Divya Darshan Pant (1919–2001)

Divya Darshan Pant (Figs. 7.49 and 7.50) was born in Ranikhet in the Indian state of Uttarakhand on October 18, 1919, the youngest of four sons of a local much-respected physician and writer. Growing up in this hill paradise in the Kumaon Himalayas, the young Pant went on lengthy walks with his father and rapidly developed an appreciation of nature and "the great outdoors." After school days in Ranikhet and nearby Nainital, his academic studies commenced at the University of Lucknow where he secured his first research position in 1941 under the supervision of Professor Birbal Sahni. Four years later, he joined the University of Allahabad, ascending to Reader, and ultimately to Professor and Head in 1966. Pant was an inspiring teacher who demanded high standards from his colleagues and his students. Under his 16 years of leadership, the University of Allahabad's Department of Botany emerged as a global leader in paleobotany and plant morphology. Pant's forté was in the field of paleobotany, and his knowledge of Indian plant fossils remains unsurpassed. Particularly noteworthy are his many contributions to the understanding of the Gondwanan *Glossopteris* flora from the Permian-Carboniferous of India, the Triassic flora of Nidpuri, and his associated research work on geobotanical aspects of the supercontinent of Gondwanaland. He has made profound contributions to the understanding of gametophytes of *Rhynia*, fossil conifers, and paleopalynology. His expertise in the field of epidermal, cuticular, and stomatal studies of almost all the groups of living and fossil land plants has resulted in the classification of stomata and other homologous structures based on their ontogeny and is accepted by botanists worldwide. Several obituaries (e.g., Joshi 2001; Srivastava 2001a, b; Sharma 2002) testify to his outstanding work in these areas.

Of particular interest to readers of this piece is that Pant's research contributions extended to extant cycads of all genera. His work focused in particular on cuticular, epidermal, and stomatal anatomy, an area in which he produced numerous

Fig. 7.49 Professor D. D. Pant, immaculately dressed as always in his three-piece suit with leather briefcase, inspecting a specimen of *Encephalartos lehmannii* in the grounds of the famous Jardin Villa Thuret at Cap d'Antibes on the French Riviera, April 1987. (Photo: Roy Osborne)

Fig. 7.50 Photo from Encephalartos 16 (December 1988) announcing honorary life memberships of the Cycad Society of South Africa for Dr. Knut Norstog (left) and Professor D. D. Pant (right). (Reproduced by permission of the Editor of Encephalartos)

publications and established himself as a world leader. Pant authored and co-authored more than 300 papers, reviews, monographs, books, and popular articles during his career; most relate to his paleobotanical work, while many others concern extant plants other than cycads. His two most important general books on cycads and gymnosperms were entitled "Cycas and the Cycadaales" (Pant 1962, 1973; Fig. 7.51) and "An Introduction to Gymnosperms, *Cycas*, and the Cycadales" (Pant 2002; Fig. 7.52), the latter being published posthumously in 2002 by the Birbal Sahni Institute of Palaeobotany.

Pant's extramural interests led to him studying and collecting fountain pens. In addition, he had an assemblage of over 10,000 "botanical" postage stamps. Appropriately, the fossil cycad that bears his given name is the specific epithet, *Birbalsahnia divyadarshanii*, which is one of the four Indian fossil stamps issued in 1977 (Fig. 7.53) to celebrate the Golden Jubilee of the Birbal Sahni Institute of Palaeobotany. A deeply spiritual man, Pant was a student of the Late Vedic Sanskrit texts known as the Upanishads and the Hindu religious teachings of Bhagwat Gita. Apart from that, he was thoroughly conversant with the nuances of English literature.

Cycas and the Cycadales

DIVYA DARSHAN PANT
MSc, DPhil, FLS, FASc, FNA,
Professor of Botany, University of Allahabad

CENTRAL BOOK DEPOT
ALLAHABAD

Fig. 7.51 Probably the best known of D. D. Pant's cycad books is his "*Cycas* and the Cycadales." First edition 1962, second edition 1972. Central Book Depot, Allahabad

Fig. 7.52 Evolving from his earlier cycad book, Pant's final volume "An Introduction to Gymnosperms, *Cycas*, and the Cycadales"

Fig. 7.53 Four Indian fossil stamps issued on the 1977 Golden Jubilee of the Birbal Sahni Institute of Palaeobotany. The 2.00-rupee (200 piase) stamp recognizes Professor Pant's name in the fossil *Birbalsahnia divyadarshanii*; other species represented are *Glossopteris*, *Pentoxylon*, and *Williamsonia sewardiana*

Those of us who met Pant at the various International Conferences on Cycad Biology remember him always appearing immaculately dressed in his three-piece suit and always willing to chat on a wide range of subjects without ever appearing pedantic or over-authoritative. A true gentleman indeed, Divya Darshan Pant passed away at the age of 81, on May 9, 2001, and was survived by his wife Radha, his son Priya, and two daughters, Vijaya and Kusum.

Professor Pant was a recipient of an astonishing number of awards and honors. These include the Medallion of the Birbal Sahni Institute of Palaeobotany (1971), the Birbal Sahni Gold Medal of the Indian Botanical Society (1976), the Biennial International Gold Medal of the Palaeobotanical Society (1987), and the Birbal Sahni Centenary Gold Medal (1991). His lectureship presentations include the Founders Day Lecture to the Maharashtra Association for the Cultivation of Science, the Panchanan Maheshwari Memorial Lecture (1971), the Birbal Sahni Memorial Lecture (1973), the Lyengar Endowment Lecture, the JBS Haldane Oration Award and Medal (1988), and the T. S. Sadasivan Lecture to the Indian National Science Academy (1990). He was elected Fellow of the Indian Academy of Sciences, Bangalore, the Indian National Academy of Sciences, the Linnean Society of London, the Indian Botanical Society, and the Palaeobotanical Society. He was an Emeritus Member of the International Society for Plant Taxonomists, an Honorary Member of the International Organization of Palaeobotany, and an Honorary Life Member of the Cycad Society of South Africa (Osborne 1988). Pant also served on the Committee for the International Code of Botanical Nomenclature and the IUCN Cycad Specialist Group.

Fifteen plant taxa recognize Pant's name in either the genus or the species epithets. Among the fossil genera are *Pantophyllum* and *Pantopteris*, while the fossil species include *Birbalsahnia divyadarshanii*, *Brachyphyllum pantii*, a fossil coconut fruit named *Cocos pantii*, *Dischotomopteris pantii*, *Glossopteris pantii*, *Hepaticites pantii*, *Podocarpoxylon pantii*, *Pseudoctenis pantii*, *Sagittophyllum pantii*, and *Scutum pantii*. Extant species named for Pant are two lycophytes, *Isoetes divyadarshanii* and *I. pantii*, and *Cycas divyadarshanii* (Fig. 7.54), recently separated from *C. pectinata* and now known to occur in China, India, Laos, Myanmar, Thailand, and Vietnam (Khuraijam and Singh 2020).

Pant's Cycad Publications

Pant, D. D. 1953. Notes on *Cycas* plants growing at Allahabad. Journal of the Indian Botanical Society 32(3): 145–156.

Pant, D. D. 1957. The classification of gymnospermous plants. Palaeobotanist 6: 65–70.

Pant, D. D. 1958. The structure of some leaves and fructifications of the *Glossopteris* flora of Tanganyika. Bulletin of the British Museum of Natural History 3(4): 125–175.

Pant, D. D. 1962 & 1973. *Cycas* and the Cycadales. First edition 1962. Second Edition 1972. Central Book Depot, Allahabad.

Pant, D. D. 1965. On the ontogeny of stomata and other homologous structures. Plant Science Series, Allahabad 1: 1–24.

Fig. 7.54 *Cycas divyadarshanii* Khuraijam, a cycad from Manipur State in north-eastern India, as named for Professor Pant in 2020. (Photos: Jiban Khuraijam)

Pant, D. D. 1987. Fossil history and phylogeny of the Cycadales. Geophytology 17: 125–162.

Pant, D. D. 1990. On the genus *Glandulataenia* nov. from the Triassic of Nidhpuri, India. Pp. 186–199 in D. W. Stevenson (ed.), The Biology, Structure and Systematics of the Cycadales. Proceedings of the First International Conference on Cycad Biology, Beaulieu-sur-Mer, France, 1987. Memoirs of the New York Botanical Garden 57.

Pant, D. D. 1997. Conservation of cycads with special reference to Indian cycads. Encephalartos 52: 7–10.

Pant, D. D. 1997. The aerophoric nature of the pinna base humps of *Macrozamia*. Phytomorphology 47(4): 345–351.

Pant, D. D. 1999. A study in contrasts: the present and past distribution and form of certain cycads. Pp. 1–23 in C. J. Chen (ed.), Biology and Conservation of Cycads. Proceedings of the Fourth International Conference on Cycad Biology, Panzhihua, China, 1996. International Academic Publishers, Beijing.

Pant, D. D. 1999. Some unusual species of *Cycas*. The Cycad Newsletter 22(3): 24–29.

Pant, D. D. 2000. Occurrence of unusual branching in the female cones of some *Zamia* plants and their implications on the branching of cycads. The Cycad Newsletter 23(1): 6–9.

Pant, D. D. 2002. An Introduction to Gymnosperms, *Cycas*, and the Cycadales. Monograph 4. Birbal Sahni Institute of Palaeobotany, Allahabad.

Pant, D. D. & K. Das. 1990. Occurrence of non-coralloid aerial roots in *Cycas*. Pp. 56–62 in D. W. Stevenson (ed.), The Biology, Structure and Systematics of the Cycadales. Proceedings of the First International Conference on Cycad Biology, Beaulieu-sur-Mer, France, 1987. Memoirs of the New York Botanical Garden 57.

Pant, D. D. & K. Das. 1993. Cuticle, epidermis and anatomy of pinnae in some species of *Cycas* Linn. Pp. 169–195 in D. W. Stevenson & K. J. Norstog (eds), The Biology, Structure and Systematics of the Cycadales. Proceedings of the Second International Conference on Cycad Biology, Townsville, Australia, 1990. Palm & Cycad Societies of Australia Ltd., Milton, Queensland.

Pant, D. D. & K. Das. 1995. Cuticle, epidermis and anatomy of some species of *Macrozamia*. Pp. 259–279 in Sahni Centenary International Conference on Global Environment and Diversification Plants through Geologic Time, Allahabad.

Pant, D. D. & R. Joshi. 1995. On *Cycadites jabalpurensis* sp. nov. from the Lower Cretaceous beds of Jabalpur Series, India. Birbal Sahni Centenary Volume 305–310.

Pant, D. D. & B. Mehra. 1962. Studies in gymnospermous plants —*Cycas*. Central Book Depot, Allahabad.

Pant, D. D. & B. Mehra. 1963. Development of stomata in leaves of three species of *Cycas* and *Ginkgo biloba*. Journal of the Linnean Society of London, Botany, 58: 491–497.

Pant, D. D. & B. Mehra. 1963. On a cycadophyte leaf *Pteronilssonia gopalii* gen. et sp. nov. from the Lower Gondwanas of India. Palaeontographica B113: 126–134.

Pant, D. D. & D. D. Nautiyal. 1960. Some seeds and sporangia of *Glossopteris* flora from Raniganj coal field, India. Palaeontographica B107: 41–64.

Pant, D. D. & D. D. Nautiyal. 1962. Seed cuticles in some modern cycads. Current Science India 31(2): 75–76.

Pant, D. D. & D. D. Nautiyal. 1963. Cuticle and epidermis of recent Cycadales: leaves, sporangia and seeds. Senckenbergiana Biologica 44(4): 257–348.

Pant, D. D. & D. D. Nautiyal. 1963. On the anatomical structure of the pinnae of some species of *Cycas* L. (Cycadaceae). Current Science India 32(5): 232–234.

Pant, D. D. & D. D. Nautiyal. 1963. Epidermal structure of pinnae of *Macrozamia* Miq. Current Science India 32(6): 280–281.

Pant, D. D. & D. D. Nautiyal. 1963. The epidermal structure of the sporangia of some modern cycads. Current Science India 32(12): 559–561.
Pant, D. D. & D. D. Nautiyal. 1963. Cuticular and epidermal structure of sporangia of modern cycads. Pp. 374–375 in Proceedings of the Indian Science Congress, Delhi.
Pant, D. D. & D. D. Nautiyal. 1982. Studies on living and fossil plants. Society of Indian Plant Taxonomists, Allahabad.
Pant, D. D., D. D. Nautiyal & S. K. Chaturvedi. 1981. Microsopic examination of fecal pellets of insect larvae feeding on leaves of some cycads. Proceedings of the Indian Academy of Plant Science 90: 508–514.
Pant, D. D. & R. Singh. 1990. Preliminary observations on insect-plant relationships in Allahabad plants of *Cycas*. Palms & Cycads 28: 10–14.
Pant, D. D. & R. Singh. 1991. Unusual orthotropous germination in seeds of *Cycas rumphii* Miq. and the morphological nature of cycad coleorrhiza. Plant Science Series Bulletin 2 (Sahni Central Issue): 9–14.
Pant, D. D. & R. Singh. 1993. Preliminary observations on insect/plant relationships in Allahabad plants of *Cycas*. Abstract in D. W. Stevenson & K. J. Norstog (eds), The Biology, Structure and Systematics of the Cycadales. Proceedings of the Second International Conference on Cycad Biology, Townsville, Australia, 1990. Palm & Cycad Societies of Australia Ltd., Milton, Queensland.
Pant, D. D., R. Singh & D. K. Chauhan. 1994. On *Cycas pectinata* Hamilton from northeast India. Encephalartos 38: 17–30.
Pant, D. D., R. Singh & P. C. Srivastava. 1995. On the occurrence of two cycadophylls: *Chiguites mamalensis* gen. et sp. nov. and *Cycadites meyenii* (Kapoor) comb. nov. in the Lower Gondwana (Permian) of Kashmir, India. Birbal Sahni Centenary Volume 311–316.
Pant, D. D. & P. C. Srivastava. 2000. Branched female cones of *Zamia*. Palms & Cycads 67: 28–30.
Pant, D. D. & B. K. Verma. 1963. On the structure of leaves of *Rhabdotaenia* Pant from the Raniganj coal field, India. Palaeontology 6(2): 301–314.
Singh, R. & D. D. Pant. 1997. On the occurrence of lenticels in seeds of *Cycas* L. Phytomorphology 47(3): 325–331.

Sergio Sabato (1941–1991)

Sergio Sabato (Fig. 7.55) was born in Catania, the ancient port city on Sicily's east coast, on April 30, 1941. His father was an agricultural engineer. The family became separated during WW II, and Sergio remembers that it took his father over a week to walk home when the war was over.

He attended a classical high school in Rome and then graduated in Agricultural Science from the University of Florence (Firenze) in 1964. He then accepted a research position at the University's Institute of Agricultural and Forestry Botany, making two field trips to Latin America in the early 1970s. In 1975, Paolo De Luca recruited him to join the team of botanists working at the Department of Plant Biology (*Dipartimento di Biologia Vegetale*) and the associated Botanical Garden (*Orto Botanico*) of the University of Naples (*Università degli Studi di Napoli Federico II*). During his 11-year tenure with the Naples University and the Garden, he and Neapolitan colleagues Paolo De Luca, Aldo Moretti (Fig. 7.56), Aldo Merola,

Fig. 7.55 An image of Sergio Sabato taken from his official church mass card

Gesualdo Siniscalco Gigliano, and Luciano Gaudio, accompanied on occasions by botanists and students from other Italian and international institutions (Fig. 7.57), undertook several expeditions to Latin America, studying cycads in habitat and collecting living and dried material (De Luca et al. 1995). Some of these trips were in the company of Professor Luigi Califano, whom we treat separately in this volume. It is largely thanks to these field collections that the Botanical Garden of Naples has one of the largest ex situ collections of authenticated cycad specimens in the world (De Luca 1990).

Fig. 7.56 Sergio Sabato with Aldo Moretti at Fairchild Tropical Gardens in Florida, U.S.A., in 1981. (Photo: Knut Norstog)

In 1989, Sergio married Silvana, and they moved to her hometown of Lecce in the Puglia region, where Sergio was appointed as Professor at the University of Lecce in 1990. In addition to his teaching and research duties, he planned and promoted a botanical garden for the university. This was a section of the local *Ecotekne* university campus, associated with another botanical garden near Masseria Sant'Angelo. It later became the Botanical Garden of Salento also home to *Herbarium Lupiense*, hosting a collection of about 20,000 dried flora specimens from Italy and Salento in particular.

Sergio's publications are numerous and diverse (De Luca et al. 1991), generally co-authored with colleagues from the Naples Botanical Garden. Most deal with descriptions of New World cycads, particularly *Dioon* species, but also cover more general aspects of cytology, taxonomy, ultrastructure, phytochemistry, and in vitro cultivation. He began summarizing data on New World cycads in later publications on *Dioon* (Sabato and De Luca 1985) and *Zamia* (Sabato 1990). His interests also included a study of bromeliads, and he was responsible for naming a new *Tillandsia* species for the late Professor Tomaselli.

Sergio passed away from a sudden coronary on his fiftieth birthday, on April 30, 1991, shortly after a dinner event with colleagues after a faculty meeting. He was survived by his wife Silvana, their son Gianluca, and his surgeon brother (De Luca et al. 1991).

Fig. 7.57 Sergio Sabato, Dennis Stevenson, Knut Norstog, and Knut's son Erik collecting herbarium material of *Zamia manicata* in Mutata, Antioquia, Colombia, in 1987. (Photo: Rodrigo Bernal)

The species *Ceratozamia sabatoi* (Vovides et al. 1993; Fig. 7.58), found in pine/oak forest and mixed oak forest in the Mexican states of Hidalgo and Querétaro, was named for him by Andrew Vovides and colleagues. The 1993 holotype, *A. P. Vovides 1205* (Fig. 7.59), is in the herbarium of the Instituto de Ecología in Xalapa. The Proceedings of the Second International Conference on Cycad Biology, held in Australia in 1990, were dedicated to Sergio's memory, with the dedication in part below.

> Sergio was an outstanding cycadologist who made significant contributions to cycad biology. He understood the neotropical Zamiaceae better than any of his predecessors. But Sergio was more than just a cycad biologist, he was also a scholar. His knowledge of botany, science in general, art, literature and music was both thorough and fundamental. He was a unique colleague who touched the lives of many of us very deeply (Stevenson and Norstog 1993).

Names of Cycads with Sabato's Authorship
Data from the *World List of Cycads*

Ceratozamia euryphyllidia Vázq.Torres, Sabato & D.W.Stev. (1986).
Dioon califanoi De Luca and Sabato (1979).
Dioon caputoi De Luca, Sabato & Vázq.Torres (1980).

Fig. 7.58 *Ceratozamia sabatoi* in the transition zone between oak forest and cloud forest in San Joaquín, Querétaro. (Photo: Lilí Martínez-Domínguez)

Dioon holmgrenii De Luca, Sabato & Vázq.Torres (1981).
Dioon merolae De Luca, Sabato & Vázq.Torres (1981).
Dioon rzedowskii De Luca, A.Moretti, Sabato & Vázq. Torres (1980).
Dioon tomasellii De Luca, Sabato & Vázq.Torres (1984).
Dioon tomasellii var. *sonorense* De Luca, Sabato & Vázq. Torres (1984) ≡ *D. sonorense* (De Luca, Sabato & Vázq.Torres) Chemnick, T.J.Greg. & Salas-Mor.
Zamia disodon D.W.Stev. & Sabato (2001).
Zamia pseudomonticola L.D.Gómez *ex* D.W.Stev. & Sabato (1986).
Zamia vazquezii D.W.Stev., Sabato & De Luca (1998).

Sabato's Cycad Publications

Caputo, P., R. Nazzaro & S. Sabato. 1985–86. Observations on seeds in the Cycadales. Delpinoa 27–28: 29–36.
De Luca, P., V. La Valva & S. Sabato. 1980. Spermatogenesis and tissue formation in cycad pollen grains. Caryologia 33(2): 261–265.
De Luca, P., P. Medeghini Bonatti, A. Moretti, S. Sabato & G. Siniscalco Gigliano. 1981. *Ultrastruttura dei canali mucipari delle Cycadales e dati chemiotassonomici sulla composizion e delle mucillagini.* Giornale Botanico Italiano 115: 397–398.
De Luca, P., A. Moretti & S. Sabato. 1979. Regeneration in megagametophytes of cycads. Giornale Botanico Italiano 113: 129–143.

Fig. 7.59 Image of the holotype of *Ceratozamia sabatoi* Vovides, Vázq.Torres, Schutzman, and Iglesias as collected by Andrew Vovides and Priscilla Fawcett in San Joaquín, Querétaro, in April 1991. (Photo: Lilí Martínez-Domínquez)

De Luca, P., A. Moretti & S. Sabato. 1980. *Distribuzione et assonomia del genere Dioon (Zamiaceae)*. Giornale Botanico Italiano 114(3–4): 129–130.

De Luca, P., A. Moretti, S. Sabato & G. Siniscalco Gigliano. 1980. The ubiquity of cycasin in cycads. Phytochemistry 19(10): 2230–2231.

De Luca, P., A. Moretti, S. Sabato & G. Siniscalco Gigliano. 1982. A comparative study of cycad mucilages. Phytochemistry 21(7): 1609–1611.

De Luca, P., A. Moretti, S. Sabato & M. Vázquez Torres. 1980. *Dioon rzedowskii* (Zamiaceae), a new species from Mexico. Brittonia 32(2): 225–229. [*Dioon rzedowskii* De Luca, A.Moretti, Sabato & Vázq.Torres].

De Luca, P. & S. Sabato. 1978. Nomenclature and taxonomy of *Dioon mejiae* Standl. & Williams (Zamiaceae). Delpinoa 20: 45–50.

De Luca, P. & S. Sabato. 1979. In vitro spermatogenesis of *Encephalartos* Lehm. Caryologia 32(2): 241–245.

De Luca, P. & S. Sabato. 1979. *Dioon califanoi* (Zamiaceae), a new species from Mexico. Brittonia 31(1): 170–173. [*Dioon califanoi* De Luca & Sabato].

De Luca, P. & S. Sabato. 1980. *Osservazioni sur adici coralloidi apogame*. Giornale Botanico Italiano 114: 112.

De Luca, P. & S. Sabato. 1980. Regeneration of coralloid roots on cycad megagametophytes. Plant Science Letters 18(1): 27–31.

De Luca, P., S. Sabato, A. Balduzzi & R. Nazzaro. 1980. Coralloid root regeneration on *Macrozamia* megagametophytes. Giornale Botanico Italiano 114(6): 271–275.

De Luca, P., S. Sabato & D. W. Stevenson. 1984. Comments opposed to the proposal to conserve the spelling *Dioon* against *Dion* (Zamiaceae). Taxon 33(4): 728–729.

De Luca, P., S. Sabato & M. Vázquez Torres. 1970. *Dioon purpusii* Rose (Zamiaceae), a mis-known species. Delpinoa n.s. 20: 31–35.

De Luca, P., S. Sabato & M. Vázquez Torres. 1980. *Dioon caputoi* (Zamiaceae), a new species from Mexico. Brittonia 32(1): 43–46. [*Dioon caputoi* De Luca, Sabato & Vázq. Torres].

De Luca, P., S. Sabato & M. Vázquez Torres. 1981. *Dioon merolae* (Zamiaceae), a new species from Mexico. Brittonia 33(2): 179–185. [*Dioon merolae* De Luca, Sabato & Vázq. Torres].

De Luca, P., S. Sabato & M. Vázquez Torres. 1981. *Dioon holmgreniii* (Zamiaceae), a new species from Mexico. Brittonia 33(4): 552–555. [*Dioon holmgrenii* De Luca, Sabato & Vázq.Torres].

De Luca, P., S. Sabato & M. Vázquez Torres. 1982. Distribution and variation of *Dioon edule* (Zamiaceae). Brittonia 34(3): 355–362.

De Luca, P., S. Sabato & M. Vázquez Torres. 1984. *Dioon tomasellii* (Zamiaceae), a new species with two varieties from western Mexico. Brittonia 36(3): 223–227. [*Dioon tomasellii* De Luca, Sabato & Vázq.Torres, *Dioon tomasellii* var. *sonorense* De Luca, Sabato & Vázq.Torres].

Medeghini Bonatti, P. & S. Sabato. 1984. Plastid ultrastructure in *Macrozamia moorei* (Zamiaceae). Caryologia 37: 133–139.

Moretti, A. & S. Sabato. 1984. Karyotype evolution by centromeric fission in *Zamia* (Cycadales). Plant Systematics and Evolution 146: 215–223.

Moretti, A. & S. Sabato. 1988. Systematics and evolution of *Dioon* and *Ceratozamia*. Fairchild Tropical Garden Bulletin 43(1): 22–28.

Moretti, A., S. Sabato & G. Siniscalco Gigliano. 1981. Monosaccharide composition of the mucilages in *Encephalartos* Lehm. (Zamiaceae). Giornale Botanico Italiano 115: 291–297.

Moretti, A., S. Sabato & G. Siniscalco Gigliano. 1981. Distribution of macrozamin in Australasian cycads. Phytochemistry 20(6): 1415–1416.

Moretti, A., S. Sabato & G. Siniscalco Gigliano. 1983. Taxonomic significance of methyl-azoxymethanol glycosides in the cycads. Phytochemistry 22(1): 115–117.
Moretti, A., S. Sabato & M. Vázquez Torres. 1980. The distribution of *Ceratozamia* Brong. (Zamiaceae). Delpinoa 22: 13–21.
Moretti, A., S. Sabato &M. Vázquez Torres. 1982. The rediscovery of *Ceratozamia kuesteriana* (Zamiaceae) in Mexico. Brittonia 34(2): 185–188.
Sabato, S. 1990. West Indian and South American Cycads. Pp. 173–185 in D. W. Stevenson (ed.), The Biology, Structure and Systematics of the Cycadales. Proceedings of the First International Conference on Cycad Biology, Beaulieu-sur-Mer, France, 1987. Memoirs of the New York Botanical Garden 57.
Sabato, S. & P. De Luca. 1985. Evolutionary trends in *Dion* (Zamiaceae). American Journal of Botany 72(9): 1353–1363.
Sabato, S., A. Moretti & V. La Valva. 1978. *Dioon purpusii Rose e nuovi taxa erroneamente attribuitia questa specie*. Delpinoa 18–19: 107–116.
Stevenson, D. W. & S. Sabato. 1986. Typification of names in *Zamia* L. and *Aulacophyllum* Regel (Zamiaceae). Taxon 35(1): 134–144. [*Zamia pseudomonticola* L.D.Gómez *ex* D.W.Stev. & Sabato].
Stevenson, D. W. & S. Sabato. 1986. Typification of names in *Ceratozamia* Brong., *Dioon* Lindl. and *Microcycas* A.DC. (Zamiaceae). Taxon 35(3): 578–584.
Stevenson, D. W. & S. Sabato. 2001. *Zamia disodon* D.W.Stev. & Sabato. Pp. 38–40 in D. W. Stevenson (ed.), Cycadales, Flora of Colombia 21. Universidad Nacional de Colombia, Bogotá, Colombia. [*Zamia disodon* D.W.Stev. & Sabato].
Stevenson, D. W., S. Sabato, A. Moretti & P. De Luca. 1998. What is *Zamia fischeri* Miquel? Delpinoa 37–38: 9–17. [*Zamia vazquezii* D.W.Stev., Sabato & De Luca].
Stevenson, D. W., S. Sabato & M. Vázquez Torres. 1986. A new species of *Ceratozamia* (Zamiaceae) from Veracruz, Mexico, with comments on species relationships, habitats and vegetative morphology in *Ceratozamia*. Brittonia 38(1): 17–26. [*Ceratozamia euryphyllidia* Vázq.Torres, Sabato & D.W.Stev.].

Oliver Wolf Sacks (1933–2015)

Oliver Wolf Sacks (Fig. 7.60) was born on July 9, 1933, in Kilburn, northwest London, the youngest of four children of Samuel Sacks and Muriel Landau, both medical doctors. During WW II, Oliver and his brother Michael were evacuated to the English midlands where they "subsisted on meagre rations of turnips and beet root and suffered cruel punishments at the hands of a sadistic headmaster" (Sacks 2001). After the war, Sacks returned to London to enroll at Saint Paul's School and later Queen's College in Oxford, where he graduated in 1956 with a BA in physiology and biology and two years later graduated with a MA and MD in medicine.

Sacks' interest in botany arose from his friendship with Marie Stopes who would take him to Kew to teach him about ferns and cycads. She was one of the world's leading paleobotanists, until changing midcareer to championing the cause for women's reproductive rights. She knew the Sacks family through the fact that Oliver's parents were both obstetrician-gynecologist physicians. Her dissertation, done in Germany with the famous morphologist Karl van Goebel, was on the anatomy of the seeds of extant cycad genera.

Fig. 7.60 Outstanding neurologist and author, Oliver Sacks (1933–2015), seated below the foliage of an *Encephalartos altensteinii* at the *Hortus Botanicus* Amsterdam. (Photo: Bill Hayes)

Sacks took a great interest in plants generally. He served on the board of the New York Botanical Garden and could regularly be seen on Wednesdays walking the grounds, botanizing in the conservatory, visiting the fern and cycad collections, and attending monthly meetings of the New York Chapter of the American Fern Society. He also brought patients to the Garden as portrayed in the film "Awakenings."

Frustrated with British conservatism, Sacks decided in 1960 to make a move to the New World, firstly completing an internship at Mount Zion Hospital in San Francisco and then a residency in neurology and neuropathology at the University of California, Los Angeles (UCLA). His somewhat unusual extramural activities included experimentation with mind-altering drugs prompting his book "Hallucinations" (Sacks 2012), sexual forays, fast motorcycles, scuba diving, and bodybuilding. In 1966, he headed east, taking a post as instructor and later Professor of Neurology at the Albert Einstein College of Medicine. In July 2007, he joined Columbia University as a Professor of Neurology and Psychiatry, as well as holding the title of Columbia University Artist. In 2012, he moved to the New York University School of Medicine. He maintained a busy nursing-home-based practice at the Beth Abraham Hospital in the Bronx, New York, and also accepted a limited number of private patients.

Fig. 7.61 A selection of leather-bound copies of some of Oliver Sacks' books. (Reproduced by courtesy of the Oliver Sacks Foundation)

In addition to his medical career, Sacks was a talented and prodigious author and a polymath as evidenced by his personal library (Fig. 7.61). Apart from 16 published books and a series of significant lectures, essays, and journal submissions, he produced letters, notes, marginalia, and audio recordings. Much of this work was based on his own life experiences, and much of it was unpublished. He became widely known for writing best-selling case histories about both his patients' disorders and his own unusual life experiences, with many of his essays inspiring plays, films, opera, dance, fine art, and musical works. The New York Times called him a "poet laureate of contemporary medicine," and "one of the great clinical writers of the 20th Century."

Of particular relevance to the cycad community was his 1997 two-part book (Fig. 7.62) "The Island of the Colorblind and Cycad Island," the second part of which tells of the high incidence of Lytico-Bodig neurodegenerative disease—an insidious combination of amyotrophic lateral sclerosis (ALS), dementia, and Parkinsonism—that appeared during the post-World War II years in the indigenous Chamorro people of Guam. Following up on a report from Leonard Kurland (1988), Sacks believed that the cause for the disease complex was most probably associated with ingestion of the toxin beta-methylamino-L-alanine (BMAA) that occurs in the local cycad, *Cycas micronesica*. Furthermore, along with Paul Cox, he postulated that it might not so much be the direct consumption of cycad seed, but rather the biomagnified amount of toxin absorbed when the Chamorros ate flesh of the fruit bat, *Pteropus mariannus*, known to feed on the fleshy sarcotesta of the cycad seeds (Cox and Sacks 2002). This was followed by a short paper on the occurrence of BMAA in patients in Guam with ALS (Murch et al. 2004).

Of cycads, he wrote in "The Island of the Colorblind and Cycad Island" (1996)

> It was not only the sheer gigantism that seemed characteristic of the group – the biggest spermatozoids, the biggest egg cells, the biggest growing apices, the biggest cones, the biggest everything in the vegetable world. It was rather the sense that cycads were brilliantly adaptable and resourceful life-forms, full of unusual capacities and developments which had enabled them to survive for a quarter of a billion years, when so many of their contemporaries had fallen by the way.

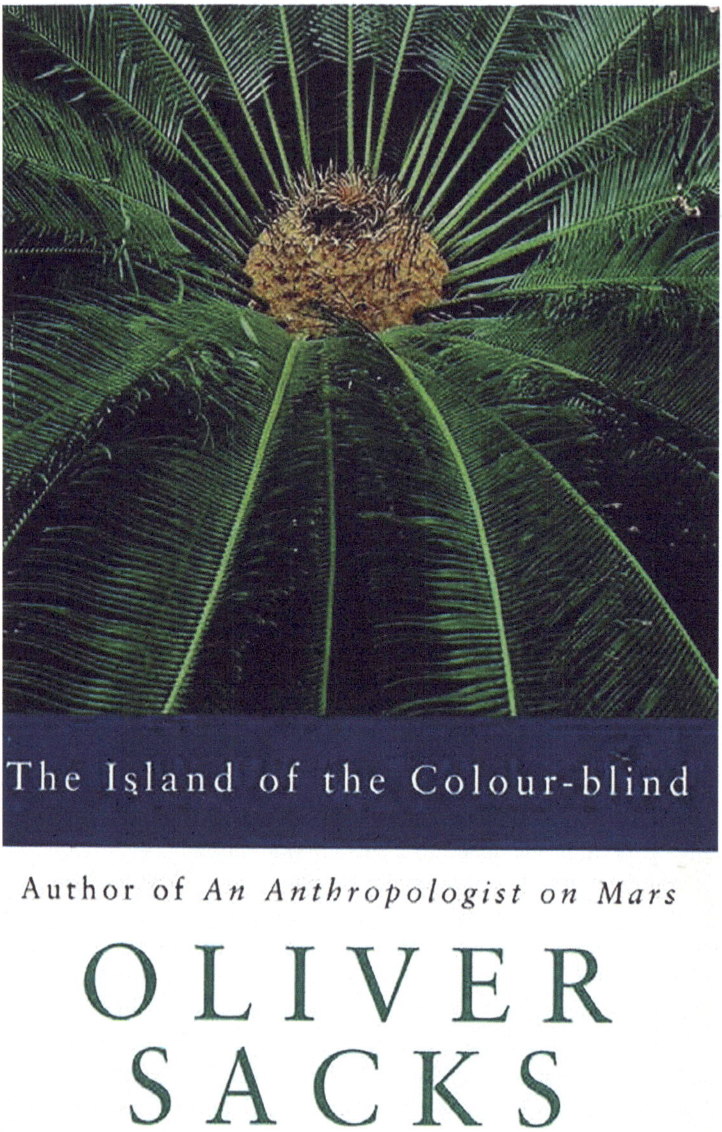

Fig. 7.62 The front cover of the Picador Pan Macmillan Sacks' double volume "The Island of the Colour-blind" and "Cycad Island" as published in 1996

Interestingly, this book was printed under two titles, one in English entitled "The Island of the Colour-blind" and "Cycad Island" and another in American English entitled "The Island of the Colorblind and Cycad Island."

Another botanical work was his "Oaxaca Journal" (2002), which tells how ferns lured Sacks to Oaxaca, but he then weaves the strands of biology, history, and culture into a tapestry of Mexico and relates how a group of fern-seekers become united in a common passion.

Among his many awards and honors, Sacks was a Member of the American Academy of Arts and Letters, a Fellow of the New York Academy of Sciences, a Fellow of the American Academy of Arts and Sciences, and an Honorary Fellow of the Royal College of Physicians. His work was recognized by the Music has Power Award from the Institute for Music and Neurologic Function, a Golden Plate Award of the American Academy for Achievement, Rockefeller University's Lewis Thomas Prize for scientific writing, and an astonishing total of 12 honorary doctorates from a dozen major international universities. In 2007, Sacks received the position of "Columbia Artist" from Columbia University, a post that gave him unconstrained access to the university, regardless of department or discipline. In 2008, he was appointed Commander of the Order of the British Empire (CBE) for services to medicine. The cycad community recognized Sacks' contributions to cycad biology by dedicating the Proceedings of the Ninth International Conference on Cycad Biology (Li et al. 2018), held in Shenzhen, China in 2011, to his memory, and by naming a new cycad from Western Oaxaca as *Ceratozamia oliversacksii* (Fig. 7.63) in Kew Bulletin (Martínez-Domínguez et al. 2022).

On the personal front, Sacks was somewhat clumsy in habit and severely shy as a youth, a passionate collector (he collected samples of each element in the periodic table), an insomniac, and unable to recognize faces. He experienced erotic fantasies and had bouts of deep anxiety. Some thought of him as unconventional and

Fig. 7.63 The recently described *Ceratozamia oliversacksii* from western Oaxaca. (Photo: Lilí Martínez-Domínguez)

eccentric. Yet he portrayed a deep humanity and sincere empathy with his patients and was brutally honest. Sacks wrote about his boyhood love for chemistry in "Uncle Tungsten" (2001) and always carried a small periodic table in his wallet. Sacks explained the reason in Ric Burns' (2019) American biographical documentary "Oliver Sacks: His Own Life," and we quote

> Oh yes, I carry a periodic table in my wallet. I love it very much. It stands for order, stability – but it also stands for imagination and mystery. And some of the elements get very, very complicated, as you go above 92 and sort of relativistic and other considerations come in. For example, you cannot understand, on the basis of the periodic table, why gold is gold. It's a very simple question. In fact, it turns out to be a very deep question. And I am not mathematical enough to tell you the answer. But it involves both quantum physics and relativity.

In 2009, at the age of 75, Sacks met a writer named Bill Hayes who later became his partner for the rest of his days. One can only imagine that the arrangement brought peace and comfort after living most of his life when homosexuality was unaccepted, declared illegal in Britain, and regarded as an abomination against God and nature in traditional Jewish families. It may explain why he was a lifelong atheist (he was named in 2010 as a distinguished achiever by the American Freedom from Religion Foundation).

Sacks suffered from bouts of ill health, including loss of sight in his right eye that prompted his book "The Mind's Eye"(Sacks 2010) attributed to an ocular tumor that ultimately metastasized to his liver and from which, despite radiation therapy, he passed away in his Manhattan home, surrounded by close friends, on August 30, 2015, at the age of 82. Sacks' books continue to be read by millions around the world and have inspired thousands of young readers to dedicate their lives to science and medicine.

Some of Oliver Sacks' Popular Books

"Awakenings" (1973) [The inspiration for the 1990 film starring Robert De Niro and Robin Williams as Dr. Sacks.

"A Leg to Stand On" (1984)

"The Man who Mistook his Wife for a Hat" (1985)

"An Anthropologist on Mars" (1995)

"The Island of the Colorblind and Cycad Island" (1996)

"Uncle Tungsten: Memories of a Chemical Boyhood" (2001)

"Musicophilia: Tales of Music and the Brain" (2007)

"The Mind's Eye" (2010)

"Hallucinations" (2012)

"On the Move" (2015)

"The River of Consciousness" (2017).

Julius Schuster (1886–1949)

Julius Schuster (Fig. 7.64), son of a tobacco wholesaler, was born in Munich, Germany, on April 7, 1886. He attended school at the Wilhelms gymnasium before studying geology and botany at the Ludwig Maximilians University in Munich. He graduated with a doctorate in 1909 after submitting a thesis on the morphology of grass flowers and was then appointed to the Research Institute for Biology. In 1911, he completed his "habilitation" in botany and paleontology but renounced the submission after allegations of scientific fraud.

Fig. 7.64 Julius Schuster in his early 20 s. (From the library of the Botanical Institute of Munich University)

He then had an eclectic series of postings in Berlin: research assistant at the Botanical Museum, assistant at the Prussian Artillery Examination Commission, employee of the German Pharmacopoeia at the Reich Health Office, research assistant and library councilor at the Berlin State Library, assistant at the Paleontological Museum and at the University's Geological-Paleontological Institute, and Head and Professor of Organic Natural Sciences at the Institute for the History of Medicine and Natural Sciences. He published numerous papers on a diversity of subjects, ranging from botany and paleobotany to science history and sex psychology. In 1922, he received the Lorenz Oken Medal from the Society of German Natural Scientists and Doctors.

In 1932, Schuster published his taxonomic work on the Cycadaceae as part of Engler's comprehensive work *Das Planzenreich* (Fig. 7.65). This monograph was intended to be a complete synthesis of the family but has been more or less condemned by all cycad biologists who have followed. In the words of Paolo De Luca (1990) in his opening address at the First International Conference on Cycad Biology

> Unfortunately, Schuster has no familiarity with living cycads, consulted a few herbaria, and was superficial with regards to nomenclature. Moreover, he did not understand the nomenclatural type concept in the modern sense and never annotated the specimens cited as types in his work, creating great confusion and instability in the nomenclature of cycads. All these deficiencies constrain and confound the contemporary botanist, complicating nomenclatural revision.

Further, although Schuster recognized traditional two subfamilies, he placed each genus in its own tribe with a consequent lack of grouping that resulted in a meaningless nomenclature. Only one of Schuster's numerous cycad names remains as accepted on the World List of Cycads, his *Cycas nathorstii* from India and Sri Lanka (1932), reference to the "other valid names" section of the current World List reveals Schuster's authorship in bewildering selection comprising a single *Ceratozamia*, 23 *Cycas*, one *Dioon*, five *Encephalartos*, and 16 *Zamia* taxa. One positive aspect was the completeness and accuracy of the non-taxonomic sections and the bibliography, providing a useful reference for cycad biologists even to this day. All of Schuster's cited specimens were filed at the Berlin herbarium and appear to have been lost during a bombing raid in March 1943 during World War II (Fig. 7.66).

Politically, Schuster was close to the National Socialists and was implicated in the robbery of cultural assets from the German-occupied territories in the1940s. On one occasion, he was supposedly involved in a beer hall incident. In 1945, he was dismissed from his academic post due to his political affiliations. Four years later, he committed suicide by drowning himself in the Müggelsee, a large lake near Berlin (Junker and Landsberg 1994).

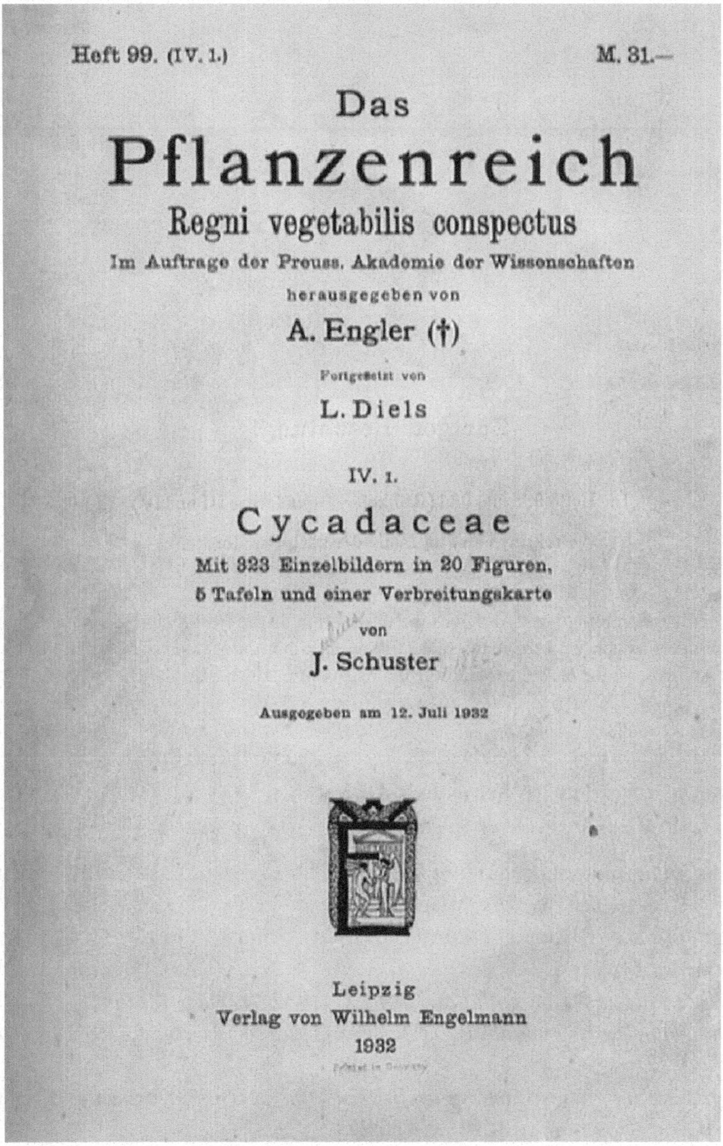

Heft 99. (IV. 1.) M. 31.–

Das

Pflanzenreich

Regni vegetabilis conspectus

Im Auftrage der Preuss. Akademie der Wissenschaften

herausgegeben von

A. Engler (†)

Fortgesetzt von

L. Diels

IV. 1.

Cycadaceae

Mit 323 Einzelbildern in 20 Figuren,

5 Tafeln und einer Verbreitungskarte

von

J. Schuster

Ausgegeben am 12. Juli 1932

Leipzig

Verlag von Wilhelm Engelmann

1932

Fig. 7.65 The front cover of Schuster's 1932 Cycadaceae volume in Engler and Diels' *Das Pflanzenreich*

Fig. 7.66 A catastrophe to botanists was the destruction of the Berlin Herbarium when nearly all cycad and other vouchers were incinerated following a bombing raid on March 1, 1943. (Photo from the Archives of the Botanical Garden and Museum of Berlin-Dahlem)

Names of Cycads with Schuster's Authorship
Data from the *World List of Cycads*

Ceratozamia mexicana var. *longifolia* forma *fuscoviridis* (W. Bull) J. Schust. (1932) ≡ *C. fuscoviridis* W.Bull.

Cycas circinalis forma *glauca* (Miq.) J. Schust. (1932) ≡ *C. glauca* Hort. *ex* Miq.

Cycas circinalis forma *gothanii* J.Schust. (1932), *nomen dubium.*

Cycas circinalis forma *undulata* (Desf. *ex* Gaudich.) J.Schust. (1932) = *C. circinalis* L.

Cycas circinalis subsp. *madagascariensis* forma *trigonocarpoides* J.Schust. (1932) = *C. thouarsii* R.Br. *ex* Gaudich.

Cycas circinalis subsp. *madagascariensis* (Miq.) J. Schust. (1932) = *C. thouarsii* R.Br. *ex* Gaudich.

Cycas circinalis subsp. *papuana* var. *scratchleyana* (F. Muell.) J. Schust. (1932) ≡ *C. scratchleyana* F. Muell.

Cycas circinalis subsp. *papuana* (F. Muell.) J.Schust. (1932) ≡ *C. papuana* F.Muell.

Cycas circinalis subsp. *riuminiana* var. *curranii* forma *chamberlainii* (W.H.Br. & Kienholz) J. Schust. (1932) = *C. riuminiana* Porte *ex* Regel.

Cycas circinalis subsp. *riuminiana* var. *curranii* forma *graminea* J. Schust. (1932) = *C. wadei* Merr.
Cycas circinalis subsp. *riuminiana* var. *curranii* forma *maritima* J. Schust. (1932) = *C. edentata* de Laub.
Cycas circinalis subsp. *riuminiana* var. *curranii* J.Schust. (1932) ≡ *C. curranii* (J.Schust.) K.D.Hill.
Cycas circinalis subsp. *riuminiana* (Porte *ex* Regel) J.Schust. (1932) ≡ *C. riuminiana* Porte *ex* Regel.
Cycas circinalis subsp. *seemannii* (A. Braun) J.Schust. (1932) ≡ *C. seemannii* A.Braun.
Cycas circinalis subsp. *vera* var. *beddomei* (Dyer) J. Schust. (1932) ≡ *C. beddomei* Dyer.
Cycas circinalis subsp. *vera* var. *pectinata* (Griff.) J. Schust. (1932) ≡ *C. pectinata* Buch.-Ham.
Cycas circinalis subsp. *vera* J. Schust. (1932) = *C. circinalis* L.
Cycas media var. *basaltica* (C.A. Gardner) J. Schust. (1932) ≡ *C. basaltica* C.A.Gardner.
Cycas media var. *furfuracea* (W. Fitzg.) J. Schust. (1932) ≡ *C. furfuracea* W.Fitzg.
Cycas media var. *lanepoolei* (C.A. Gardner) J. Schust. (1932) ≡ *C. lanepoolei* C.A.Gardner.
Cycas nathorstii J.Schust. (1932).
Cycas revoluta var. *taiwaniana* (Carruth.) J. Schust. (1932) ≡ *C. taiwaniana* Carruth.
Cycas rumphii subsp. *normanbyana* (F.Muell.) J.Schust. (1932) = *C. media* R.Br.
Cycas rumphii subsp. *zeylanica* J. Schust. (1932) ≡ *C. zeylanica* (J.Schust.) A.Lindstr. & K.D.Hill.
Cycas siamensis subsp. *balansae* (Warb.) J. Schust. (1932) ≡ *C. balansae* Warb.
Dioon edule forma *lanuginosum* (Wittm.) J. Schust. (1932) = *D. edule* Lindl.
Encephalartos horridus var. *latifrons* (Lehm.) J. Schust. (1932) ≡ *E. latifrons* Lehm.
Encephalartos horridus var. *nanus* (Lehm.) J. Schust. (1932) = *E. horridus* (Jacq.) Lehm.
Encephalartos horridus var. *van-hallii* (de Vriese) J. Schust. (1932) = *E. horridus* (Jacq.) Lehm.
Encephalartos lehmannii forma *dentatus* Regel *ex* J. Schust. (1932) = *E. lehmannii* Lehm.
Encephalartos lehmannii forma *spinulosus* (Lehm.) J. Schust. (1932) ≡ *E. spinulosus* Lehm.
Zamia chamberlainii J. Schust. (1932) = *Z. pygmaea* Sims.
Zamia floridana forma *silvicola* (Small) J. Schust.(1932) = *Z. integrifolia* L.*fil.*

Zamia floridana var. *purshiana* forma *silvicola* (Small) J.Schust. (1932) = *Z. integrifolia* L.*fil.*
Zamia floridana var. *purshiana* J. Schust. (1932) = *Z. integrifolia* L.*fil.*
Zamia loddigesii var. *longifolia* J. Schust. (1932) = *Z. loddigesii* Miq.
Zamia media var. *commeliniana* forma *silicea* (Britton) J. Schust. (1932) = *Z. pygmaea* Sims.
Zamia media var. *commeliniana* J. Schust. (1932) = *Z. pumila* L.
Zamia media var. *gutierrezii* forma *calcicola* J. Schust. (1932), *nomen dubium.*
Zamia media var. *gutierrezii* (Sauvalle) J. Schust. (1932) = *Z. muricata* Willd.
Zamia media var. *jacquiniana* forma *brevipinnata* J. Schust. (1932), *nomen dubium.*
Zamia media var. *jacquiniana* J. Schust. (1932) = *Z. integrifolia* L.*fil.*
Zamia media var. *portoricensis* (Urb.) J. Schust. (1932) ≡ *Z. portoricensis* Urb.
Zamia media var. *tenuis* (Willd.) J. Schust. (1932) = *Z. integrifolia* L.*fil.*
Zamia pygmaea var. *kickxii* (Miq.) J. Schust. (1932) = *Z. pygmaea* Sims.
Zamia pygmaea var. *ottonis* (Miq.) J.Schust. (1932) = *Z. pygmaea* Sims.
Zamia wielandii J. Schust. (1932) = *Z. poeppigiana* Mart. & Eichler.

Schuster's Cycad Publications

Schuster, J. 1932. Cycadaceae. Pp. 1–168 in A. Engler (ed.), *Das Pflanzenreich* 99; also as *Die Natürlichen Pflanzenfamilien* 4(1), Wilhelm Engelmann,Leipzig. [*Cycas nathorstii* J.Schust. and numerous synonyms as listed above].
Schuster, J. 1935. *Zur Ehrenrettung der Cycas.* Gartenflora 84(9): 259–264.

Loran Martin Whitelock (1930–2014)

Loran Whitelock (Fig. 7.67) was born on April 21, 1930, in Oakland, California, son to an engineer, Loran M. Whitelock Sr., and Katherine (née Land). He studied botany, zoology, and microbiology at California State University, Los Angeles, and he worked initially as Health Inspector for the City and later the County of Los Angeles but then switched his career to become a landscape designer. His first design project was for the Newporter Hotel in Newport Beach.

Pam Ward and Ed Medlock, close friends of Loran's, told us a number of interesting and lesser-known facts about the young Whitelock. We summarized some of these below, but not necessarily in chronological order:

As a young child, his mother would walk him to MacArthur Park in Los Angeles. He knew all the plant names, and the gardeners in the park referred to him as "the Professor."

Fig. 7.67 Loran Whitelock inspecting *Ceratozamia euryphyllidia* in his home greenhouse, November 2010. (Photo: Chip Jones)

He also was an enthusiastic square dancer in Las Vegas, thanks to his mother's love of dance and her coaxing him to join a group called the Y-Knot dancers.

During his college days, Loran earned pocket money in various ways. One was doing night-shift janitorial duties for a biological laboratory; another was working at the Page Museum classifying fossil bones, mostly saber-toothed cats, from the La Brea Tar Pits at Hancock Park. In his early years, he joined a martial arts dojo and earned a black belt in judo. Loran was an avid and eclectic collector throughout his life. One of his first childhood hobbies was that of acquiring assorted paint cans. Another collection that Loran assembled was that of premier Winchester rifles.

He also gathered an extensive collection of reptiles, especially rattlesnakes. Each snake had its own cage, and all were kept in a closet in the apartment that he shared with his mother. When their landlord got wind of this, they were forced to move, but Loran said that his mom never blinked an eye! In the 1960s, he was a "sometimes guest" of the new Arizona-Sanora Desert Museum and on one visit met the well-known American herpetologist, Laurence Klauber, and together, they collected a specimen of the rare Tiger Rattlesnake, *Crotalus tigris.*

Loran also had an interest in vampire bats. To get to know bats better, he would stay overnight in Griffith Park, quite against all the city rules. Further, to get familiar with bats' habitats, he and a friend spent about three months in a cave in Texas. He also assisted the producers of a Hollywood movie that featured bats and had to scale cliff faces at Vasquez Rocks in the Sierra Pelona Mountains during the film production.

During a visit to Mexico in the 1960s, Loran became enamored with the cycad flora of that country that set him off on a lifelong quest to study all aspects of these plants. He traveled extensively throughout the world collecting plants and seeds and taking photographs at every opportunity. Many of the plants he gathered were prize specimens from habitat, but we must remember that his early collections were done prior to the enactment of the CITES legislation (Convention on International Trade in Endangered Species of Wild Fauna and Flora) that came into effect on July 1, 1975.

At his hillside property in Eagle Rock, northeast of Los Angeles, Loran then established one of the most comprehensive and best-displayed private cycad collections in the world. He registered a business called Cycad Gardens that traded vigorously on an international basis; a great many of the current private cycad collections globally were from Loran's garden.

His long-awaited book "The Cycads" (Fig. 7.68) was the result of many years' worth of effort and was received with acclaim in 2002. An updated second edition, in two volumes, was co-authored by Duke Benadom in 2015 and issued shortly after Loran's death. Loran attended and presented material at several International Conferences on Cycad Biology and a cycad workshop at Montgomery Botanical Center in 2002.

Quite apart from his avid interest in cycads, Loran's garden and home had an extensive representation of other plants, including palms, cacti and succulents, ferns, anthuriums, bromeliads, and other unusual and often quite rare plants many of which he had collected personally. His animal interests extended to birds, beetles, butterflies, and reptiles, while memorabilia from his travels extended to Asian artwork, screens, bronzeware, tapestries, wooden figures, sculptures, Chinese foo dogs, Tiffany glass, meteorites, and fossils of all kinds. He married Eva White in 1962.

Loran passed away on May 27, 2014, at the age of 84. Californian palm expert Don Hodel named *Chamaedorea whitelockiana* for him, Jeff Chemnick and Tim Gregory (1996) acknowledged him in *Ceratozamia whitelockiana* (Fig. 7.69) from Oaxaca, and Johan Hurter (1995) recognized his name in the Ugandan cycad *Encephalartos whitelockii* (Fig. 7.70). He bequeathed his cycad collection of nearly 500 specimens to The Huntington Botanical Gardens, near Pasadena in the city of San Marino, along with an endowment to ensure their ongoing care. The gargantuan task of moving the plants, some of them weighing more than 900 kgs, fell to Gary Roberson, the lead project gardener of the Garden's cycad and palm collections, and took some four months to complete. Incredibly, all but ten plants survived.

Fig. 7.68 The front cover of Loran Whitelock's popular volume "The Cycads" published in 2002. The feature plant is a fine specimen of *Encephalartos heenanii* in a habitat near Barberton on the South Africa/Swaziland (now Eswatini) border

Fig. 7.69 *Ceratozamia whitelockiana* in habitat in Oaxaca, Mexico. (Photo: Chip Jones)

Fig. 7.70 The Ugandan cycad, *Encephalartos whitelockii*, named for Loran by Johan Hurter in 1995. (Photo: Duke Benadom)

Names of Cycads with Whitelock's Authorship
Data from the *World List of Cycads.*

Ceratozamia hondurensis J.L. Haynes, Whitelock, Schutzman & R.S. Adams (2008).

Zamia bussellii Schutzman, R.S.Adams, J.L.Haynes & Whitelock (2008) = *Zamia onan-reyesii* C.Nelson & Sandoval.

Whitelock's Cycad Publications

Haynes, J. L., L. M. Whitelock, B. Schutzman & R. S. Adams. 2008. A new endemic *Ceratozamia* from Honduras (Cycadales: Zamiaceae). Cycad Newsletter 31(2–3): 16–21, figs. 1–2. [*Ceratozamia hondurensis* J.L.Haynes, Whitelock, Schutzman & R.S.Adams].

Schutzman, B., R. S. Adams, J. L. Haynes & L. M. Whitelock. 2008. A new endemic *Zamia* from Honduras (Cycadales: Zamiaceae). Cycad Newsletter 31(2–3): 22–26, figs. 1–16, 18–19. [*Zamia bussellii* Schutzman, R.S.Adams, J.L.Haynes & Whitelock].

Whitelock, L. M. 1975. Cycads — age and growth rate. Cactus and Succulent Journal (U.S.) 47(2): 71–81.

Whitelock, L. M. 1978. The twilight of the cycads. Garden 2(5): 6–10.

Whitelock, L. M. 1993. Field studies of the genus *Cycas* in Australia. Pp. 300–316 in D. W. Stevenson & K. J. Norstog (eds), The Biology, Structure and Systematics of the Cycadales. Proceedings of the Second International Conference on Cycad Biology, Townsville, Australia, 1990. Palm & Cycad Societies of Australia Ltd., Milton, Queensland.

Whitelock, L. M. 1995. Cycad conservation in the past and the need for improvements in the future. Pp. 1–8 in P. Vorster (ed.), Proceedings of the Third International Conference on Cycad Biology, Pretoria, South Africa, 1993. Cycad Society of South Africa, Stellenbosch.

Whitelock, L. M. 2002. The Cycads. Timber Press, Portland, Oregon.

Whitelock, L. M. 2004. Classification concepts in *Ceratozamia* (Zamiaceae). Pp. 95–108 in T. Walters & R. Osborne (eds), Cycad Classification: Concepts and Recommendations. CABI Publishing, Wallingford, UK, pp. 95–108.

Whitelock, L. M. 2004. Range and variation in the genus *Ceratozamia* (Zamiaceae). Proceedings of the Fifth International Conference on Cycad Biology, Miami, U.S.A., 1999. The Botanical Review 70(2): 235–239.

Whitelock, L. M. & D. Benadom. 2015. The Cycads. Vols.1–2. Superb Succulents, Simi Valley, California.

Correction to: The Cycad Botanists

Roy Osborne and Dennis Wm. Stevenson

Correction to:
R. Osborne, D. Wm. Stevenson, *The Cycad Botanists*, Memoirs of The New York Botanical Garden 123, https://doi.org/10.1007/978-3-031-79082-9

The original version of the book was inadvertently published with the incorrect volume number in the front matter and cover and also with the incorrect surnames of the author on the spine. This has been corrected now.

The updated version of this book can be found at
https://doi.org/10.1007/978-3-031-79082-9

R. Osborne, D. WM. Stevenson, *The Cycad Botanists*, Memoirs of The New York Botanical Garden 123, https://doi.org/10.1007/978-3-031-79082-9_8

Literature Cited

Aiton, W. 1789. Palmae pennatifoliae in Hortus Kewensis 3: 477–479. [*Zamia debilis* L.*fil*., *Z. furfuracea* L.*fil*., *Z. integrifolia* L.*fil*., *Z. pungens* L.*fil*.].

Allan, M. 1967. The Hookers of Kew, 1785–1911. M. Joseph, London.

André. 1873. *L'Illustration horticole* 20: 126–127 & Plates 133–134.

André. 1875. *L'Illustration horticole* 22: 23 & Plates 195.

Anonymous. 1836. Beschreibung von *Encephalartos lehmanni* Ecklon. Allgemeine Gartenzeitung 4: 217-218, pl. 1.

Anonymous. 1877. Alexander Braun. Nature 15: 490. https://doi.org/10.1038/015490a0. Accessed 25 February 2022.

Anonymous. 1888. Memoirs: August Wilhelm Eichler. Proceedings of the American Academy of Arts and Sciences 23(2): 355–356. http://www.jstor.org/stable/20021536. Accessed 25 February 2022.

Anonymous. 1902. Obituary, William Bull. The Gardeners' Chronicle 31, series 3: 381.

Aran, S, 2020. The sacred geometry of French botanist Charles Lemaire. https://bonjourparis.com. Accessed 25 February 2022.

Arbose, J. 1974. To the rescue of endangered plants. The New York Times, December 8.

Artabe, A., A. B. Zamunder & D. W. Stevenson. 2010. *Neochamberlainia*, a new name for *Chamberlai*nia Artabe, Zamuner & D.W.Stev. *non Chamberlainia* Grout (Brachytheciaceae). Brittonia 62(1): 95.

Aymonin, G. 1983. Le Professeur Jacques Leandri (1903–1982). *Bulletin du museum national d'historie naturelle* (ser. 4), Section B, Botanique: 3–10.

Baillon, H. E. 1894. *Histoire des plantes* 12: 68. [*Stangeria eriopus* (Kunze) Baill.]

Balfour, I. B. 1887. August Wilhelm Eichler, biography and list of publications. Annals of Botany 1: 400–403.

Banks, J.1962. The *Endeavour* Journal 1768–1771. J. C. Beaglehole (ed.). Angus & Robertson Ltd., Sydney.

Banks, J., M. Gooding, D. Mabberley & J. Studholme. 2019. Florilegium. Compact Edition. Thames & Hudson, London.

Beekman, E. 2003. Georgius Everhardus Rumphius: The Ambonese Herbal. Yale University Press, New Haven.

Beentje, H. J. 1998. J.M. Hildebrandt (1847–1881): notes on his travels and plant collections. Kew Bulletin 53(4): 835–856.

Benson, D. 1996. L. A. S. Johnson: taxonomist, ecologist, conservationist… botanist *sens. lat.* Telopea 6: 521–526.

R. Osborne, D. Wm. Stevenson, *The Cycad Botanists*, Memoirs of The New York Botanical Garden 123, https://doi.org/10.1007/978-3-031-79082-9

Benson, D. 2020. Joseph Banks and Daniel Solander at Botany Bay: how serious was the science in 1770; where did they actual go? Australian Plants 30(243): 258–273.

Benson, D. & G. Eldershaw. 2007. Backdrop to encounter; the 1770 landscape of Botany Bay, the plants collected by Banks & Solander and rehabilitation of natural vegetation at Kurnell. Cunninghamia10(1): 113– 137.

Bentham, G. & J. D. Hooker. 1862–1883. *Genera Plantarum*. 3 volumes. Lovell, Reeve & Co., London.

Bernal, R., Galeano, G. & Henderson, A. 1989. Neotypification of Colombian palms collected by W. Kalbreyer. Taxon 38: 98–107.

Bery, M. 2021. Thomas Moore – gardener and writer of horticulture. The Streatham Society. https://streathamsociety.org.uk. Accessed 25 February 2022.

Best, H. 1952. Memories of the kindly Baron. The Victorian Naturalist 68(1): 179–180.

Boulger, G. S. 1897. Seemann, Berthold Carl. Dictionary of National Biography, 1885–1900. Vol. 51: 1–6.

Braun, A. 1849–1850. *Betrachtungen über die Erscheinung det Verjüngung in der Natur, inbesondere in der Lebens- und Bilduingsgeschichte der Planze*. Poppen, Freiburg. English translation in 1853 by A. Henfrey, London.

Braun, A. & C. D. Bouché. 1874. *Plantae novae a cl. Hildebrandt ex Africa orientali allatae, a. 1874 in horto Berolinensi cultae. Index seminum in Horto Botanico Berolinensi collectorum* 18

Braun A. 1875a. *Die Frage nach der Gymnospermie der Cycadeen erläutert durch die Stellung dieser Familie im Stufengang des Gewächsreichs Monatsberichte der Königlichen Preussische Akademie des Wissenschaften zu Berlin*: 289–377

Braun, A. 1875b. *Ueber Lepidozamia. Sitzungsberichte der Gesellschaft Naturforschen der Freunde zu Berlin* 1875: 29–37.

Braun, A. 1876. *Ueber zwei von Hildebrandt angeführte Cycadeen nebst Bermerkungen über einige Cycadeen. Sitzungsberichte der Gesellschaft Naturforschen der Freunde zu Berlin* 1876: 113–125.

Brenan, J. P. M. 1974. Dr John Hutchinson,1884–1972. Kew Bulletin 29(1): 1–14. [Includes a detailed bibliography by Carolyn Pope.]

Briggs, B. G. 1996. L. A. S. Johnson – a botanical career. Telopea 6: 511–520.

Briggs, B. G. 2000. Lawrence Alexander Sidney Johnson 1925–1997. Historical Records of Australian Science 13(4): 475–494.

Briggs, B. G. & L. A. S. Johnson. 1979. Evolution in the Myrtaceae —evidence from inflorescence structure. Proceedings of the Linnean Society of New South Wales 102: 157–256.

Briquet, J. 1940. Biographies des botanistes à Genève de 1500 à 1931.Bulletin de la Société botanique Suisse 50a: 1-494.

Britten, J. 1897. Ferdinand von Mueller. Obituary. Journal of Botany 35: 272–278.

Britton N. L. 1923. Scientific Survey of Porto Rico & Virgin Islands 5: 141.

Brongniart, A. T. 1822. *Sur la classification et la distribution des végétaux fossiles*. A. Belin, Paris.

Brongniart, A. T. 1828. *Prodrome d'une histoire des végétaux fossiles*. F. G. Levrault, Paris & Strasbourg.

Brongniart, A. T. 1828–1837. *Histoire des végétaux fossiles, ou recherches botaniques et géologiques sur les végétaux renfermés dans les diverses couches du globe*, 2 vols. Paris.

Brongniart, A. T. 1829. *Recherches sur l'organisation des tiges des cycadées. Annales des sciences naturelles* 16: 389–402.

Brongniart, A. T. 1830. *Recherches sur la structure et les fonctions des feuilles. Annales des sciences naturelles* 21: 420–458.

Brongniart, A. T. 1843. *Énumération des genres de plantes cultivées au museum d'histoire naturelle de Paris*. Fortin, Masson & Cie, Paris & Leipzig.

Brongniart, A. T. 1846. *Note sur un nouveau genre de cycadées du Mexique. Annales des sciences naturelles, botanique, série* 3, 5: 5–9, Plate 1.

Brongniart, A.T. 1881 (posthumous): *Recherches sur les graines fossiles silicificées. Description des grains fossils silicifiées de terrain Houiller de Saint-Étienne*. G. Masson, Paris.

Brown, N. E. 1915. John Medley Wood. 1827–1915. Bulletin of Miscellaneous Information 10: 417–419.

Brown, R. 1810. *Prodromus florae Novae Hollandiae et insulae Van- Diemen.* 1830. *Supplementum primum prodromiflorae Novae Hollandiae.* Richard Taylor, London. Reprinted in 1960 by J. Cramer, Weinheim, Germany.

Brown, R. 1814. General remarks, geographical and systematical, on the botany of Terra Australis. In: M. Flinders, Voyage to Terra Australis. Two vols, G. & W. Nicol, London.

Brown, R. 1828. A brief account of microscopical observations made in the months of June, July and August 1827 on the particles contained in the pollen of plants and on the general existence of active molecules in organic and inorganic bodies. Philosophical Magazine (Series 2) 4: 21, 161–173.

Brown, R. 1844. On the plurality and development of the embryos in the seeds of Conifers. Annals and Magazine of Natural History 44: 1–7.

Brummitt, R. K. & C. E. Powell. 1992. Authors of Plant Names. Royal Botanic Gardens, Kew.

Bull, W. 1861–1914. A Retail List of New Beautiful and Rare Plants. Liberty Hyde Bailey Hortorium. Ethel Z. Bailey Horticultural Catalogue Collection. Digitised. https://archive.org/stream/WilliamBullFirm00WillI/WilliamBullFirm00W illI_djvu.txt. Accessed 25 February 2022.

Burbidge, N. T. 1966. Brown, Robert (1773–1858). Australian Dictionary of Biography, https://adb.anu.edu.au/biographybrown-robert-1835/text2113. Accessed 24 February 2022.

Burkill, I. H. 1944. David Prain (1857–1944). Obituary Notices of Fellows of the Royal Society 4(13): 746–770.

Burns, R., Director, 2019. "Oliver Sacks: His Own Life," documentary film by Zeitgeist Films.

Burtt Davy, J. 1926–1932. Manual of the Flowering Plants and Ferns of the Transvaal with Swaziland, South Africa, Part 1, Pteridophyta to Bombacaceae (1926), Part 2, Malvaceae to Umbelliferae (1932), Parts 3–4 unpublished.

Brunner, J. 1989. The Bundesgärten Schönbrunn. Encephalartos19: 18–21.

Calonje, M. & Meerman, J. 2009. What is *Zamia prasina*? Journal of the Botanical Research Institute of Texas 3(1): 43–49.

Calonje, M., G. Morales, C. López-Gallego & F. J. Roldan. 2015. A taxonomic revision of *Zamia montana* and *Zamia oligodonta*, with notes on their conservation status. Phytotaxa 192(4): 279–289.

Calonje, M., D. W. Stevenson & R. Osborne. 2024.The World List of Cycads, online edition, http://www.cycadlist.org.

Calonje, M., A. S. Taylor Blake, D. W. Stevenson, G. Holzman & Y. A. Ramos. 2012. *Zamia lindleyi*: a misunderstood species from the highlands of western Panama. Memoirs of the New York Botanical Garden 106: 419–437.

de Candolle, A. P. 1855. *Géographie botanique raisonnée.* Two vols. V. Masson, Paris.

de Candolle, A. P. 1868. *Prodromus systematis naturalis regni vegetabilis.* Vol. 16(2): 522–547.

de Candolle, A. P. 1873. *Histoire des sciences et des savants depuis deux siècles. H. Georg, Geneva.*

de Candolle, A. P. 1880. *La phytogeographie; ou l'art de décrire les végétaux considéres sous différents point de vue*. G. Masson, Paris.

de Candolle, A. P. 1883. *Origine des plantes cultivées.* Germer Baillière et Cie, Paris.

de Candolle, A. P. 1886. *Origin of Cultivated Plants.* Germer Baillière et Cie, Paris.

de Candolle, A. L., A. P. & A. C. 1824–1873. *Prodromus systematis naturalis regni vegetabilis. Seventeen vols, different publishers, Paris.*

Chamberlain, C. J. 1919. The Living Cycads. University of Chicago Press, Chicago.

Chambers, N. 2016. Endeavouring Banks: Exploring Collections from the Endeavour Voyage 1768–1771. Paul Holberton Publishing, London

Chemnick, J. & T. J. Gregory. 1996. A new species of *Ceratozamia* (Zamiaceae) from Oaxaca, Mexico, with comments on distribution, habitat and relationships. Phytologia 70(1): 51–57. [*Ceratozamia whitelockiana* Chemnick & T.J.Greg.].

Chisholm, H.1911. Brongniart, Adolphe Théodore. Encyclopædia Britannica, 11th edition, Cambridge University Press.

Churchill, D. M., T. B. Muir and D. M. Sinkora. 1978. The published works of Ferdinand J.H. Mueller (1825–1896). Muelleria 4(1): 1–124. The National Herbarium of Victoria.

Codd, L. E. 1989. Obituary: Robert Allen Dyer, FRSSAf (1900–1987). Transactions of the Royal Society of South Africa 47(1): 109–110.

Costantino, G. n.d. The roots of palaeobotany: Brongniart and fossil plants.Biodiversity heritage library, https://blog.biodiversitylibrary.org. Accessed 24 February 2022.

Cox, P. A., and O. W. Sacks. 2002. Cycad neurotoxins, consumption of flying foxes, and ALS-PDC disease in Guam. Neurology 58: 956–959.

Curtis, W. M. 1915. *Encephalartos hildebrandtii* Plate 8582 & 8693.

Dammer, U. 1892. [Protest about Herder's obituary of Regel in *Botanisches Zentralblatt* 50(11): 323. 1892.] Gartenflora 41: 609.

De Luca, P. 1990. A historical perspective on cycads from antiquity to the present. Pp. 1–7 in D. W. Stevenson (ed.), The Biology, Structure and Systematics of the Cycadales. Proceedings of the First International Conference on Cycad Biology, Beaulieu-sur-Mer, France, 1987. Memoirs of the New York Botanical Garden 57: 1–7.

De Luca, P., A. Moretti & A. Balduzzi. 1995. Twenty-five years of field studies at the Botanical Garden of Naples, Italy. Encephalartos 42: 11–18.

De Luca, P., A. Moretti & G. Siniscalco Gigliano. 1991. *Ricordo di Sergio Sabato. Informatore Botanico Italiano* 23(2–3): 111–115.

De Luca, P. & S. Sabato. 1979. *Dioon califanoi* (Zamiaceae), a new species from Mexico. Brittonia 31(1): 170–173. [*Dioon califanoi* De Luca & Sabato].

de Vriese, W.H. 1841. Herinneringen aan Franzjulius Ferdinand Meyen. Tijdschrift voor Natuurlijke Geschiedenis en Physiologie 8(2): 3–38.

Desmond, R. G. C. 1975. The Hookers and the development of the Royal Botanic Gardens, Kew. Biological Journal of The Linnean Society of London 7:173–182.

Desmond, R. G. C. 2004, Prain, Sir David (1857–1944), Oxford Dictionary of National Biography, online edition. www.oxforddnb.com/view/article/35602. Accessed 27 February 2022.

Dorr, L. J. & D. H. Nicolson. 2008–2009. Taxonomic Literature. A selective guide to botanical publications and collections with dates, commentaries and types. Suppl. 7: F–Frer, Suppl 8: Fres–G. A. R. G. Gantner Verlag K. G., Königstein.

Dumas, J.-B. 1881. *Notice sur les derniers travaux d'Adolphe Brongniart, relatifs aux v.g.taux fossiles.* pp. v–xiv in A.T. Brongniart *Recherches sur les graines fossiles silicifices. Description des grains fossils silicifiées de terrain Houiller de Saint-Étienne.* G. Masson, Paris.

Duperry, L. I. 1828–1829. *Voyage au tour du monde, execute par ordre du Roi, sur la corvette de sa majesté, la Coquille, pendant les années 1822, 1823, 1824 et 1825. Botanique,* two vols, Arthus Bertrand, Paris.

Dyer, R. A. 1965a. New species and notes on type specimens of South African *Encephalartos.* Journal of South African Botany 31(2): 111–121.

Dyer, R. A. 1965b. The Cycads of Southern Africa. Bothalia 8(4): 405–515.

Dyer, R. A. 1975–1976. The genera of southern African flowering plants. Vol. 1, Dicotyledons. Vol. 2, Monocotyledons, with A. Mauve, A. E. Loxton & P. Goldblatt. Department of Agricultural Technical Services, Pretoria.

Dyer, R. A. & I. C. Verdoorn. 1966. Zamiaceae. Pp. 3–34 in L. E. Codd, B. De Winter & H. B. Rycroft (eds), Flora of Southern Africa Vol.1.

Ecklon, C. F. 1833. Nachricht ben Pflanzen, welche Herr Ecklon aus Sudafrica mit gerbract hat. Allegemeine Gartenzeitung 1: 158.

Edwards, I. Benedikt Roezl (1823–1885). http://www.pacsoa.org.au. Accessed 25 February 2022.

Ehret, G. D., C. J. Trew & J. J. Haid. 1752. *Plantae selectae quarum imagines ad exemplaria naturalia Londini in hortis curiosum nutrit.* Vol. 3, Plate 26.

Eichler, A. W. 1863. Cycadaceae. Pp 410–416, Plates 108–109 in C.F.P. de Martius (ed.), *Flora Brasiliensis,* Frid. Fleischer, Leipzig.

Eichler, A. W. 1875–1878. *Blütendiagramme*. Wilhelm Engelmann, Leipzig.
Eichler, A. W. 1883a. Ein neues *Dioon*. Gartenzeitung 2: 411–413, 438–439, Plate 80.
Eichler, A. W. 1883b. *Lepidozamia peroffskyana Rgl.* Gartenzeitung 2(1): 38–42.
Eichler, A. W. 1887a. Cycadaceae, Sago- oder Farnpalmen. Pp. 6–23 in: A. Engler & K. Prantl (eds), *Die natürliche Pflanzenfamilien nebst ihren Gattungen und wichtigeren Arten.* Vol 2. Wilhelm Engelmann, Leipzig.
Eichler, A. W. 1887b. Fossile Gattungen der Cycadaceae. Pp. 24–26 in: A. Engler & K. Prantl (eds), *Die natürliche Pflanzenfamilien nebst ihren Gattungen und wichtigeren Arten.* Vol. 2. Wilhelm Engelmann, Leipzig.
Erickson, R. F. 2011. Nikolaus Joseph, Freiherr von Jacquin 1727–1817. Missouri Botanical Garden's Rare Books.
Fawcett, P. K. S., K. J. Norstog & A. P. Vovides. 1995. Pollination of two species of *Zamia* in South Florida and its implication for the evolution of cycad strobili. Pp. 111–121 in D. D. Pant et al. (eds), Proceedings of the International Conference on Global Environment and Diversification of Plants though Geological Time. South Asian Publishers, Allahabad.
Farlow, W. G. 1893. Alphonse de Candolle. Proceedings of the American Academy of Arts and Sciences 28: 406–411.
Fernández, T. & E. Tamaro. 2004. *Biografia de José Celestino Mutis. Biografias y Vidas, La enciclopedia biográfia en línea.* Barcelona. https://www.biografiasyvidas.com/biografia/m/mutis.htm.
FitzGerald, S. 2009. Hooker, Sir William Jackson (1785–1865). The Oxford Dictionary of National Biography. https://www.oxforddnb.com. Accessed 25 February 2024.
Gaerdt, H. 1868. *Le Chevalier Joseph de Rawicz-Warszewicz, esquisse biographique.* Pp. 39–45 in E. Morren (ed.) La Belgique Horticole, Annales d'Horticulture Belge et Étrangère. Liége.
Garside, S. 1942. Baron Jacquin and the Schönbrunn Gardens. Journal of South African Botany 8: 199–224.
Gaudichaud, C. 1829. *Voyage au tour du monde, entrepris par ordre du roi. Exécuté sur les corvettes de S.M. l'Uranie et la Physicienne. Botanique.* Pp. 432–441. [*Cycas thouarsii* R.Br. *ex*Gaudich.].
Geison, G. L. n.d. Thiselton-Dyer, William Turner. https://www.encyclopedia.com. Accessed 25 February 2022.
Giddy, C. 1984. Cycads of South Africa. Purnell & Sons, Cape Town. Second revised edition 1984, C. Struik Publishers, Cape T
Giddy, C. 1990. Focus on Encephalartos ngoyanus I.Verd. Encephalartos 23: 3–9.
Giddy, C. 1995. Cycad conservation — a third world perspective. Pp. 19–25 in P. Vorster (ed.), Proceedings of the Third International Conference on Cycad Biology, Pretoria, South Africa, 1993. Cycad Society of South Africa, Stellenbosch.
Gregersen, E. 2021. Alexander Braun. Encyclopedia Britannica. https://www.britannica.com/biography/Alexander-Braun. Accessed 25 February 2024.
Grobbelaar, N. 2002a. Broodbome, met special verwysing na die Suider-Afrikaanse soorte (Afrikaans version). Published by the author, Pretoria.
Grobbelaar, N. 2002b. Cycads – with special reference to the southern African species. (English version edited by R. Osborne). Published by the author, Pretoria.
Gunn, M. & L. E. Codd.1981. Botanical exploration of southern Africa. A. A. Balkema, Rotterdam.
Hemsley, W. B. 1886–1888. *Biologia centrali-Americana.* Contributions to the knowledge of fauna and flora of Mexico and Central America. Vol. 4, Botany. R. H. Porter & Dulau & Co., London.
Hemsley, W. B. 1904. In memoriam, Berthold Seemann. Journal of the Kew Guild 1(3): 31–32.
Hill, K. D. 1992. A preliminary account of *Cycas* (Cycadaceae) in Queensland. Telopea 5: 177–206.
Hill, K. D. 1994. The *Cycas rumphii* complex (Cycadaceae) in New Guinea and the Western Pacific. Australian Systematic Botany 7(6): 543–567
Hill, K. D. 1995. The genus *Cycas* (Cycadaceae) in the Indian region, with notes on the application and typification of the name *Cycas circinalis.* Taxon 44: 23–31. [*Cycas circinalis* L.].

Hill, K. D. 1996. A taxonomic revision of the genus *Cycas* (Cycadaceae) in Australia. Telopea 7(1): 1–64.
Hill, K. 1998. Cycas thouarsii. https://web.archive.org/web/20191228031243/http://plantnet.rbg-syd.nsw.gov.au/cgi-bin/cycadpg?taxname=Cycas+thouarsii
Hill, K. D. & L. A. S. Johnson. 1995. Systematic studies in the eucalypts. 7. A revision of the bloodwoods, genus *Corymbia* (Myrtaceae). Telopea 6(2–3): 185–504.
Hill, K. D., H. T. Nguyen & P. K. Lôc. 2004. The genus *Cycas* (Cycadaceae) in Vietnam. Proceedings of the Fifth International Conference on Cycad Biology, Miami, U.S.A., 1999. Botanical Review 70(2): 134–193.
Hill, K. D. & R. Osborne. 2001. Cycads of Australia. Kangaroo Press, Sydney, Australia.
Hinterding M. 1995. The cycad collection at the botanical garden of Amsterdam. Encephalartos 42: 18–23.
Home, R. W. 1997. Ferdinand Mueller: Migration and the Sense of Self. Historical Records of Australian Science 11(3): 311–323.
Hooker, W.J. 1830. Botanical Miscellany. 1: 74-77.
Hooker, J. D. 1872. *Macrozamia corallipes*, native of New South Wales. Curtis's Botanical Magazine 98: Plate 5943 and text.
Hooker, J. D. 1890. *Zamia wallisii*. Plate 7103 in Curtis's Botanical Magazine 116.
Hooker, J. D. 1897. *Zamia obliqua*, native of New Grenada. Curtis's Botanical Magazine 123: Plate 7542 and text
Hooker WJ. 1851. Observations upon the elevated temperature of the male inflorescence of Cycadaceous plants; communicated by Dr. W.H. de Vriese, Professor of Botany and Director of the Royal Garden of the University of Leyden. Journal of Botany and Kew Garden Miscellany 3: 186–188.
Hooker WJ. 1857. Mission of Dr and Prof de Vriese to Dutch East India. Journal of Botany and Kew Gardens Miscellany 9: 346.
Hooker, W. J. 1859. *Stangeria paradoxa*. Plate 5121 in Curtis's Botanical Magazine 85.
Hooker, W. J. 1863a. *Encephalartos horridus* var. *trispinosus*, three-toothed *Encephalartos*. Curtis's Botanical Magazine 89: Plate 5371 and text.
Hooker, W. J. 1863b. *Bowenia spectabilis,* Australian *Bowenia*. Curtis's Botanical Magazine 89: Plate 5398 and text.
Howard, R. A.1973. The *Enumeratio* and *Selectarum* of Nicolaus von Jacquin. Journal of the Arnold Arboretum 54: 435–436.
Hull, D. L. 1996. Rainbows in retrospect: L. A. S. Johnson's contributions to taxonomic philosophy. Telopea 6: 527–539.
Hurter, P. J. H. 1995. Two newly described cycads from Africa. Phytologia 78(6): 409–416. [*Encephalartos nubimontanus* P.J.H.Hurter and *E. whitelockii* P.J.H.Hurter].
Hutchinson, J. 1926–1934. The Families of Flowering Plants, arranged according to a new system based on their probable phylogeny. Vol. 1, Dicotyledons. Vol. 2, Monocotyledons. Macmillan & Co., London.
Hutchinson, J. 1932. *Encephalartos kosiensis* Hutch. Bulletin of Miscellaneous Information 1932: 512.
Hutchinson, J. 1946. A Botanist in Southern Africa [with numerous photographs, drawings, and maps]. P. R. Gawthorne, London.
Hutchinson, J. 1964–1967. The Genera of Flowering Plants. Vols.1–2, Vol. 3 published posthumously. Oxford University Press.
Hutchinson, J. & J. M. Dalziel.1927–1928. Flora of West Tropical Africa. First edition: two vols. Second edition, with R. W. J. Keay & F. N. Hepper, three vols, 1954–1972.
Hutchinson, J. & R. Melville. 1948. The Story of Plants and their Uses to Man. Gawthorn, London.
Hutchinson, J. & G. Rattray. 1933. Cycadaceae. Pp. 24–44 in *Flora Capensis* 5(2).
Jacot-Guillardmo, A. 1958. Temperature Variations in Male Cones. Nature 182: 474.
Jacquin, N. J. 1760. *Enumeratio systematica plantarum.*
Jacquin, N. J. 1762–1763. *Selectarum stirpium Americanarum historia.*

Jacquin, N. J. 1791. *Collectanea ad botanicam, chemiam, et historiam naturalem spectantia*. Vol. 3: 261–265. [*Zamia angustifolia* Jacq.].
Jacquin, N. J. 1801. *Fragmenta botanica, figures coloratis illustrata,* Nos 91–94: 27–28 & Plates 25–31. [*Zamia cycadifolia* Jacq., *Z. horrida* Jacq., *Z. lanuginosa* Jacq., *Z. longifolia* Jacq.].
Johnson, L. A. S. 1959. The families of cycads and the Zamiaceae of Australia. Proceedings of the Linnaean Society of New South Wales (Series 2) 84: 64–117.
Jones, D. L. & K. D. Hill. 1992. *Macrozamia johnsonii*, a new species of *Macrozamia* section *Macrozamia* (Zamiaceae) from northern New South Wales. Telopea 5(1): 31–34. [*Macrozamia johnsonii* D.J.Jones & K.D.Hill].
Jones, W. G., K. D. Hill & J. M. Allen. 1995. *Wollemia nobilis*, a new living Australian species and genus in the Araucariaceae. Telopea 6(2–3): 173–176.
Joshi, L. M. 2001. Divya Darshan Pant — Obituary. The Guardian, online edition.
Junker, T. & H. Landsberg. 1994. The two deaths of a naturalist. Julius Schuster's (1886–1949) path from botany to the history of biology. Medical History Journal, International Quarterly Journal for the History of Science 29: 149–170.
Karsten, G. K. 1857. *Organographische Betrachtung des* Zamia muricata *Willd. Ein Beitrag zur Kenntnifs der Organisations-Verhältnisse der Cycadeen und der en Stellung im natürlichen Systeme*. Pp. 193–219 in: *Adhandlungen der Königlichen Akademia der Wissenschaften zu Berlin.*
Khuraijam, J. S. & R. Singh. 2020. Taxonomic status of the *Cycas pectinata* complex. Pp. 59–76 in C. López-Gallego, M. Calonje, M. Patrick Griffith & J. S. Khuraijam (eds), Proceedings of the Tenth International Conference on Cycad Biology, Medellín, Colombia, 2015. Cycads 5(1): 59–76. [*Cycas divyadarshanii* Khuraijam & Rita Singh].
Kieza, G. 2020. Banks. HarperCollins, Sydney.
King, C.J. 1974. Moore, Charles (1820–1905). Australian Dictionary of Biography 5: 60–132.
Knapp, J. A. 1892. Geheimrath Dr. Eduard August v. Regel, Nachruf, *Verhandlungen der Kaiserlich-Königlichen Zoologisch-Botanischen Gesellschaft in Wien* 42: 260–304.
Köhler, P. 2014. The Life of Józef Warszewicz (1812–1866): the Kraków Period. *Acta Baltica Historiae et Philosophiae Scientarum* 2(1): 18–26.
Kunze, G. *Additamentum enumerationis filicum mexacanarum, partim a B. Schiedeo, partima C. Car. Ehrenbergio aliisque collectarum*. Linnaea 13: 129–153 1839.
Kurland, L. T. 1988. Amyotropic lateral sclerosis and Parkinson's disease complex on Guam linked to an environmental neurotoxin. Trends in Neuroscience 11: 51–54.
Lavranos, J. J. & D. L. Goode. 1988. Notes on southern African Cycadales. Bulletin du Jardin Botanique National de Belgique 58(1–2): 219–224. [*Encephalartos dolomiticus* Lavranos & D.L.Goode, *E. dyerianus* Lavranos & D.L.Goode, *E. eugene-maraisii* I.Verd. subsp. *middelburgensis* Lavranos & D.L.Goode].
Lehmann, J. G. C. 1821–1824. *Icones et descriptions novarum et minus cognitarum stirpium.* Five parts of 10 plates each.
Lehmann, J. G. C. 1822–1840. *Semina in Horto Botanico Hamburgensi.*
Lehmann, J. G. C. 1830–1840, 1849–1852. *Delectus seminum quae in Horto Hamburgensium Botanico e collectione anni 1834 mutuae commutation.*
Lehmann, J. G. C.1834. *De plantis cycadeis praesertim Africae Australis. Novarum et minus cognitarum stirpium. Pugillus* 6: 1–66, Tabs. 1–5. Meissner, Hamburg.
Lehmann, J. G. C. 1838. *Tijdschrift voor natuurlijke geschiedenis en physiologie* 4(4): 414–425.
Lehmann, J. G. C. 1844. *Novarum et minus cognitarum stirpium. Pugillus* 8: 31. Meissner, Hamburg.
Lehmann, J. G. C. 1844–1847. *Plantae preissianae sive enumeratio plantarum quas in Australasia occidental et meridionale-occidentali annis 1838–1841 collegit Ludovicus Preiss.*
Lemaire C. A. 1838. *Cactearum aliquot novarum AC insuetarum in Horto Monvilliano cultarum accurata descriptio*, Vol. 1. F.G. Levrault, Paris.
Lemaire C. A. 1839. *Cactearum genera nova speciesque novæ et omnium in Horto Monvilliano cultarum ex affinitatibus naturalibus ordinatio nova indexque methodicus.* J. Loss, Paris.

Lemaire C. A. 1841–1847. *Iconographie descriptive des cactées, ou, essais systématiques et raisonnés sur l'histoire naturelle, la classification et la culture des plantes de cette famille.* H. Cousin, Paris.
Lemaire C. A. 1855. *Dioon edule.* L'illustration Horticole 2: 91–94. With 3 Icones.
Lemaire C. A. 1867. Deux espéces nouvelles D'*Encephalartos*. L'illustration Horticole 14: 79-81.
Lemaire C. A. 1868. *Encephalartos ghellinckii.* L'illustration Horticole 15: 79-80. Planche 567.
Li, N., D. W. Stevenson & M. Patrick Griffith (eds), 2018. Cycad Biology and Conservation. Proceedings of the Ninth International Conference on Cycad Biology. Shenzhen, China, 2011. Memoirs of the New York Botanical Garden 117.
Linden, J. 1873. "Zamia Roezlei". *Catalogue des plantes nouvelles* 90: 10. [*Zamia roezlii* Linden].
Linden, J. 1875. Botanical collectors. Gardeners' Chronicle n.s. 3: 279–280.
Linden, L. & E. Rodigas. 1885. *Zamia tonkinesis. L'Illustration horticole* 32: 27, Plate 547. [*Z. tonkinesis* L.Linden & Rodigas].
Lindley J. 1843. Edwards's Botanical Register 28: ad t.10.
Lindstrom, A. J. 2002. Circumscription and lectotypification of *Cycas rumphii* (Cycadaceae). Brittonia 54: 305–309.
Lindström, A. J. 2009. Typification of some species names in *Zamia* L. (Zamiaceae), with an assessment of the status of *Chigua* D.Stev. Taxon 58(1): 265–270.
Linnaeus, C.1753. *Species plantarum: exhibentes plantas rite cognitas, ad genera relatas, cum differentiis specificis, nominibus trivialibus, synonymis selectis, locis natalibus, secundum systema sexual e digestas.* Laurentius Salvius, Stockholm.
Mabberley, D. J. 1985. *Jupiter Botanicus*–Robert Brown of the British Museum. J. Cramer, Braunschweig.
Mabberley, D. J. 2019. Botanical Revelation: European Encounters with Australian Plants before Darwin. The Peter Cross Collection. New South Publishing, Sydney.
Mabberley, D. J. 2020. Journey to Australia, then New Holland, by Joseph Banks, leads to the publication of hundreds of drawings of new plants and animals. ABC Science Show broadcast, 28 Nov 2020, directed by Robyn Williams.
Maconochie, J. R. 1968. Studies on the leaf production of selected plant species in the temperate, southern arid and central arid regions of South Australia. M.Sc. thesis, University of Adelaide.
Maconochie, J. R. 1978. Two new species of *Cycas* from Northern Australia. Journal of the Adelaide Botanical Gardens 1(3): 175–178.
Maconochie, J. R. 1980. Key to Cycas. In H. T. Clifford & J. Constantine (eds), Ferns, Fern Allies and Conifers of Australia. University of Queensland Press, St. Lucia, Queensland.
Madriñán, S. 2013. Nikolaus Joseph Jacquin's American Plants. Botanical Expedition to the Caribbean (1754–1759) and the Publication of the *Selectarum stirpium Americanarum historia*. Brill, Leiden.
Maiden, J. H. 1908. Records of Australian botanists. Journal and Proceedings of the Royal Society of New South Wales 42: 60–132.
Malaisse, F., J. P. Sclavo & C. Crosiers. 1992. *Recherches sur les Encephalartos Lehm. (Zamiaceae) d'Afrique centrale.* 1. *Un nouvel Encephalartos de Tanzanie. Annales de Gembloux* 98(2): 153–157
Margulis, L. & P. H. Raven. 2009. The Herbal of Rumphius. American Scientist 97(1): 7–9.
Martínez-Domínguez, L., F. Nicolalde-Morejón, F. Vergara-Silva & D. W. Stevenson. 2017. Qualitative and quantitative morphological evidence for recognition of a new species within *Ceratozamia* (Zamiaceae) from Mexico, Phytotaxa 317(1): 17–28. [*Ceratozamia chamberlainii* Mart.-Domínguez, Nic.Mor. & D.W.Stev.].
Martínez-Domínguez, L., F. Nicolalde-Morejón, F. Vergara-Silva & D. W. Stevenson. 2022. *Ceratozamia oliversacksii* (Zamiaceae) a new species of gymnosperm from western Oaxaca, Mexico. Kew Bulletin 77: 211–219.
McCracken, P. 1989. The Great Tree Robbery. "Personality" Magazine. 23 October. Pages 20–21.
Melville, R. 1957. *Encephalartos* in Central Africa. Kew Bulletin 12(2): 237–257.
Melville, R. 1960. A new theory of the angiosperm flower. Nature 188: 14–18.

Melville, R. 1969. Leaf venation patterns and the origin of the angiosperms. Nature 224: 121–125.
Melville, R. 1970. Red Data Book of Threatened Plant Species, volume 5, Angiospermae. International Union for Conservation of Nature, Morges.
Mendelson, W. 2024. From Despair to Discovery: The Botanical Odyssey of Matthias Jakob Schleiden and the Dawn of Cell Theory. Pythagoras Press.
Merola, A. 1976–1977. *Le funzioni degli orti botanici e dei musei naturalistici nel pensiero di Luigi Califano. Breve raccolta antologica con catalogo* delle *specie donate al l'Orto Botanico di Napoli*. Delpinoa 18–19: 177–193.
Meyer, M. & K. Eloff. 2015. Obituary: Prof Nathanaël Grobbelaar (1928–2015). South African Journal of Botany 97: 182–183.
Miquel, F. A. W. 1839. *Cycas rumphii. Bulletin des sciences physiques et naturelles en Néerlande* 2: 45. [*Cycas rumphii* Miq.].
Miquel, F. A. W. 1842. Pp.1–82 in *Monographia cycadearum*, R. Nathan, Utrecht.
Miquel, F. A. W. 1843. De Cycadeis Loddigesianis. Tijdschrift voor natuurlijke geschiedenis en physiologie 10(1): 68–74.
Miquel, F.A.W. 1851a. *Cycadaceae quaedam Americanae* 9. *Nieuwe verhandelingen van dee erste klasse van het koninklijke Nederlandsche instituut* 3(4): 181–189.
Miquel, F. A. W. 1851b. *Over de rangschikking der fossiele cycadeae. Tijdschrift voor de wis- en natuurkundige wetenschappen.*4: 205–227.
Miquel, F. A. W. 1852. *Sur une espèce nouvelle de Zamia des Indes occidentales, introduite dans l'establissement L. van Houtte à Gand. Flore des serres, et des jardins de l'Europe* 7(6): 141–142.
Moore, C. 1884. Notes on the genus *Macrozamia*. Journal and Proceedings of the Royal Society of New South Wales 17: 115–122.
Moore, T. 1851. A popular history of the British ferns and the allied plants. Reeve & Benham, London.
Moore, T. 1853. *Stangeria paradoxa*. Hooker's Journal of Botany and Kew Garden, Misc. 2: 228.
Moore, T. 1855. The ferns of Great Britain and Ireland. Bradury & Evans, London.
Moore, T. 1857–1862. *Index filicum.* In 20 Parts. William Pamplin, London.
Moore, T. 1861. British ferns and their allies. Routledge, Warne, and Routledge, London.
Moore, T., R. Warner & B. S. Williams. 1882–1897. The Orchid Album. B. S. Williams, London. Pates by John Nugent Fitch.
Moretti, A. & S. Sabato. 1984. Karyotype evolution by centromeric fission in *Zamia* (Cycadales). Plant Systematics and Evolution 146: 215–223.
Morren, E. 1879. *A la mémorie Gustave Wallis. La Belgique Horticole, annales de botanique er d'horticulture*: 5–19. A translation of the German obituary by Eduard Otto.
Morris, D. 1974. Mueller, Sir Ferdinand Jakob Heinrich von (1825–1896). Australian Dictionary of Biography 5.
Mueller, F. J. H. 1853. *Breviarum plantarum ducatas slesvicensis austro-occidentalis.* (Mueller's first botanical paper, completed in 1847 as part of his doctoral thesis under the name Müller but published only in 1853.)
Mueller, F. J. H. 1858. An historical review of the explorations of Australia. Philosophical Institute, Melbourne.
Mueller, F. J. H. 1881. Australasian Chemist and Druggist 4: 84. [*Macrozamia moorei* F.Muell.]. See also *Fragmenta phytographiæ Australiæ* 11: 125. [*Encephalartos moorei* (F.Muell.) F.Muell.].
Murch S. J., P. A. Cox, S. A. Banack, J. C. Steele, & O. W. Sacks. 2004. "Occurrence of beta-methylamino-l-alanine (BMAA) in ALS/PDC patients from Guam". Acta Neurologica Scandinavica **110** (4): 267–269.
Musgrave, T. 2020. The Multifarious Mr. Banks. From Botany Bay to Kew. The Natural Historian who shaped the World. Yale University Press, New Haven.
Mutis, J. C. 1783–1816. *Flora de la Real Expedición Botánica del Nuevo Reino de Granada.* Volumes published sequentially since 1953 by Ediciones Cultural Hispánica, Madrid.

Nakai, T. 1943. *Ordines, familiae, tribi, genera, sectiones, species, varietates, formae et combinationes novae a Prof. Nakai-Takenosin adhuc ut novis edita.* Hokuryukan, Tokyo. [*Dyerocycas micholitzii* (Dyer) Nakai].

Nelson, D. 2012. Working with John Maconochie (1941–1984). Australasian Systematic Botany Society Newsletter 150: 36–44.

Nicolalde-Morejón, F., A. P. Vovides & D. W. Stevenson. 2009. Taxonomic revision of *Zamia* in Mega-Mexico. Brittonia 61(4): 301–335.

Norstog, K. J. 1967. Fine structure of the spermatozoid of Zamia with special reference to flagellar apparatus. American Journal of Botany 54(7): 831–840.

Norstog, K. J. 1974. Fine structure of the spermatozoid of Zamia: the vierer Grüppe. American Journal of Botany 61(5): 449–456.

Norstog, K.J. 1985. Exploring for *Zamia wallisii.* Fairchild Tropical Garden Bulletin 40(1): 6–18.

Norstog, K. J. 1986. *Zamia chigua*, a case of mistaken identity? Fairchild Tropical Garden Bulletin 41(1): 6–13.

Norstog, K. J. 1990. The spermatozoid of Microcycas calocoma: ultrastructure. Botanical Gazette 151(3): 275–284.

Norstog, K. J. 1995. C. J. Chamberlain and the ladies. Unpublished manuscript and correspondence with Selma Voth.

Norstog, K. J. 2003. Foreword in Cycads: status survey and conservation action plan. IUCN/SSC Cycad Specialist Group. IUCN, Gland, Switzerland and Cambridge, UK.

Osborne, R. 1985. From the bookshelf. "The Living Cycads," C. J. Chamberlain, book review. Encephalartos 2: 30.

Osborne, R. 1988. Honorary Life Members [Dr. K. J. Norstog and Professor D. D. Pant]. Encephalartos 16: 33.

Osborne, R. 1998. Obituary, Cynthia Giddy. Including tributes from Avis Meresman and Cynthia Giddy's daughters, Caitlin and Janet. Encephalartos 55: 14–18.

Osborne, R. 2003. In memoriam: Knut Norstog (1921–2003). My memories of Knut. The Cycad Newsletter 26(3): 10–11.

Osborne, R. 2018. From Russia with Love. Encephalartos 134: 18–19.

Osborne, R. & H. Baijnath. 1995. Morphometric analysis of vegetative characters of *Encephalartos woodii*, *E. natalensis*, and an apparent intermediate. In P. Vorster (ed.), Proceedings of the 3rd International Conference on Cycad Biology (Pretoria, South Africa, 1993). Cycad Society of South Africa, Stellenbosch, pp. 241–244.

Osborne, R., B. Dalziel & Y. Dalziel. 2019. Focus on...*Cycas angulata* R.Br. Encephalartos 136: 3–9.

Osborne, R. & R. Gorelick. 2007. Sex change in cycads — cases, causes and chemistry. In A. P. Vovides, D. W. Stevenson & R. Osborne (eds), Proceedings of the Seventh International Conference on Cycad Biology (Xalapa, Mexico, 2005). Memoirs of the New York Botanical Garden 97: 335–345.

Osborne, R., A. Grove, P. Oh, T. J. Mabry, J. C. Ng & A. A. Seawright. 1994.The magical and medicinal usage of *Stangeria eriopus* in South Africa. Journal of Ethnopharmacology 43(2): 67–72.

Pahud C. H., de Vriese W. H. & Mieling C. W. 1854. Illustrations d'Orchidées des Indes Orientales Néerlandaises. C.W. Mieling, The Hague.

Pant, D. D. 1962. Cycas and the Cycadales. First edition. Central Book Depot, Allahabad.

Pant, D. D. 1973. Cycas and the Cycadales. Second Edition. Central Book Depot, Allahabad.

Pant, D. D. 2002. An Introduction to Gymnosperms, Cycas, and the Cycadales. Monograph 4. Birbal Sahni Institute of Palaeobotany, Allahabad.

Pearson, R. H. (ed.). 1908 (25 April & 2 May). New plants at Ghent: *Encephalartos woodii*, Hort. Sander. The Gardeners' Chronicle, series 3, 43: 257 & Supplement.

Peters, N. 2021. Rumphius and his Ambonese Herbal, a series of unfortunate events. Leiden special collections blog, University of Leiden.

Pilet, P. E. n.d. Candolle, Alphonse de. Encyclopedia.com. https://www.encyclopedia.com/science/dictionaries-thesauruses-pictures-and-press-releases/candolle-alphonse-de. Accessed 25 February 2022.

Pilger, R. K. F. 1926a. *Phylogenie und Systematik der Coniferae*. In: A. Engler & K. Prantl (eds). *Die natürlichen Pflanzenfamilien* Vol. 13. Wilhelm Engelmann, Leipzig.

Pilger, R. K. F. 1926b. Cycadaceae. Pp. 44–82 in: A. Engler & K. Prantl (eds), *Die natürliche Pflanzenfamilien nebst ihren Gattungen und wichtigeren Arteninsbesondere den Nutzpflanzen*. Vol. 13. Wilhelm Engelmann, Leipzig.

Poissan MJ. 1878. Du dégagement de chaleur qui accompagne l'épanouissement des inflorescences males de *Dioon edule*. Société Botanique de France 25: 253–254.

Prain, D. (ed.). 1914. *Encephalartos woodii*. Bulletin of Miscellaneous Information 7: 250–251 plus plate.

Prain, D. 1915. *Encephalartos hildebrandtii*, East Africa. Curtis's Botanical Magazine 141: Plates 8592–8593 plus text.

Prain, D. 1917. Cycadaceae. Pp. 344–354 in D. Prain (ed.), Flora of Tropical Africa Vol. 6, Part 2.

Regel, E. A. 1876. *Cycadearum generum specierumque revisio. Acta Horti Petropolitani* 4(4):275-320.

Rheede tot Draakestein, H. A. van. 1678–1693. *Hortus Indicus Malabaricus, Continens regni Malabarici apud Indos celeberrimi omnis generis plantas rariores.*12 vols. J. van Someren & J. van Dyck, Amsterdam.

Robyns, W. 1928. *Tentamen monographiae Vanguerieae generum que affinium. Bulletin du Jardin Botanique del'état Bruxelles* 11(1): 1–359.

Rodionenko, G.I.. 1986. Eduard Regel. Floriculture 2: S41, kindly translated by Dr. K. Tkachenko.

Rumphius, G. E. 1705. *D'Amboinsche Rariteitkamer*, Halma, Amsterdam. "The Ambonese Curiosity Cabinet". English translation 1999 by E. M. Beekman, Yale University Press, New Haven.

Rumphius, G. E. 1742. *Herbarium Amboinense* (Latin), *Amboinsche Kruidboek* (Dutch). J. Burman, Amsterdam. English translation 2011 by E. M. Beekman, 6 vols. Yale University Press, New Haven.

Sabato, S. 1990. West Indian and South American Cycads. Memoirs of the New York Botanical Garden 57: 173–185.

Sabato, S. & P. De Luca. 1985. Evolutionary trends in *Dion* (Zamiaceae). American Journal of Botany 72(9): 1353–1363.

Sacks, O. W. 1997. The Island of the Colorblind and Cycad Island. Alfred Knopf, New York.

Sacks, O. W. 2001. Uncle Tungsten: Memories of a Chemical Boyhood. Vintage Books, London.

Sacks, O. W. 2002. The Oaxaca Journal. National Geographic Society, Washington, D.C.

Sacks, O. W. 2010. The Mind's Eye. Alfred Knopf, New York

Sacks, O. W. 2012. Hallucinations. Knopf/Picador, New York

Sands, M. J. S. 1986. Obituary: Ronald Melville I.S.O., B.Sc., Ph.D., M.P.S., Ph.C., F.L.S. (1903–1985). Kew Bulletin 41(4): 761–768.

Savage, J. M. 1970. On the trail of the golden frog: with Warszewicz and Gabb in Central America. Proceedings of the California Academy of Sciences, Series 4, 38(14): 273–288.

Schätti, B., I. Ineich & C. Kucharzewski. 2018. Auguste Ghiesbrecht's natural history explorations in Oaxaca and other parts of Mexico until 1854. Acta zoológica Mexicana 34: 1–22.

Schleiden, M. J. 1839. Beiträge zur Phytogenesis. Archiv für Anatomie, Physiologie und wissenschaftliche Medicin. **1838**: 137–176.

Schultz, M. & H-H. Poppendieck. 2019. *Bildnis Johann Georg Christian Lehmann. Kunstschätze und Wissensdinge, Eine Geschichte der Universität Hamburg in 100 Objekten*, No. 19, pp. 72–75. Michael Imhof Verlag, Hamburg.

Schuster, J. 1932. Cycadaceae. Pp. 1–168 in A. Engler (ed.), *Das Pflanzenreich* 99; also *Die Natürlichen Pflanzenfamilien* 4(1), Wilhelm Engelmann, Leipzig.

Schwann, T. 1839. Mikroskopische Untersuchungen über die Uebereinstimmung in der Struktur und dem Wachsthum der Thiere und Pflanzen. Berlin: Sander.

Seemann, B. C.1853. Narrative of the voyage of HMS *Herald* during the years 1845–1851.Two vols. London, with a simultaneous German edition.
Seemann, B. C. 1854. Julius van Warscewicz. Bonplandia 8: 95–96.
Sharma, M. 2002. Obituary. Professor Divya Darshan Pant, Oct 18, 1919– May 8, 2001. Palaeobotanist 51: 169–178.
Sibilio, G. 2014–2015. The greenhouses of the Botanical Garden of Naples, Italy. Delpinoa 56–57: 60–103.
Smuts, J. C. 1946. Forward in A Botanist in Southern Africa P. R. Gawthorne, London.
Srivastava, P. C. 2001a. Biographical Memoirs of INSA Fellows, Divya Darshan Pant (1919–2001): 99–127. http://www.insa.ndl.iitkgp.ac.in. Accessed 28 February 2022.
Srivastava, P. C. 2001b. A link with the past. Divya Darshan Pant 1919– 2001. Current Science India 81(1): 116–117.
Stanberg, L. 2010. Hill, Kenneth (Ken) D. Biographical Notes. The Australian National Herbarium. Website: cpbr-info@anbg.gov.au. Accessed 28 February 2022.
Stanberg, L. 2012. Ken D. Hill, bibliography. Pp. xvii-xxv in D. W. Stevenson, R. Osborne & A. S. Taylor Blake (eds), Proceedings of the Eighth International Conference on Cycad Biology, Panama City, Panama, 2008. Memoirs of the New York Botanical Garden 106.
Stapf, O. 1889. *Die Arten der Gattung Ephedra. Denkschriften der Kaiserlichen Akademie der Wissenschaften, Mathematisch-Naturwissenschaftliche Classe* 56(2): 1–112.
Stapf, O. 1914. *Encephalartos hildebrandtii*, Kew Bulletin of Miscellaneous Information 10: 386–392.
Stafleu, F. A. 1966. F. A. W. Miquel, Netherlands Botanist. Wentia 16: 1–95.
Stafleu, F. A. 1972. The Willdenow Herbarium. Taxon 21(5–6): 685–692.
Stafleu, F. A. 1978. Dedication to the Memory of F. A. W. Miquel. Flora Malesiana Series 1, Vol. 8, Part 3: 6–16.
Stafleu, F. A. 1971. Jacquin and his American plants, an introduction to facsimile reprint of N. J. Jacquin, *Selectarum stirpium Americanarum historia*: F7–F17.
Stafleu, F. A. & R. S. Cowan. 1976–1988. Taxonomic Literature. A selective guide to botanical publications and collections with dates, commentaries and types. Vol. 1: A–G, Vol. 2: H–Le, Vol. 3: Lh–O, Vol. 5: Sal–Ste, Vol. 7: W–Z. Second Edition, Bohn, Scheltema & Holkema, Utrecht.
Stafleu, F. A. & E. A. Mennega. 1992–2000. Taxonomic Literature. A selective guide to botanical publications and collections with dates, commentaries and types. Suppl. I: A–B, Suppl. 2: Be–Bo, Suppl. 3: Br–Ca, Suppl. 4: Ce–Cz, Suppl. 5: Da–Di, Suppl. 6: Do–E. Koeltz Scientific Books, Königstein.
Stevenson, D. W. 1982. A new species of *Ceratozamia* (Zamiaceae) from Chiapas, Mexico. Brittonia 34(2): 181–184. [*Ceratozamia norstogii* D.W.Stev.].
Stevenson, D. W. 1992. A formal classification of the extant cycads. Brittonia 44(2): 220–223.
Stevenson, D. W. 1993. Lectotypification of *Cycas circinalis* L. in C. E. Jarvis, F. R. Barrie, D. M. Allan & J. L. Reveal (eds), A list of Linnaean generic names and their types. *Regnum Vegetabile* 127: 40.
Stevenson, D. W. 2001. *Zamia encephalartoides* D.Stev., *sp. nov.* Pp.40–43 *Flore de Colombia, Monografía 21, Instituto de Ciencias Naturales, Facultad de Ciencias, Universidad Nacional de Colombia, Bogotá.*
Stevenson, D. W. 2004a. Cycads of Colombia. Botanical Review 70(2): 194–234.
Stevenson, D.W. 2004b. In memoriam, Knut Norstog (1921–2003). The Botanical Review 70(1): 1–4.
Stevenson, D. W. & S. Sabato.1986a. Typification of names in *Zamia* L. and A*ulacophyllum* Regel (Zamiaceae). Taxon 35(1): 134–144.
Stevenson, D.W. & S. Sabato. 1986b. Typification of names in C*eratozamia* Brong., *Dioon* Lindl. and *Microcycas* A.DC. (Zamiaceae). Taxon 35(3): 578–584.
Stevenson, D. W. & K. J. Norstog. 1993. Sergio Sabato, dedication. Frontispiece and Introduction. Pp. 1–2 in D. W. Stevenson & K. J. Norstog (eds), The Biology, Structure and Systematics

of the Cycadales. Proceedings of the Second International Conference on Cycad Biology, Townsville, Australia, 1990. Palm & Cycad Societies of Australia Ltd, Milton, Australia.

Stevenson, D. W., K. J. Norstog & P. K. S. Fawcett. 1998. Pollination biology of cycads. Pp. 277–294 in S. Owens & P. Rudall, Reproductive Biology: in systematics, conservation and economic botany. Royal Botanic Gardens, Kew.

Stevenson, D. W., R. Osborne & A. S. Taylor Blake. 2012. In memoriam, Ken D. Hill. Pp. xvi-xvii in D. W. Stevenson, R. Osborne & A. S. Taylor Blake (eds), Proceedings of the Eighth International Conference on Cycad Biology. Panama City, Panama, 2008. Memoirs of the New York Botanical Garden 106.

Strey, R. G. 1977. The Father of Natal Botany, John Medley Wood. Natalia 7: 43–45.

Svedelius, N. 1944. Carl Peter Thunberg (1743–1828) on his bicentenary. Isis 35(2): 128–134.

Tang W. 1987. Heat Production in Cycad Cones. Botanical Gazette 148: 165–174.

Taylor, J. 2018. Picturing the Pacific. Joseph Banks and the shipboards artists of Cook and Flinders. Adlard Coles, London.

Teysman, J.E. 1850. Over eene verhoogde temperatur bij den mannelijken bloei van Cycas circinalis, waargenomen in 's Lands Plantetuin te Buitenzorg. Door J.E. Teysman, Medegedeeld door W.H. de Vriese. Nederlandsch Kruidkundig Archief 2(2): 172–180.

Thiselton-Dyer, W. T. 1882. New garden plants: *Zamia montana*. Gardeners' Chronicle 17: 460.

Thiselton-Dyer, W.T. 1884. Cycadaceae. Pp. 190–195 in W.B. Hemsley, *Biologia Centrali-Americana*, Contributions to the Knowledge of the Flora and Fauna of Mexico and Central America, Vol. 3, Botany. R.H. Porter, London

Thiselton-Dyer, W.T. 1893. Alphonse de Candolle. Nature, No. 1238, Vol. 48: 269–271.

Thunberg, C. P. 1784a. *Curculio zamiae, nova et singularis insecti species e Capite Bonae Spei, descripta. Nova acta regiae societatis scientiarum Upsaliensis* 4: 29–30.

Thunberg, C. P. 1784b. *Flora Japonica: sistens plantas insularum Japonicarum*. Leipzig.

Thunberg, C. P. 1793. *Resa uti Europa, Africa, Asia förrättad åren 1770–1779*. W. Richardson & J. Egerton, London.

Thunberg, C. P. 1794–1800. *Prodromus plantarum capensium*. Vols1–2. Uppsala.

Thunberg, C. P. 1807–1813. *Flora Capensis: Sistens plantas promontorii Bonae Spei Africes*. Vols 1 & 2. Uppsala. Complete edition by J. A. Schultes, 1823. Stuttgart.

Tomlinson, P. B. 2012. Rescuing Robert Brown — The Origins of Angio-Ovuly in the Seed Cones of Conifers. Botanical Review 78: 310–334.

Trew, 1752. C. J. Plantae Selectae, Tab. 26. 1752.

Turrill, W. B. 1963. Joseph Dalton Hooker, botanist, explorer and administrator. The Scientific Book Club, London.

Vázquez Torres, M., A. Moretti & C. Carvajal-Hernanez. 2013. A new species of *Ceratozamia* from Veracruz, Mexico. Delpinoa 50–51: 129–133.

Veldkamp, J. F. 2011. Georgius Everhardus Rumphius (1627–1702), the blind seer of Ambon. The Gardens' Bulletin, Singapore 63(1–2): 1–15. de Vriese WH. 1841. Eenige opmerkingen aangaande den stam van *Cycas circinalis*, vooral met betrekking tot de afbeeldingen daarvan, voorkomende in het III^{de} Deel van den Hortus Malabaricus. Het Instituut of Verslagen en Mededeelingen, uitgegeven door de vier Klassen van het Koninklijk-Nederlandsch Instituut van Wetenschappen, Letterkunde en Schoone Kunsten 1: 217–226.

Von Mueller, F. 1825–1896. Regardfully yours: selected correspondence of Ferdinand von Mueller. Three vols, edited by R.W. Home et al. Bern, New York, Peter Lang 1998–2006.

Vorster, P. 1987. R. A. Dyer. Encephalartos 12: 12–15.

Vorster, P. 2004. Zamiaceae. Typification of *Encephalartos*. Bothalia 34(2): 112–113.

Vorster, P & E. Vorster. 1985. Focus on … *Stangeria eriopus* (Kunze) Baill. Encephalartos 2: 8–17. Reprinted from Excelsa 4 of 1974.

Vovides, A. P., M. Vázquez Torres, B. Schutzman & C. G. Iglesias. 1993. A new species of *Ceratozamia* (Zamiaceae) from Querétaro and Hidalgo, Mexico. Novon 3(4): 502–506.

de Vriese WH. 1842. Berigt aangaande een' onlangs uit Java ontvangen' *Cycas circinalis*, L. Gekweekt en thans bloeijende in den kruidentuin der Stad Amsterdam. Bij A. Zweesaardt, Amsterdam.

de Vriese WH. 1843. De *Encephalarto lehmanni* Eckl. ad virum clarissimum Fr. Ant. Guil. Miquel animadversiones mjttit Guil. Henr. de Vriese Tijdschrift voor Natuurlijke Geschiedenis en Physiologie 10: 57–67.

de Vriese WH. 1845a. Nog een woord over den stam van *Cycas circinalis* en dien van *C. revoluta*, naar aanleiding van onlangs uit Java ontvangene exemplaren. Tijdschrift voor Natuurlijke Geschiedenis en Physiologie 12: 158–162.

de Vriese WH. 1845b. *Zamia Galeotti* eene nieuwe soort uit de groep der Cycadeen, gekweekt in den Hortus Botanicus te Amsterdam. Tijdschrift voor Natuurlijke Geschiedenis en Physiologie 12: 23–24.

de Vriese WH. 1846a. Nadere toelichting omtrent *Zamia muricata* Willd., naar aanleiding van eene bloeijende plant in de kassen van den Heer A. van der Hoop, ip Spaa. Nederlandsch Kruidkundig Archief 1: 162-167.

de Vriese WH. 1846b. Aanteekeningen omtrent eenen bloeijenden *Encephalartos altensteinii* Lehm., eene bijdrage tot de nadere kennis der Cycadeen. Nederlandsch Kruidkundig Archief 1: 168–180.

de Vriese WH. 1846c. Note sur le *Zamia muricata* Willd. Annales des Sciences Naturelles ser.3, t.6: 358–359.

de Vriese, W. H. 1847. *Descriptions et figures des plantes nouvelles et rares du jardin botanique de l'université de Leide et des principaux jardins du Royaume des Pay-bas.* Chez A. Arnz & comp., Leiden.

de Vriese WH. 1848. De palmen van Suriname: beschouwd in betrekking tot derzelver kruidkundige kermerken, kultuur en nut voor nijverheid en handel. Uitgegeven voor Rekening van de Maatschappij, Leiden.

de Vriese WH. 1855. Tuinbouw-Flora van Nederland en zijne overzeesche bezittingen: bevattende de geschiedenis en afbeeldingen van nieuwe of merwaardige planten, bloemen en vruchten. A.W. Sythoff, Leiden.

de Vriese, W. H. 1856. *Plantae Indiae Batavae Orientis quas in itinere per insulas archipelagi indici Javam, Amboinam, Celebem, Ternatum, aliasque annis1815-r821 exploravit Casp. Georg. Carol/ Reinwardt.* E. J. Brill, Lugduni-Batavoru.

Wang, D. Y. 1996. *Cycas elonga* (Leandri) D.Y.Wang, P. 51 in F. X. Wang & H. B. Liang (eds), Cycads of China. Guangdong Science and Technology Press, Guangzhou.

Warszewicz, J. 1845. *Correpondenz: Nachricht.* Allgemeine Gartenzeitung 13: 252–253.

Weddell, H. A. 1849. *Additions a la flore de l'Amérique du sud. Annales de Sciences Naturelles*, sér. 3, 13: 249 & Plate 4. [*Zamia brongniartii* Wedd.].

White, A., R. A. Dyer & B. L. Sloane. 1941. The Succulent Euphorbieae. Vols 1–2, Abbey Garden Press, Pasadena.

Whitelock, L. M. 2002. The Cycads. Timber Press, Portland, Oregon.

Whitelock, L. M. & D. Benadom. 2015. The Cycads. Vols.1–2. Superb Succulents, Simi Valley, California.

Willdenow, C. L. 1787. *Florae Berolinensis*. Wilhelmi Viewegii, Berlin.

Willdenow, C. L. 1792. *Grundriss der Kräuterkunde zu Vorlesungen*. Haude & Spener, Berlin (translated into English as "Principles of Botany", 1805).

Willdenow, C. L. (ed.). 1793. Juan de Loureiro's *Flora Cochinchinensis*. Haude & Spener, Berlin.

Willdenow, C. L. 1806. *Species Plantarum*, Fourth Edition. *Dioecia polyandra* 4: 845–849

Willdenow, C. L. 1822. *Anleitung zum Selbststudium der Botanik*. Third Edition, Oehmigke, Berlin.

Willis, J. H. 1949. Botanical pioneers in Victoria. Victorian Naturalist 66(5–7): 83–89, 103–109, 123–128.

Index

R. Osborne, D. Wm. Stevenson, *The Cycad Botanists*, Memoirs of The New York Botanical Garden 123, https://doi.org/10.1007/978-3-031-79082-9

The manufacturer's authorised representative in the EU is Springer Nature Customer Service Centre GmbH, Europaplatz 3, 69115 Heidelberg, Germany. If you have any concerns regarding our products, please contact ProductSafety@springernature.com

Printed and bound by CPI Group (UK) Ltd, Croydon, CR0 4YY
30/09/2026
02225562-0013